T0135156

Fireworks Algorithm

Ying Tan

Fireworks Algorithm

A Novel Swarm Intelligence
Optimization Method

 Springer

Ying Tan
Peking University
Beijing
China

ISBN 978-3-662-51618-8   ISBN 978-3-662-46353-6 (eBook)
DOI 10.1007/978-3-662-46353-6

Springer Heidelberg New York Dordrecht London

Parts of the edition are translations from the Chinese language edition: 烟花算法引论 by Ying Tan
© Science Press 2015. All rights reserved
© Springer-Verlag Berlin Heidelberg 2015
Softcover re-print of the Hardcover 1st edition 2015

Printed on acid-free paper

Springer-Verlag GmbH Berlin Heidelberg is part of Springer Science+Business Media
(www.springer.com)

*Dedicated to my wife Chao Deng
and my daughter Fei Tan for their
unconditional love and support!*

Preface

The swarm intelligence optimization method is used to study bio- or non-bio-inspired and population-based iterative algorithms for seeking the intrinsic cooperative mechanism in a swarm. It has recently attracted a great deal of attention from researchers from different fields and diverse domains. Many novel algorithms and their efficient improvements have been proposed continuously. The swarm intelligence optimization method is increasingly becoming one of the hottest and most important paradigms under the big umbrella of evolutionary computation (EC).

Inspired by the fireworks explosion in the night sky, fireworks algorithm, abbreviated as FWA, was proposed by the author in 2010. FWA is a swarm intelligence optimization algorithm, which seems effective at finding a good enough solution to the global optimum of a complex optimization problem. In FWA, as a firework explodes, a shower of sparks is shown in the adjacent area. These sparks explode again and generate other showers of sparks in a smaller area. Gradually, the sparks search the whole solution space in a fine structure and focus on a small region to eventually find (a) good enough solution(s).

To my memory, on the night of the Eve of the 2006 Chinese Lunar Year, as the municipality authorities of Beijing lifted the ban on fireworks during that Spring Festival, people in Beijing set off a large amount of fireworks with sparks of diverse colors which lighted up the dark sky in a variety of beautiful patterns.

While I stared at the glorious scene and colorful patterns for a long time, suddenly an idea came to my mind that the way fireworks explode may be an efficient and effective way or strategy to search for a potential good solution in a vast solution space. Such a search strategy would be different from the established ones in the EC community. Since then I began to study this explosion-like search method.

Like other practical optimization algorithms, FWA is able to fulfill three user requirements given by Storn and Price in 1997. First of all, FWA can process linear, nonlinear, and multi-model test functions. Second, FWA can be parallelized for tackling complicated real-world problems. Third, FWA has good convergence properties and can always find good enough solution(s) for a global minimization problem.

As mentioned above, the motivation for studying FWA is to seek an efficient and effective method with a novel searching mechanism for addressing a variety of complex optimization problems, especially multi-modal optimization problems.

There are three reasons that drove me to publish such a monograph based on our latest research work. The first is that the FWA, with characteristics of explosive ability, randomness, implicit parallelism, instantaneity, and diversity, is a new explosive searching manner to find the global optimum of a complex problem, which considerably enriches and promotes the study of swarm intelligence. The second reason is that the FWA and its variants show their stable convergence and superior performance compared to the standard particle swarm optimization (SPSO) and the latest version of SPSO (SPSO2011) in terms of extensive experiments on the 28 benchmark functions at IEEE CEC2013. The third reason is that the FWA is pretty suitable for multiple-modal optimization functions which will find a wide range of applications in the real world. In this regard, a number of practical applications (including, NMF solving, Spam detection and filtering, JSP, to name a few) based on the FWAs are listed in application chapters of the monograph. Therefore, the FWA provides a brand new way to search the global optimum of complex optimization problems. The current FWA and its applications prove that it can be used to solve many complex optimization problems effectively. Furthermore, the FWA can be also parallelized and is thus suitable to deal with big data problems. Whether for theoretical or applied research, the FWA is worth researching on and can bring great scientific and economic benefits.

This book is the first monograph focused on FWA based on a number of academic papers published primarily by the author and his guided students and team members, and is intended to systematically and completely summarize the most important research work on FWA till date, including FWA's basic principle and implementation, its modeling and theoretical analysis, its most important improvements, and several successful applications.

I hope this book would shape the research on FWA appropriately and can show a complete picture of FWA for interested readers and newcomers who may find many algorithms in the book that can be directly used in their projects on hand without any modification; furthermore, some algorithms can also be viewed as a start point for some active researchers to work on.

This book is primarily intended for researchers, engineers, graduates, and senior undergraduates with interests in novel swarm intelligence algorithms such as fireworks algorithms and its applications. The structure of the contents of this book is organized in a manner from bottom to top or from simple to complex. In order to understand the contents of this book, the readers must have the fundamental skills of digital iterative computing, artificial intelligence, computational intelligence as well as pattern recognition.

In addition, the author presents many newly proposed FWAs in didactic approach with detailed materials and shows their excellent performance using a number of complete experiments and comparisons with state-of-the-art swarm-based algorithms. Furthermore, a collection of resources, source codes, and references of FWA is also provided in the book or accompanied with some webpages that are available and ready for readers to download freely in an easy-to-use manner at http://www.cil.pku.edu.cn/research/fwa/resource.html.

Specifically, this monograph is organized into four parts for easy reference.

Part I is the fundamental and basic theory, which is consisted of three chapters, forming the most substantial theoretical analysis part of this book, including introduction, basic principle, theories and implementation of FWA, modeling and theoretical analysis of FWA, as well as studying the effects of different types of random number generators on the performance of FWA.

Part II is FWA variants, which groups together the most important FWA variants so far, consisting of seven chapters each of which describes one kind of recent significant improvements in FWA. Since the invention of the Fireworks Algorithm (FWA) in 2010, it has attracted a lot of attention in the community of swarm intelligence (SI). All the improvement work on FWA can be classified into two aspects, one, based on working on the limitations of FWA, and the other, its hybridization with other algorithms.

In Part III on advanced topics on FWA, four chapters are included to present some advanced topics in the research on FWA recently, including FWA for multi-objective optimization (MOO), discrete FWA (DFWA) for combinatorial optimization, and GPU-based FWA for parallel implementation of FWA.

Part IV shows how fireworks algorithms can be applied to various applications in different areas. These applications include traditional pattern recognition problems (nonnegative matrix factorization, document clustering, spam detection, and image recognition), complex model estimation problem (seismic inversion), and emerging swarm robotics searching problem. These applications sit in areas that differ greatly from each other and have different requirements for optimization algorithms. The fireworks algorithm can solve these problems successfully, which illustrates that the algorithm can be adapted to different requirements in real-life applications. These applications are reported for instances that might bring insights into more and more real-world applications of FWA in the future.

Due to my limited specialty knowledge and capability, a few errors, typos, inappropriateness, and inadequacy are present in the book, for which critical reviews, constructive comments, and valuable suggestions from interested readers and active researchers are warmly welcome. All comments and suggestions should be sent to ytan(AT)pku.edu.cn for justifying whether or not they will be adopted in a future revision according to their validations and correctness. Finally, I would here like to deliver my faithful and heartfelt thanks to all who gave and will give me help in improving the quality of this book in advance.

Beijing, China   Ying Tan
October 2014

Acknowledgments

I would also like to thank my past and present students who gave me strong assistance in research for such an innovative idea and amazing work.

I am grateful to my students who took part in the research work on FWA under my guidance at Computational Intelligence Laboratory at Peking University (CIL@PKU) (http://www.cil.pku.edu.cn). Almost the whole content of this book is from the research works achieved by, or academic papers published by myself and my supervised postdoctoral fellows, PhD and Master's students, who are Dr. Yuanchun Zhu, Dr. Andreas Janecek, Dr. Jianhua Liu, Dr. Zhongyang Zheng, Mr. Zhongmin Xiao, Mr. You Zhou, Miss Wenrui He; my PhD candidates who include Shaoqiu Zheng, Ke Ding, Chao Yu, Junzhi Li, Guyue Mi and Weiwei Hu; as well as my master's graduates Xiang Yang, Lang Liu, Yang Gao, and Xiaolin Zhang. I would like to deliver my special thanks to all of them. Without their hard work and unremitting efforts, it would be impossible for FWA research to be considerably focused with the current prospects of today.

In addition, in the process of preparing the manuscript for this volume, the following persons gave me efficient help and support: Shaoqiu Zheng, Chao Yu, Ke Ding, Zhongyang Zheng, Junzhi Li, Weidi Xu, Lang Liu, Xiaolin Zhang, Guyue Mi, and Weiwei Hu. It is due to all their cooperative efforts and hard work that the FWA research works scattered in a variety of academic journal papers and international conference papers were well organized in such a tight form. I would like to give my special thanks to them.

I am graceful to Dr. Pei Yan for his outstanding work on accelerating FWA by approximating fitness landscape, as my visiting student from Kyushu University, I here want to thank his supervisor Prof. Hideyuki Takagi from Kyushu University for his helpful suggestions.

Next, I also want to thank Prof. Yujun Zheng and his team from Zhenjiang University of Science and Technology for unselfishly providing and granting me use of their latest research works on FWA required for the completion of this book.

My special thanks go to Prof. Xingui He. I am deeply indebted to him for his encouragement and support over the years and also for this book. His comments on the general organization of the book greatly improved it in both content and form.

Thanks to Prof. Yuhui Shi, my good friend, for his efficient help and effort in advertising the FWA as well as for his invaluable feedback provided after reading parts of the book.

I owe my gratitude to all colleagues and students who have collaborated directly or indirectly in collaboration with research on FWA and in writing this monograph.

While working on this book, I was supported by the Natural Science Foundation of China (NSFC) under grant no. 61375119, 61170057, 60875080, 60673020, 61110306107 and 61010306001. This work was also partially supported by the National Key Basic Research Development Plan (973) Project of China with grant no. 2015CB352302. I thank the Natural Science Foundation of China (NSFC) and the Ministry of Science and Technology of China greatly for its generous financial support.

I want to thank Dr. Chang, Celine, a senior editor from Springer publisher, for her kind coordination and help in reviewing the manuscript of this book. I am also grateful to Miss Jane Li, an editorial assistant of Springer publisher, for her assistance in editing the manuscript and formatting this book. I also want to thank Springer International publishing for their commitment to publish and stimulate innovative ideas and for giving me the opportunity to publish this book so quickly.

Anyway, I want to thank honestly everyone who give me help and support in any form at any time.

This book is dedicated to my wife Dr. Chao Deng and to my daughter Fei Tan, for their unconditional love; without their unselfish and uninterrupted salient support and encouragement, it would be impossible to make this book a reality.

Contents

About the Author

**Dr. Ying Tan** is a full professor and Ph.D. advisor at School of Electronics Engineering and Computer Science of Peking University, and director of Computational Intelligence Laboratory at Peking University (CIL@PKU). He received his BEng from the Electronic Engineering Institute, MSc from Xidian University, and PhD from Southeast University, in 1985, 1988, and 1997, respectively. From 1997, he was a postdoctoral fellow, then an associate professor at University of Science and Technology of China (USTC), served as director of Institute of Intelligent Information Science, and a full professor since 2000. He worked with the Chinese University of Hong Kong (CUHK) in 1999 and 2004–2005. He was an electee of 100 talent programs of the Chinese Academy of Science (CAS) in 2005.

He serves as the Editor-in-Chief of International Journal of Computational Intelligence and Pattern Recognition (IJCIPR), an Associate Editor of IEEE Transactions on Cybernetics (Cyb), IEEE Transactions on Neural Networks and Learning Systems (TNNLS), International Journal of Artificial Intelligence (IJAI), International Journal of Swarm Intelligence Research (IJSIR), International Journal of Intelligent Information Processing (IJIIP), IES Journal B, Intelligent Devices and Systems, and Advisory Board of International Journal on Knowledge Based Intelligent Engineering (KES), and The Editorial Board of Journal of Computer Science and Systems Biology (JCSB), Journal of Applied Mathematics (JAM), Applied Mathematical and Computational Sciences (AMCOS), Immune Computing (ICJ), Defense Technology (DT), CAAI Transactions on Intelligent Systems. He also served as an Editor of Springer Lecture Notes on Computer Science (LNCS) for 15 volumes, and Guest Editor of several referred Journals, including Information Science, Softcomputing, Neurocomputing, International Journal of Artificial Intelligence (IJAI), International Journal of Swarm Intelligence Research (IJSIR), BB, CJ, IEEE/ACM Transactions on Computational Biology and

Bioinformatics (IEEE/ACM TCBB), etc. He is a member of Emergent Technologies Technical Committee (ETTC), Computational Intelligence Society of IEEE since 2010. He is a senior member of IEEE and ACM and a senior member of the CIE. He is the general chair of joint conference ICSI'2015 in conjunction with 2nd BRICS CCI'2015, and He was the general chair of the International Conference on Swarm Intelligence (ICSI'2010-2014) and one of joint general chairs of 1st BRICS CCI'2013, program committee co-chair of IEEE CEC'2014 at IEEE WCCI'2014, 2013 IEEE International Conference on Intelligent Science and Technology, 2012 International Conference on Advanced Computational Intelligence (ICACI'2012), 2008 International Symposium on Neural Networks (ISNN'2008) and so on.

His research interests include computational intelligence, swarm intelligence, data mining, machine learning, pattern recognition, intelligent information processing for information security, fireworks algorithm, etc. He has published more than 280 papers in refereed journals and conferences in these areas, and authored/co-authored 6 books and 10+ chapters in book, and received 3 invention patents.

For details, please visit the official website of Computational Intelligence Laboratory at Peking University (CIL@PKU) at http://www.cil.pku.edu.cn.

Abbreviations

ABC	Artificial bee colony
ACO	Ant colony optimization
AFWA	Adaptive fireworks algorithm
ALSPG	ALS using projected gradient
AR-Mutation	Attract-repulse mutation
BBO	Biogeography-based optimization
BSO	Brain storm optimization
CCPSO	Co-evolutionary particle swarm optimization algorithm
CEC	IEEE congress on evolutionary computation
CF	Core firework
CFWA	Cultural firework algorithm
CI	Computational intelligence
CMR	Combined multiple recursive generator
CNMOIA	Constrained nonlinear multi-objective optimization immune algorithm
CoFWA	Cooperative firework algorithm
CPSO	Clonal particle swarm optimization
CPU	Central processing unit
CUDA	Compute unified device architecture
DE	Differential evolution
DFWA	Discrete fireworks algorithm
DPSO	Discrete particle swarm optimization
DTEA	Dominating-tree based multi-objective evolutionary algorithm
dynFWA	Fireworks algorithm with dynamic search
EC	Evolutionary computation
EFWA	Enhanced fireworks algorithm
FIR	Finite impulse response
FSS	Fish school search
FWA	Fireworks algorithm
FWA-DE	Fireworks algorithm with differential evolution
FWA-DM	Fireworks algorithm with differential mutation

GA	Genetic algorithm
GES	Group explosion strategy
GFSR	Generalized feedback shift register
GPGPU	General-purpose computation on GPU
GPU	Graphics processing unit
GPU-FWA	GPU Fireworks Algorithm
HPC	High performance computing
IDF	Inverse document frequency
IFWABS	Improved FWA with best fitness selection and random mutation
IFWAFS	Improved FWA with the fitness selection using the roulette and random mutation
IIR	Infinite impulse response
LC	Local concentration
LCG	Linear congruential generator
LLE	Locally linear embedding
MEAC	Minimal explosion amplitude check
MKL	Math kernel library
MOEA	Multi-objective evolutionary algorithms
MOEPSO	Multi-objective endocrine particle swarm optimization
MOFWA	Multi-objective FWA
MOO	Multi-objective optimization
MOP	Multi-objective optimization problem
MOPSO	Multi-objective particle swarm optimization
MRG	Multiple recursive generator
MT	Mersenne twister
MU	Multiplicative update
NFL	No free lunch
NMF	Non-negative matrix factorization
NNDSVD	Nonnegative double singular value decomposition
NSGA	Non-dominated sorting genetic algorithm
NSPSO	Non-dominated sorting particle swarm optimizer
PAES	Pareto archived evolution strategy
PCA	Principal component analysis
PDE	Pareto differential evolution algorithm
POEM	Probabilistically oriented explosion mechanism
PRNGs	Pseudorandom random number generators
PSO	Particle swarm optimization
QAP	Quadratic assignment problem
QRNGs	Quasi random number generators
RDPSO	Robotic darwinian particle swarm optimization
RNGs	Random number generators
RPSO	Robotic particle swarm optimization
SI	Swarm intelligence
SIA	Swarm intelligence algorithm
SIO	Swarm intelligence optimization

S-MOFWA	S-metric based multi-objective fireworks algorithm
SPEA	Strength pareto evolutionary algorithm
SPSO	Standard particle swarm intelligence
SVD	Singular value decomposition
SVM	Support vector machine
TF	Term frequency
TRNGs	True random number generators
TSP	Traveling salesman problem
UDM	Unified distance measure
VFR	Variable-rate fertilization

Symbols

$f()$	Optimization function f
D	Dimension
x_i	Variable
x_{ij}	The jth dimension value of variable x_i
x_{UB}	The upper bound of variable x
x_{LB}	The lower bound of variable x
$x_{UB,k}$	The kth dimension upper bound value of variable x
$x_{LB,k}$	The kth dimension lower bound value of variable x
$evals_{max}$	The maximum evaluation times
$maxIter$	The maximum iteration number
N	Fireworks number
m	Explosion sparks number
$\hat{m}$	Gaussian mutation sparks number
X_i	The location of the ith firework in the fireworks swarm
X_{ik}	The kth dimension value of firework X_i
$f(X_i)$	The function fitness of firework X_i
A_i	The explosion amplitude of the ith firework
$\hat{A}$	Coefficient in the calculation of explosion amplitudes
S_i	The explosion sparks number of the ith firework
$U(a,b)$	A random value generated between (a,b) with mean distribution
$\mathcal{N}(\mu,\sigma)$	A random value generated with Gaussian distribution-with mean μ and variance σ
g	$\mathcal{N}(1,1)$
e	$\mathcal{N}(0,1)$
K	Candidates set, in general, it contains fireworks, explosion sparks and Gaussian mutation sparks
$d(.,.)$	Distance measure

ε,	Machine smallest value
x_j	The position of jth individual in candidates set
$x_{j,k}$	The kth dimension value of x_j
$R(x_i)$	The sum distances between individual x_i and the rest individuals in candidates K
$p(x_i)$	The probability that individual x_i is selected as firework
s^*	The best spark generated by firework X
X_B	The best firework
X_{elite}	The generated elite solution
$\hat{x}_b$	The explosion sparks with minimal fitness among all the explosion sparks
CF	The firework with best function fitness
$nonCF$	The fireworks except for CF
X_{CF}	The location of CF
$X_{CF,k}$	The kth dimension value of X_{CF}
A_{CF}	The explosion amplitude of CF
C_a	Amplification factor for explosion amplitude of CF
C_r	Reduction factor for explosion amplitude of CF
A_{init}	The initial explosion amplitude
A_{final}	The final explosion amplitude
$\|\cdot\|_F$	F norm
$\|\cdot\|_\infty$	Infinite norm
SI	Shift index
SV	Shift value
round(.)	Round operation
E_t	The current evaluation times
$o(.)$	Low order
$\%$	Modular arithmetic operation
C'	Coverage measure
A	Objective matrix
W	Reduced rank nonnegative factors matrix
H	Reduced rank nonnegative factors matrix
D	Distance matrix
w_i^r	The ith row of W
h_j^c	The jth column of H
a_i^r	The ith row of A
a_j^c	The ith column of A
L	The length of the root
p_a	The probability that a solution is accepted
θ	Parameter which controls the p_a
Ω_S	The super volume of region S

F_{Max}	Maximum fitness
r_t	Radius of the target
R_i	The ith robot
$P_i(t)$	Robot i's position
$V_i(t)$	Robot i's velocity
$G_i(t)$	Grouping component
$H_i(t)$	History component

List of Figures

List of Tables

Part I
Fundamentals and Basic Theory

This part describes the principle, fundamentals, and basic theory of fireworks algorithm (FWA). The first chapter introduces the study of the origins and motivations, research areas, problems to be solved, features, as well as future research directions of fireworks algorithms. The second chapter gives the details of fireworks algorithms, including FWA's components and framework, characteristics, and comparisons with other SI algorithms. The third chapter gives a stochastic model of fireworks algorithm, proves its global convergence, discusses and analyzes its time complexity, and finally studies the effects of different types of random number generators on the performance of FWA.

Chapter 1
Introduction

This chapter presents the motivation of when, why, and how the fireworks algorithm (FWA), as a novel swarm intelligence optimization algorithm, came out. After a concise review on swarm intelligence domain, a brief introduction to FWA is presented with primary focuses on four aspects of theoretical analysis, algorithm study, problem solving, and applications. The characteristics and advantages of FWA are also described. Finally, overviews of FWA research are detailed with completed reference citations.

1.1 Motivations

During my childhood in Sichuan, in southwest China, it was a great pleasure for me to set off fireworks, or firecrackers, or sparkler, or Yan Hua in Chinese, with my friends during the Spring Festival, the most important traditional festival in China. Now and then, we competed for the title of master who could ignite fireworks that could fly the highest into the sky and make the loudest noise. Though the time has long passed, the splendid sparks showering the night sky is imprinted in my memory.

By the Spring Festival of 2006, I had worked at Peking University for one year. During that time, I devoted to the in-depth study of evolutionary computation (EC) in computational intelligence. Therefore, I tried to relate any novel phenomena I met to EC during that time. Just in that year, Beijing authorities relaxed fireworks ban to restriction during the Spring Festival. After many years' ban on firework celebrations, citizens of the Capital were eager for the arrival of the Chinese Lunar New Year's Eve and looked forward to a livelier and jollier Spring Festival.

On this year's Eve Day, people set off a large amount of fireworks as if it could relieve the whole year's stress. The sparks of diverse colors lighted up the dark sky and made various beautiful patterns. The glorious scene, reminiscent of childhood memories, brought much joy and comfort to me.

An idea suddenly came to my mind that the way fireworks explode may be an efficient strategy for searching the solution space. Such a search strategy would be

© Springer-Verlag Berlin Heidelberg 2015

Y. Tan, *Fireworks Algorithm*, DOI 10.1007/978-3-662-46353-6_1

different from the established ones. So I began to study this explosion-like search method, and named it Fireworks Algorithm, i.e., FWA, for short.

Although the name "Fireworks Algorithm" sounds intuitive and compact, since this name does not directly relate to optimization problems it is good at, some other names are also used by other researchers, such as "Fireworks Optimization," "Fireworks Explosion Algorithm," "Fireworks Explosion Optimization Algorithm," "Fireworks Explosion Search Algorithm," "Explosion Search Method," and so on. In order to avoid any possible confusion, only the original name "Fireworks Algorithm," abbreviated by FWA, is used hereafter in this book.

The motivation of studying FWA is to seek a simple and efficient method dealing with complex optimization problems, especially multimodal optimization problems.

In the summer semester of 2006, Yuanchun Zhu from Jilin University was admitted by Peking University as a PhD student under my supervision and I arranged for him to do his graduate project in my laboratory in the early months of the following year. We discussed the idea about FWA and set the project content as the study and implementation of Fireworks Algorithm. Together, we started a thorough study of FWA.

After half a year's exploration and study, FWA's main components and overall framework were completely established. Finally, the conventional FWA was designed based on the explosive operator. The pioneering work was finished by May 2007.

Unfortunately, in the following two years, due to my hosting of a project supported by the National High-tech Research and Development Program (863 Program), the FWA-related researches were suspended. Till 2010, the original work on FWA was reported at the First International Conference of Swarm Intelligence (ICSI' 2010), entitled "Fireworks Algorithm for Optimization." After then, FWA has increasingly gained more attention in the field of swarm intelligence.

1.2 Brief Introduction to Swarm Intelligence

Swarm intelligence is an active branch of Evolutionary Computation which is one of the most important research topics in Computational Intelligence (CI) Community. In the past several decades, fruitful achievements have been made on the research of Computational Intelligence, such as artificial neural networks [1–6], fuzzy logic and systems [7], evolutionary computation [8–11], chaos computation [12], simulated annealing [13], tabu search [14], and hybrid strategies, to name a few. All these methods simulate and give an insight into natural phenomena or biological processes.

Swarm intelligence (SI) is regarded as the collective behavior of decentralized, self-organized, and populated systems. A typical swarm intelligence system consists of a population of simple agents that can communicate (either directly or indirectly) locally with each other by acting on their local environment. Though the agents in a swarm follow very simple rules, the interactions between such agents can lead to the emergence of very complicated global behavior, far beyond the capability of individual agents. Examples in natural systems of swarm intelligence include bird flocking, ant foraging, and fish schooling.

Inspired by such behavior of swarms, a class of algorithms was proposed for tackling optimization problems, under the title of Swarm Intelligence Algorithm. In SI Algorithm, a swarm is made up of multiple artificial agents. The agents can exchange heuristic information in the form of local interaction directly or indirectly (via environment). Such interaction, in addition to certain stochastic elements, generates the behavior of adaptive search, and finally leads to global optimization.

The most respected and popular SI algorithms are particle swarm optimization (PSO), which is inspired by the social behavior of bird flocking or fish schooling, and ant colony optimization (ACO) which simulates the foraging behavior of ant colony. PSO is widely used in real-parameter optimization while ACO has been successfully applied to solve combinatorial optimization problems. The most well-known problems are the traveling salesman problem (TSP) and quadratic assignment problem (QAP).

Swarm intelligence algorithms can fall into two major categories, bio-inspired and non-bio-inspired [15]. The former includes ant colony optimization (ACO) [16], particle swarm optimization (PSO) [17], fish schooling search (FSS) [18], firefly algorithm-I [19], firefly algorithm-II [20], buccock algorithm [21], bat algorithm [22], artificial bee algorithm (ABC) [23–25], bacterial foraging optimization (BFO) [26], and so forth. Non-bio-inspired algorithms consist of fireworks algorithm (FWA) [27], water drops algorithm [28], brain storm optimization (BSO) [29], and magnetic optimization algorithms [30], to name a few.

Nowadays, research efforts on swarm intelligence are mainly devoted to algorithm design, problem solving, and applications. Swarm intelligence algorithms are more and more applied to large-scale problems to tackle the issues of the curse of dimension and big data. Hybrid algorithms and variants are actively proposed.

The cooperative mechanism in swarm intelligence algorithms has the capability of breaking down the curse of no free lunch (NFL) [31] which means there exist efficient intelligent algorithms that can also to be found by fully utilizing the cooperation and co-evolution among individuals in a swarm [32]. This will stimulate more and more researchers to enter the field of swarm intelligence and develop more and more efficient algorithms to solve many real-world problems for us.

1.3 Brief Introduction to FWA

Just like conventional swarm intelligence algorithms, FWA is an iterative algorithm made up of four key components or building blocks, i.e., explosive operator, mutation operation, mapping rule, and selection strategy. Explosive operator can be divided further into explosion strength, explosion amplitude, shift mutation, and other operators. Gaussian mutation is the most widely used mutation operator. Mirror mapping rule and stochastic mapping rule are two popular mapping rules for FWA. As for selection strategy, there are distance-based selection and stochastic selection.

Specifically, workflow of FWA can be stated as follows. First, N fireworks are generated randomly as the initial swarm. Then, every firework conducts explosion operation and mutation operation, and mapping rule is triggered if necessary. Finally, N fireworks are selected from all the fireworks and generated sparks according to the selection strategy. The iteration continues until particular termination criterion is satisfied. As the iteration proceeds, better optimization results can be achieved eventually.

The study of FWA mainly focuses on four aspects: theoretical analysis, algorithm study, problem solving, and applications.

1. Theoretical Analysis
   Theoretical analysis involves FWA's mechanism, convergence quality, trajectory, and parameters. It helps the design of new algorithms and provides helpful insight for improving established algorithms.
2. Algorithm Study
   By analyzing and adjusting FWA's components, people try to improve FWA's performance (convergence, solution quality, and time efficiency), and come up with improved FWA variants. In the meantime, it will be helpful to integrate FWA with other methods, ending up with efficient hybrid algorithms.
3. Problems to be Solved
   - Single-Objective Optimization Problems—SOO: it is concerned with mathematical optimization problems involving only one objective function.
   - Constrained Single-Objective Optimization Problems—CSOO: it is the SOO problem with constraint(s).
   - Multiple-Objective Optimization Problems—MOO: it is to make decision under multiple criteria, which is concerned with mathematical optimization problems involving more than one objective function to be optimized simultaneously.
   - Constrained Multiple-Objective Optimization Problems—CMOO: it is the MOO problem with constraint(s).
   - Many-Objective Optimization Problems—ManyOO: it is concerned with mathematical optimization problems involving more than three objective functions to be optimized simultaneously.
   - Combinatorial Optimization Problems—CO: it is a topic that consists of finding an optimal object from a finite set of objects
   - Dynamic Optimization Problems—DOP: it is a time-dependent optimization problem, i.e., either the objective function and constraints or the associated parameters or both vary with time.
   - Other optimization problems
4. Applications
   FWA can be directly applied to many different real-world problems from which an optimization task can be drawn.

1.4 Characteristics and Advantages of FWA

Some of FWA's characteristics and advantages are summarized as follows

1. Explosiveness: Each iteration, every firework explodes and generates a number of sparks within the radiation range. Then, selection procedure is triggered to choose a fixed number of fireworks for the next iteration.
2. Instantaneity: The sparks are instantaneous, which means that sparks not selected will vanish.
3. Simplicity: every agent has only limited capability, thus the whole algorithm is simple to be implemented.
4. Locality: For a particular firework, its explosion amplitude is smaller than the feasible bound, so it can only exploit locally in the search space.
5. Emergent properties: Though following simple rules, by cooperation and competition, the swarm, as a whole, shows complicated behaviors, far more complex than every single firework, i.e., intelligence emerges.
6. Parallelism: There is no central control mechanism among fireworks, thus the fireworks are highly independent, much suitable for parallelization.
7. Diversity: The diversity is threefold. First, fireworks are diverse. Proper selection guarantees that the selected fireworks distribute in diverse places in the search space. Second, there are diverse explosion strengths and amplitudes. According to their particular fitness, different fireworks will have various explosion strengths and amplitude. Besides, there are multiple mutation operators in the FWA. Two of the well-studied mutations are displacement mutation and Gaussian mutation. Displacement mutation is subjected to the firework's fitness while Gaussian mutation firework is independent. The two mutations of very different flavors guarantee the diversity of the mutation procedure.
8. Robustness: As the fireworks are highly independent, the swarm behaves will not degrade too much in the presence of a few individuals' failure.
9. Flexibility: The problem does not need to have an explicit expression to be optimized by FWA. Thus FWA can address a very broad range of problems for which conventional optimization procedures unfortunately fail.

1.5 Overviews of FWA Research

Since the inception of FWA introduced by the seminal paper entitled "Fireworks algorithm for optimization [27]" by the author et al., FWA has drawn much attention from both academic and industrial fields. Much effort has been devoted to analyzing the conventional FWA and many improvements have been proposed to compensate its shortcomings. Several hybrid algorithms have also been introduced. These research efforts lead to a huge leap in terms of FWA's performance. Many applications are reported, where FWA is successfully utilized to tackle diverse problems.

A brief summary of these achievements on the studies of FWAs is listed as follows.

1. Theories
   Liu et al. [33] analyzed FWA's convergence property theoretically. The authors pointed out that FWA is an absorbed Markov process. Besides, this book also studied the impact of random number generators on the performance of FWA, for the first time.
2. Algorithms
   Ding et al. [34] proposed a GPU-based parallel FWA, dubbed GPU-FWA. GPU-FWA achieved 200+ speedup compared to the CPU implementation. GPU-FWA made some major modifications on FWA to minimize the inter-firework communication and thus maximize the utilization of GPU.

   Pei et al. [35] proposed a method to accelerate FWA's convergence by appropriating the objective function. The experiments show that optimal performance is achieved when a quadratic model is utilized with random selection strategy. The proposed method outperforms the conventional FWA significantly.

   Zheng S et al. [36] observed FWA extensively and proposed several improvements accordingly. The authors proposed Enhanced FWA (EFWA), which modifies all the key components of the conventional FWA.

   In [37] and [38], Li et al. and Zheng S et al. devoted to studying the adaptation in FWA's explosion amplitude, as a result, dynamic search FWA and adaptive FWA were proposed, respectively.

   Besides, some researchers studied the combination of FWA with other SI algorithms and proposed several hybrid FWAs. In [39] and [40], Zheng Y.J. and Yu et al. devoted to the hybrid of FWA and DE. In [39], the proposed FWA-DE shows advantage over both FWA and DE on several benchmark test functions. In [41], Gao et al. combined FWA with cultural algorithm and then applied it to the optimization of digital filter design. Gao et al. compared the proposed algorithm with two PSO variants [42, 43]. The experimental results show that the new algorithm achieved the best performance of the three. Zhang et al. proposed BBO FWA, which outperformed BBO and FWA with clear advantages [44].

   Very recently, Dr. James McCaffrey presented "Test Run—Fireworks Algorithm Optimization" on the latest issue of MSDN magazine [45] which is the Microsoft journal for developers. His article describes a relatively new (first published in 2010 [27]) optimization technique called fireworks algorithm optimization (FAO). The technique does not explicitly model or mimic the behavior of fireworks, but when the algorithm is visually displayed, the resulting image resembles the geometry of exploding fireworks. In one word, he thought that the FAO is indeed an interesting heuristic optimization approach.
3. Varieties of Optimization Problems
   FWA has been used to solve single-objective and multi-objective real-valued optimization problems. Zheng Y. et al. proposed a multi-objective FWA (MOFWA) and applied it to the oil crops' fertilization problem [46]. The comparative experiments show that MOFWA could obtain better performance than other popular

swarm intelligence multi-objective algorithms. This book describes, for the first time, the applications of FWA on tackling the well-known combinatorial optimization problem of Traveling Salesman Problem (TSP).

4. Applications

FWA has been applied to diverse real-world problems, for example, equations solving [47], nonnegative matrix factorization (NMF) [48], spam detection [49], distance metric [50], filter design [41], crops fertilization [46], swarm robot [51–53], etc.

Zhen-xin et al. [54] proposed an improved FWA to solve nonlinear equations and system problems. His experiments show that the FWA has advantages over the other algorithms in solving variable-coupling equations and also gave an analysis to disclose why the improved FWA is more effective. Recently, Sujin Bureerat, and Nantiwat Pholdee, in [55], systematically compared and studied the optimizing performance of 24 metaheuristic algorithms for mass minimization of trusses with dynamic constraints, and gave an objective ranking among which FWA is above the average with a promising performance. Mohamed Imran, Kowsalya and Kothari, in [56] and [57], presented a novel integration technique for optimal network reconfiguration and distributed generation (DG) placement in distribution system with an objective of power loss minimization and voltage stability enhancement. They used fireworks algorithm (FWA) to simultaneously reconfigure and allocate optimal DG units in a distribution network. Six different scenarios are considered during DG placement and reconfiguration of network to assess the performance of the proposed technique [56]. Simulations were carried out on well-known IEEE 33- and 69-bus test systems at three different load levels to demonstrate well the performance and effectiveness of the proposed FWA method. Recently, in [58], R. Rajaram et al. presented their latest work of selective harmonic elimination in PWM inverter using fireworks algorithm in which a slightly modified FWA and firefly were used to solve the selective harmonic elimination problem in inverter output waveforms. It turns out from experiments that the FWA works faster than modified firefly algorithm as well as other popular algorithms of ant colony, PSO, and GA. Moreover, specific comparisons with other methods in the literature like GA [59], Ant colony [60], and Bees intelligence [61] suggested that Fireworks algorithm worked excellently and Firefly was also good for this application. Noora Hanni Abdulmajeed and Masri Ayob, in [62], applied the FWA to the capacitated vehicle routing problem (CVPR), which is an important problem in the industry sector with many applications in the field of transportation, distribution, and logistics. Based on the tests on 14 instances of Christofides benchmarks, the FWA is very competitive in terms of the quality of the solutions compared to other SI methods like OCGA, AGES, and AHMH [62]. The application of FWA on document clustering is also presented in this book for the first time.

For a detailed review of FWAs, readers can refer to [63, 64].

1.6 Overview of the Book

Briefly, this book is organized into four parts. In a sequel, Part A grouped together three chapters, forming the most substantial theoretical analysis part of this book, including introduction, basic principle and implementation of FWA, and modeling and theoretical analysis of FWA. Part B covers almost the most important FWA variants so far, consisting of seven chapters each of which describes one kind of recent significant improvements of FWA. In Part C, four chapters are included to present some advanced topics in the research of FWA, including two multi-objective optimization (MOO) schemes, discrete FWA (DFWA) for combinatorial optimization, and GPU-based FWA for parallel implementation of FWA. In Part D, several successful applications of FWA on nonnegative matrix factorization (NMF), text clustering, pattern recognition, and seismic inversion problem are reported for instances which might give insights on more and more real-world applications of FWA in the future, and finally, a conclusion and postscript are given. Specifically, let us briefly describe the abstract of each chapter for every part in this book as follows.

In part A of fundamental and basic theory, it describes the basic principle and theories of fireworks algorithm (FWA). The first chapter introduces the study of the origins and motivations, research areas, the problem content, and features, as well as the future research directions, of FWA. Chapter 2 describes the details of fireworks algorithm, including FWA's components and framework, characteristics, implementation, experimentation, and detailed comparisons with other SI algorithms such as particle swarm optimization (PSO), genetic algorithm (GA), etc. Chapter 3 presents modeling and theoretical analysis of FWA. It gives a stochastic model of fireworks algorithm, proves its global convergence, discusses, and analyzes its time complexity, and, finally, studies the effects of different types of random number generators on the performance of FWA.

In part B of FWA variants, it has the most important improvements of FWA till now. Since the invention of fireworks algorithm (FWA) in 2010, it has attracted a lot of attention in the community of swarm intelligence (SI) [27, 63, 64]. The improvement work on FWA is able to be classified into two categories. One is based on improving the limitations of FWA, and the other is hybridized with other SI algorithms.

Chapter 4 describes an empirical study on the influence of approximation approaches on accelerating the fireworks algorithm search by elite strategy. It uses three data sampling methods to approximate fitness landscape, i.e., the best fitness sampling method, the sampling distance near the best fitness individual sampling method, and the random sampling method. For each approximation method, this book conducts a series of combinative evaluations with the different sampling methods and sampling numbers for accelerating fireworks algorithm. The experimental evaluations on benchmark functions show that this elite strategy can enhance the fireworks algorithm search capability effectively. This chapter also analyzes and discusses the related issues on the influence of approximation model, sampling method, and sampling number on the fireworks algorithm acceleration performance.

In Chap. 5, FWA with controlling exploration and exploitation is introduced by providing a kind of new method to calculate the number of explosion sparks and amplitude of firework explosion. By designing a transfer function, the rank number of firework is mapped to scale of the calculation of scope and spark number of firework explosion. A parameter is used to dynamically control the exploration and exploitation of FWA with iteration going on.

In Chap. 6, the enhanced fireworks algorithm (EFWA) is presented, which gives the comprehensive analysis of FWA and points out five limitations in FWA which leads to the related improvement work on those limitations. The new operators which overcome the limitations are presented, which include the minimal explosion amplitude check strategy, the new mapping operator, the new operator for generating explosion sparks, the new Gaussian mutation operator, and selection operator. Experimental results suggest that the newly proposed operators are effective to deal with the limitation pointed in this chapter.

In Chap. 7, a dynamic search firework algorithm (dynFWA) is presented. The dynFWA is an improvement of the recently developed enhanced fireworks algorithm (EFWA) which uses a dynamic explosion amplitude for the core firework (CF), i.e., the firework at the currently best position. This dynamic explosion amplitude depends on the quality of the current local search around the CF. The main task for the CF is to perform a local search, while the responsibility of all other fireworks is to maintain the global search ability. Additionally, this chapter analyzes the possibility to remove the rather time-consuming Gaussian sparks operator of EFWA. Extensive experimental evaluations show that the proposed dynFWA algorithm significantly improves the results of EFWA and also reduces the runtime by more than 20%.

The explosion amplitude is a key factor influencing the performance of the Fireworks Algorithm, which needs to be controlled precisely. In Chap. 8, a new algorithm called adaptive fireworks algorithm, which replaces the explosion amplitude operator in EFWA with an adaptive method, is proposed. The distance of the best firework and a certain individual subjecting to some conditions is employed as the amplitude of the explosion. Chapter 8 analyzes the property of the adaptive explosion amplitude and come to the conclusion that the adaptive explosion amplitude for explosion is a theoretically promising operator. According to the experimental results on CEC13's 28 benchmark functions, the performance is greatly improved.

In previous FWA's studies so far, researchers ignored the cooperation and interaction between the individual fireworks in the swarm, which are the most important core for any swarm intelligence algorithm. By incorporating a probabilistically oriented explosion mechanism (POEM) into the conventional FWA, a novel cooperative fireworks algorithm (CoFWA, for short) is proposed in Chap. 9 to enhance the interactions among the individual fireworks in the swarm. In CoFWA, the POEM mechanisms of sparks generation and fireworks selection are well designed to strengthen the cooperative capability of the individual fireworks in the CoFWA. Experiments show that the CoFWA significantly outperforms two most recent variants of FWA (i.e., EFWA and dynFWA) and SPSO2011 and shows a competitive performance against the state-of-the-art swarm intelligence algorithms.

Fireworks algorithm is also suitable for combining with other EC algorithms for producing new efficient hybrid algorithms. In Chap. 10, the combinations between FWA and other SI algorithms are given out. Focus is emphasized on hybrid fireworks algorithms, including fireworks algorithm with differential mutation (FWA-DM), hybrid fireworks optimization method with differential evolution operators (FWA-DE), culture fireworks algorithm (CFWA), and hybrid biogeography-based optimization and fireworks algorithm (BBO FWA).

In Part C of advanced topics of FWA, four chapters are to present the advanced topics of FWA, including multi-objective optimization (MOO), discrete FWA (DFWA) for combinatorial optimization, and GPU-based FWA for parallel implementation of FWA.

As we know, FWA was first proposed for single-objective optimization problems and has been widely studied based on the CPU platform. Multi-objective optimization problems are universal and much more complicated than their single-objective counterparts. Chapter 11 will introduce the applications of FWA on multi-objective problems. An efficient MOFWA algorithm is proposed for oil crop VRF problems, which uses a problem-specific strategy for generating the initial population, uses the concept of Pareto dominance for individual evaluation and selection, and combines the DE operators to increase the information sharing and thus diversify the search. The MOFWA algorithm has been successfully applied to a number of VRF problems and demonstrated its efficiency and effectiveness.

Furthermore, in Chap. 12, with the help of the well-known S-metric, a kind of hyper-volume indicator, an S-metric multi-objective fireworks algorithm (MOFWA), is proposed for efficiently solving multi-objective optimization problems. The S-metric is a frequently used quality measure for solution sets' comparison in evolutionary multi-objective optimization algorithms (EMOAs), which is also used to evaluate the contribution of a single solution among the solution set. Traditional multi-objective optimization algorithms usually perform a $(\mu + 1)$ strategy and update the external archive one by one, while the proposed S-MOFWA performs a $(\mu + \mu)$ strategy, thus converging faster to a set of Pareto solutions. The comparison results with NSGA-II, SPEA2, and PESA2 demonstrate the efficiency of the proposed S-MOFWA.

Discrete FWA (DFWA) for combinatory optimization problems is just in its fiddle era, and Chap. 13 presents FWA's application in traveling salesman problem (TSP). The DFWA remains the basic framework of FWA and introduces some major changes in explosion operator, selection strategy, and mutation operator, respectively. In explosion operator, every firework is able to accept a worse solution and generate a spark with lower fitness, which refers to the mechanism of simulated annealing. However, a controlling parameter changes with the feedback of optimization process rather than time. In addition, this version of discrete firework algorithm properly changes its behavior of the local search method to suit in the framework of FWA. As the experimental results indicate, the DFWA is very effective for TSP, and potentially applied to other discrete optimization problems.

GPU is a game-changing force in the domain of high-performance computing (HPC). Thanks to GPU's parallelism and great computational power, swarm intelligence algorithms are able to fully exploit their inherent parallelism. Building swarm intelligence algorithms on the GPU platforms is an increasingly important and popular research topic. In Chap. 14, the implementation of FWA based GPU is presented. In this chapter, a very efficient FWA variant based on GPUs, dubbed GPU-FWA, is introduced. GPU-FWA modifies the original FWA to suit the particular architecture of the GPU. It does not need special complicated data structure, thus making it easy to implement; in the meantime, it can fully exploit the great computing power of GPUs. A mutation mechanism called attract-repulse mutation is introduced to guide the search process. To make the chapter self-contained, a brief introduction of general purpose computing on GPUs (GPGPU) is presented first. Then, this chapter describes GPU-FWA in detail, followed by the empirical comparison of GPU-FWA with conventional FWA as well as PSO.

In Part D, this book verifies how fireworks algorithms can be applied to various applications in different areas. These applications include traditional pattern recognition problems (nonnegative matrix factorization, document clustering, spam detection, and image recognition), complex model estimation problem (seismic inversion), and emerging swarm robotics searching problem. These applications sit in areas which differ greatly than each other and have different requirements for the optimization algorithms. Fireworks algorithm can solve these problems successfully which illustrates that the algorithm has a great adaption to different requirements in real-life applications.

Chapter 15 presents two new optimization strategies for improving the NMF using optimization algorithms based on swarm intelligence. While strategy one uses swarm intelligence algorithms to initialize the factors and prior to the factorization process of NMF, the second strategy aims at iteratively improving the approximation quality of NMF during the first iterations of the factorization. Five different optimization algorithms were used for improving NMF, including particle swarm optimization (PSO), genetic algorithms (GA), fish school search (FSS), differential evolution (DE), and Fireworks Algorithm (FWA).

Chapter 16 describes the applications of fireworks algorithm for dealing with some practical optimization problems such as document clustering, spam detection, image recognition, and seismic inversion problem. The experimental results given herein suggest that FWA is one of the most promising swarm intelligence algorithms in dealing with those real-world problems.

Swarm robotics is an emerging research area combining swarm intelligence and robotics. In Chap. 17, inspired by Fireworks Algorithm, a group explosion strategy (GES) for searching multiple targets is proposed. GES method is applied to the multiple targets searching problem on a self-built simulation platform. The swarm searches and collects targets in the environment without prior knowledge. Several tests are run to evaluate how GES performs in various aspects including stability, robustness, and flexibility. Simulation results demonstrate that GES shows great efficiency when fitness is either adequate or inadequate in the environment. GES also shows good stability in obstructive and large-scale environments.

In Backmatter, a Postscript is given for a few words I want to deliver after the accomplishment of this manuscript.

After this, several appendices are attached by giving the benchmark functions suites, web resources of FWA researches, and lists of figures, tables, and symbols appeared in the book.

Finally, indices of key words are drawn out at the end of this book.

References

1. M.T. Hagan, H.B. Demuth, M.H. Beale et al., *Neural Network Design* (Pws Pub, Boston, 1996)
2. G.C. Ruan, Y. Tan, A three-layer back-propagation neural network for spam detection using artificial immune concentration. Softcomputing **14**, 139–150 (2010)
3. X. Huang, Y. Tan, X.G. He, An intelligent multi-feature statistical approach for discrimination of driving conditions of hybrid electric vehicle. IEEE Trans. Intell. Transp. Syst. **12**(2), 453–456 (2011)
4. Y. Tan, C. Deng, Solving for a quadratic programming with a quadratic constraint based on a neural network frame. Neurocomputing **30**, 117–128 (2000)
5. Y. Tan et al., Neural network design approach of cosine-modulated FIR filter bank and compactly supported wavelets with almost PR property. Signal Process. **69**(1), 29–48 (1998)
6. Y. Tan, Z.K. Liu, On matrix eigendecomposition by neural networks. (Neural Netw. World) International Journal on Neural and Mass-Parallel Computing and Information Systems **8**(3), 337–352 (1998)
7. G.J. Klir, B. Yuan, *Fuzzy Sets and Fuzzy Logic*, vol. 4 (Prentice Hall, NewD Jersey, 1995)
8. A.E. Eiben, J.E. Smith, *Introduction to Evolutionary Computing* (springer, Berlin, 2003)
9. Y. Tan, J. Wang, Nonlinear blind separation using higher-order statistics and a genetic algorithm. IEEE Trans. Evol. Comput. **5**(6), 600–612 (2001)
10. J. Zhang, Y. Tan, L. Ni, C. Xie, Z. Tang, AMT-PSO: an adaptive magnification transformation based particle swarm optimizer. IEICE Trans. Fundam. Electron. Commun. Comput. Sci. **E94-D**(4): 786–797 (2011)
11. Y. Tan, J. Wang, A support vector network with hybrid kernel and minimal Vapnik-Chervonenkis dimension. IEEE Trans. Knowl. Data Eng. **26**(2), 385–395 (2004)
12. H.-O. Peitgen, H. Jrgens, D. Saupe, *Chaos and Fractals: New Frontiers of Science* (Springer, Berlin, 2004)
13. P.J.M. Van Laarhoven, E.H.L. Aarts, *Simulated Annealing* (Springer, Berlin, 1987)
14. F. Glover, M. Laguna, *Tabu Search* (Springer, 1999)
15. Y. Tan, S. Zheng, Research progress on swarm intelligence optimization algorithms. Commun. Chin. Autom. Soc. **34**(3), (2013)
16. M. Dorigo, M. Birattari, T. Stutzle, Ant colony optimization. IEEE Comput. Intell. Mag. **1**(4), 28–39 (2006)
17. J. Kennedy, R. Eberhart et al., Particle swarm optimization, in *Proceedings of IEEE International Conference on Neural Networks*, vol. 4(2) (Perth, Australia, 1995), pp. 1942–1948
18. C.J.A Bastos Filho, F.B. de Lima Neto, A.J.C.C. Lins, A.I.S. Nascimento, M.P. Lima, Fish school search, in *Nature-Inspired Algorithms for Optimisation* (Springer, Berlin, 2009) pp. 261–277
19. S. Ukasik, S. Ak, Firefly algorithm for continuous constrained optimization tasks, in *Computational Collective Intelligence. Semantic Web, Social Networks and Multiagent Systems* (Springer, Heidelberg, 2009), pp. 97–106
20. X.-S. Yang, Firefly algorithms for multimodal optimization, in *Stochastic Algorithms: Foundations and Applications* (Springer, Berlin, 2009), pp. 169–178
21. X.-S. Yang, S. Deb, Cuckoo search via Lvy flights, in *2009 World Congress on IEEE Nature & Biologically Inspired Computing (NaBIC)* (IEEE, 2009), pp. 210–214
22. X.-S. Yang, A new metaheuristic bat-inspired algorithm, in *Nature Inspired Cooperative Strategies for Optimization (NICSO)* (Springer, Berlin, 2010), pp. 65–74

23. D. Karaboga, An idea based on honey bee swarm for numerical optimization. Technical report-tr06, Erciyes University, Engineering Faculty, Computer Engineering Department, (2005)
24. D. Karaboga, B. Basturk, A powerful and efficient algorithm for numerical function optimization: artificial bee colony (ABC) algorithm. J. Glob. Optim. **39**(3), 459–471 (2007)
25. D. Karaboga, B. Basturk, On the performance of artificial bee colony (ABC) algorithm. Appl. Soft Comput. **8**(1), 687–697 (2008)
26. C.R. Blomeke, S.J. Elliott, T.M. Walter, Bacterial survivability and transferability on biometric devices, in *2007 41st Annual IEEE International Carnahan Conference on Security Technology* (IEEE 2007), pp. 80–84
27. Y. Tan, Y. Zhu, Fireworks algorithm for optimization, in *Advances in Swarm Intelligence* (Springer, Berlin, 2010), pp. 355–364
28. H. Shah-Hosseini, The intelligent water drops algorithm: a nature-inspired swarm-based optimization algorithm. Int. J. Bio-Inspir. Comput. **1**(1), 71–79 (2009)
29. Y. Shi, Brain storm optimization algorithm, in *Advances in Swarm Intelligence* (Springer, Berlin, 2011), pp. 303–309
30. N.M.H. Tayarani, M.R. Akbarzadeh-T, Magnetic optimization algorithms a new synthesis, in *2008 IEEE World Congress on Computational Intelligence Evolutionary Computation (CEC)* (IEEE, 2008), pp. 2659–2664
31. D.H. Wolpert, W.G. Macready, No free lunch theorems for optimization. IEEE Trans. Evol. Comput. **1**(1), 67–322 (1997)
32. D.H. Wolpert, W.G. Macready, Coevolutionary free lunches. IEEE Trans. Evol. Comput. **9**(6), 721–735 (2005)
33. J. Liu, S. Zheng, Y. Tan, Analysis on global convergence and timecomplexity of fireworks algorithm, in *IEEE Congress on Evolutionary Computation (CEC'2014)* (Beijing, China, 2014), pp. 3207–3213
34. K. Ding, S. Zheng, Y. Tan. A GPU-based parallel fireworks algorithm for optimization. In *Proceedings of the 15th Annual Conference on Genetic and Evolutionary Computation Conference* (ACM, The Netherlands, 2013), pp. 9–16
35. Y. Pei, S. Zheng, Y. Tan, H. Takagi, An empirical study on influence of approximation approaches on enhancing fireworks algorithm, in *Proceedings of the 2012 IEEE Congress on System, Man and Cybernetics* (IEEE, 2012), pp. 1322–1327
36. S. Zheng, A. Janecek, Y. Tan, Enhanced fireworks algorithm, in *2013 IEEE Congress on Evolutionary Computation (CEC)* (IEEE, 2013), pp. 2069–2077
37. J. Li, S. Zheng, Y. Tan, Adaptive Fireworks Algorithm, in *Proceedings of IEEE Congress on Evolutionary Computation (CEC'2014)* (Beijing, China, 2014), pp. 3214–3221
38. S. Zheng, A. Janecek, Y. Tan, Dynamic Search in Fireworks Algorithm, in *Proceedings of IEEE Congress on Evolutionary Computation (CEC'2014)* (Beijing, China, 2014), pp. 3222–3229
39. Y.J. Zheng, X.L. Xu, H.F. Ling, A hybrid fireworks optimization method with differential evolution. Neurocomputing **148**, 75–82 (2012)
40. C. Yu, L. Kelley, S. Zheng, Y. Tan, Fireworks Algorithm with Differential Mutation for Solving the CEC 2014 Competition Problems, in *Proceedings of IEEE Congress on Evolutionary Computation (CEC'2014)* (Beijing, China, 2014) pp. 3238–3245
41. H. Gao, M. Diao, Cultural firework algorithm and its application for digital filters design. Int. J. Model. Identif. Control **14**(4), 324–331 (2011)
42. W. Fang, J. Sun, W. Xu, J. Liu, FIR digital filters design based on quantum-behaved particle swarm optimization, in *2006 IEEE First International Conference on Innovative Computing, Information and Control (ICICIC'06)* (IEEE, 2006), vol. 1, pp. 615–619
43. W. Fang, J. Sun, W.B. Xu, FIR filter design based on adaptive quantum-behaved particle swarm optimization algorithm. Syst. Eng. Electron. **30**(7), 1378–1381 (2008)
44. M. Zhang, B. Zhang, Y. Zheng, A hybrid biogeography-based optimization and fireworks algorithm, in *Advances in Swarm Intelligence* (Springer, Berlin, 2014), pp. 1–7
45. J. McCaffrey, Fireworks algorithm optimization, MSDN Mag. **29**(12). (2014). http://msdn.microsoft.com/en-us/magazine/dn857364.aspx

46. Y-J. Zheng, Q. Song, S-Y. Chen, Multiobjective fireworks optimization for variable-rate fertil-
    ization in oil crop production. Appl. Soft Comput. **13**(11), 4253–4263 (2013)
47. J. Zhang, On fireworks algorithm for solving 0/1 knapsack problem. J. Wuhan Eng. Inst. **23**(3),
    64–66 (2011)
48. A. Janecek, Y. Tan, Swarm intelligence for non-negative matrix factorization. Intern. J. Swarm
    Int. Res. (IJSIR) **2**(4), 12–34 (2011)
49. W. He, G. Mi, Y. Tan, Parameter optimization of local-concentration model for spam detection
    by using fireworks algorithm. *Advances in Swarm Intelligence* (Springer, Berlin 2013), pp.
    439–450
50. S. Zheng, Y. Tan, A unified distance measure scheme for orientation coding in identification,
    in *2013 IEEE Congress on Information Science and Technology* (IEEE, 2013), pp. 979–985
51. Z. Zheng, Y. Tan, Group explosion strategy for searching multiple targets using swarm robotic,
    in *2013 IEEE Congress on Evolutionary Computation* (IEEE, 2013), pp. 821–828
52. Y. Tan, Swarm robotics: collective behavior inspired by nature. J. Comput. Sci. Syst. Biol.
    (JCSB)
53. Y. Tan, Z.Y. Zheng, Research advance in swarm robotics. Def. Tech. **9**(1), 31–62 (2013)
54. D.U. Zhen-xin, Fireworks algorithm for solving nonlinear equation and system. Mod. Comput.
    6(2), 18–21 (2013). doi:10.3969/j.issn.1007-1423.2013.04.005
55. N. Pholdee, S. Bureerat, Comparative performance of meta-heuristic algorithms for mass min-
    imisation of trusses with dynamic constraints. Adv. Eng. Softw. **75**(4), 1–13 (2014). doi:10.
    1016/j.advengsoft.2014.04.005
56. I.A. Mohamed, M. Kowsalya, A new power system reconfiguration scheme for power loss
    minimization and voltage profile enhancement using fireworks algorithm. Electr. Power Energy
    Syst. **63**(4), 461–472 (2014). doi:10.1016/j.ijepes.2014.04.034
57. I.A. Mohamed, M. Kowsalya, D.P. Kothari, A novel integration technique for optimal network
    reconfiguration and distributed generation placement in power distribution networks. Electr.
    Power Energy Syst. **63**(6), 461–472 (2014). doi:10.1016/j.ijepes.2014.06.011
58. R. Rajaram, K. Palanisamy, S. Ramasamy, P. Ramanathan, Selective harmonic elimination in
    PWM inverter using firefly and fireworks algorithm. Int. J. Innov. Res. Adv. Eng. (IJIRAE) **1**(8),
    55–62 (2014). doi:10.1016/j.ijepes.2014.06.011. http://www.ijirae.com/volumes/voll/issue8/
    SPEE10082.08.pdf
59. A.I. Maswood, S. Wei, M.A. Rahman, A flexible way to generate PWM-SHE switching patterns
    using genetic algorithm. IEEE SPEC **2**, 1130–1134 (2001)
60. K. Sndareswaran, K. Jayant, T.N. Shanavas, Inverter harmonic elimination through a colony
    of continuously exploring ants. IEEE Trans. Ind. Electron. **54**(10), 2558–2565 (2007)
61. K. Sndareswaran, V.T. Sreedevi, Inverter harmonic elimination using honey bee intelligence.
    Aust. J. Electr. Electron. Eng. **6**(2) (2009)
62. N.H. Abdulmajeed, M. Ayob, A firework algorithm for solving capacitated vehicle routing
    problem. Int. J. Adv. Comput. Tech. (IJACT) **6**(1), 79–86 (2014)
63. Y. Tan, S. Zheng, Research progress on fireworks algorithm. CAAI Trans. Intell. Syst. **9**(10),
    1–17 (2014)
64. Y. Tan, C. Yu, S.Q. Zheng, K. Ding, Introduction to fireworks algorithms. Int. J. Swarm Intell.
    Res. **4**(4), 39–70 (2013)

Chapter 2
Fireworks Algorithm (FWA)

Inspired by fireworks explosions in the sky at night, the fireworks algorithm (FWA) was proposed by the author in 2010, through the observation of the fact that fireworks explosion is similar to the way an individual searches for optimal solution in swarm intelligence algorithms. Recently, FWA has received extensive concerns from many active researchers in the swarm intelligence community. This chapter presents the fundamental principle, main constitution, implementation, and performance of the FWA, aiming to elaborate the FWA systematically and completely. The main contents include the key components, realization, characteristic, and impact of operations of FWA, as well as comparisons with genetic algorithm and particle swarm optimization.

2.1 Introduction

Setting off fireworks is an important creative and joyful activity during Spring Festival in China. At this time, tens of thousands of fireworks explode in the night sky and show beautiful patterns of sparks. Usually, fireworks of different prices and specifications produce entirely different patterns. For example, fireworks of lower price produce less sparks with larger amplitude compared with higher price fireworks and vice versa.

The way fireworks explode is similar to the way an individual searches the optimal solution in swarm intelligence algorithms. As a swarm intelligence algorithm, fireworks algorithm consists of four parts, i.e., the explosion operator, mutation operator, mapping rule and selection strategy. The effect of the explosion operator is to generate sparks around fireworks. The number and amplitude of the sparks are governed by the explosion operator. After that, some sparks are produced by mutation operator. The mutation operator utilizes Gaussian operator to produce sparks in Gaussian distribution. Under the effect of the two operators, if the produced spark is not in the feasible

© Springer-Verlag Berlin Heidelberg 2015
Y. Tan, *Fireworks Algorithm*, DOI 10.1007/978-3-662-46353-6_2

region, the mapping rule will map the new generated sparks into the feasible region. To select the sparks for next generation, the selection strategy is used. Fireworks algorithm runs iteratively until it reaches the termination conditions [1].

2.2 FWA Principle

2.2.1 Explosion Operator

In initialization, FWA generates N fireworks randomly. Then the N fireworks generate sparks by explosion operations. The explosion operator is a key in FWA and plays an important role. The explosion operator include explosion strength, explosion amplitude and displacement operation [1].

2.2.1.1 Explosion Strength

The explosion strength is a core operation in explosion operator. It simulates the way of explosion of fireworks in real life. When a firework blasts, the firework vanishes in one second and then many small bursts appear around it. Fireworks algorithm first determines the number of sparks, then calculates the amplitudes of each explosion.

Through the observations on the curves of some typical optimization functions, it can be seen that there are more points with good fitness values around the optima than those away from the optima. Therefore, fireworks with better fitness values produce more sparks, avoiding swing around the optima but fail to locate it. For fireworks with worse fitness values, their generated sparks are less in number and sparse in distribution, avoiding unnecessary computing. Fireworks with worse fitness values are used to explore the feasible space, preventing the algorithm from premature convergence. Fireworks algorithm determines the number and amplitude of the fireworks according to their fitness values, letting fireworks with better fitness values produce more sparks within a smaller amplitude and vice versa as shown in Fig. 2.1.

It can be seen from Fig. 2.1 that fireworks with better fitness values produce more sparks within a smaller amplitude (good explosion) than those with worse fitness values within a larger amplitude (bad explosion). After determining the number of sparks, it is needed to calculate the amplitude of the sparks in the explosion of a firework.

2.2.1.2 Explosion Amplitude

Through observation on the curves of some typical optimization functions, the points around the local optima and global optima always have better fitness values. Therefore, by controlling the explosion amplitude, the amplitude of fireworks with better

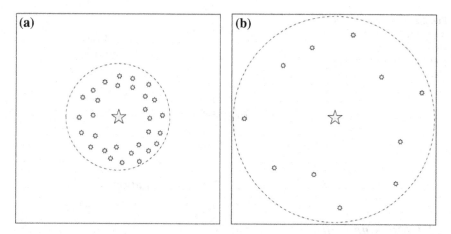

Fig. 2.1 The good and bad explosions of a firework. Reprinted from Ref. [1], with kind permission from Springer Science+Business Media. **a** Good explosion. **b** Bad explosion

fitness values gradually reduce, leading fireworks algorithm to find the local and global optima. By contrast, fireworks with worse fitness values explore the optima through a large amplitude. This is how the FWA controls the magnitude of the explosion amplitude.

2.2.1.3 Displacement Operation

After the calculation of explosion amplitude, it is necessary to determine the displacement within the explosion amplitude. FWA uses the random displacement. In this way, each firework has its own specific explosion number and amplitude of sparks. FWA generates different random displacements within each amplitude to ensure the diversity of population. Through the explosion operator, each firework generates a shower of sparks, helping to find the global optimal of an optimization function.

2.2.2 Gaussian Mutation Operator

To further improve the diversity of a population, the Gaussian mutation is introduced into FWA. The way of producing sparks by Gaussian mutation is as follows: choose a firework from the current population, then apply Gaussian mutation to the firework in randomly selected dimensions.

For Gaussian mutation, the new sparks are generated between the best firework and the selected fireworks (Fig. 2.2). Yet, Gaussian mutation may produce sparks that exceed the feasible space. When a spark lies beyond the upper or lower boundary, the mapping rule will be carried out to map the spark to a new location within the feasible space.

Fig. 2.2 Gaussian mutation.
Reprinted from Ref. [1], with
kind permission from
Springer Science+Business
Media

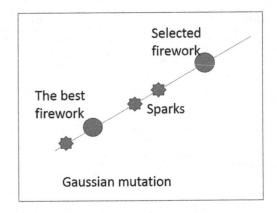

2.2.3 Mapping Rule

If a firework is near the boundary of the feasible space, while its explosion amplitude
covers both the feasible and infeasible space, the generated sparks may lie out of the
feasible space. As such, the spark beyond the feasible space is useless. Therefore,
it needs to be getting back into the feasible space. The mapping rule is used to deal
with this situation. The mapping rule ensures that all sparks are in the feasible space.
If there is any spark that is generated by a firework beyond the feasible space, it will
be mapped back to the feasible space.

2.2.4 Selection Strategy

After applying the explosion operator, the mutation operator, and the mapping rule,
some of the generated sparks need to be selected and passed down to the next generation. The distance-based strategy is used in the fireworks algorithm. In order to select
the sparks for next generation, first, the best spark is always kept for next generation.
Then, the other $(N - 1)$ individuals are selected based on distance maintaining the
diversity of the population. The individual that is farther from the other individuals
has greater chance to be selected than those individuals near the other individuals.

2.3 Implementation of FWA

FWA starts to run iteratively till the given termination conditions are met. It consists
of the explosion operator, the mutation operator, the mapping rule, and the selection strategy. There are two termination conditions, such as meeting the accuracy
requirements and reaching the maximum number of function evaluations.

The realization of FWA consists of four steps as follows.

(1) Randomly generate fireworks in the feasible space.
(2) Calculate the fitness value of each firework according to the fitness function. The number of sparks is calculated based on immune concentration theory in immunology and the fireworks with better fitness values produce more sparks.
(3) Considering the fireworks phenomena in reality and the landscape of the functions, the fireworks generate sparks within a certain amplitude in FWA. The explosion amplitude is determined by the fitness value of that firework. The explosion amplitude for the firework with better fitness value is smaller and vice versa. Each spark represents a solution in the feasible space. To keep the diversity of the population, mutation operation is needed and Gaussian mutation is one of them.
(4) Calculate the best fitness value. If the terminal condition is met, stop the algorithm. Otherwise, continue the iteration process. The best spark and the selected sparks formed a new population.

2.3.1 Explosion Operator

2.3.1.1 Explosion Strength

In the explosion strength, i.e. the number of sparks is determined as follows:

$$S_i = m * \frac{Y_{\max} - f(x_i) + \varepsilon}{\sum_{i=1}^{N} (Y_{\max} - f(x_i)) + \varepsilon}, \tag{2.1}$$

where S_i is the number of sparks for each individual or firework, m is a constant standing for the total number of sparks, and $Y_{\max}$ means the fitness value of the worst individual among the N individuals in the population. Function $f(x_i)$ represents the fitness for an individual x_i, while the last parameter ε is used to prevent the denominator from becoming zero.

The limitation of number of sparks are as follows:

$$\hat{s}_i = \begin{cases} round(a \cdot m), & \text{if } s_i < am \\ round(b \cdot m), & \text{if } s_i > bm, \ a < b < 1, \\ round(a \cdot m), & \text{otherwise} \end{cases} \tag{2.2}$$

where a and b are constant, $\hat{s}_i$ is the limitation of the number of sparks, and $round()$ is the rounding function.

2.3.1.2 Explosion Amplitude

The explosion amplitude is defined below.

$$A_i = \hat{A} * \frac{f(x_i) - Y_{\min} + \varepsilon}{\sum\limits_{i=1}^{N} (f(x_i) - Y_{\min}) + \varepsilon}, \tag{2.3}$$

where A_i denotes the amplitude of each individual, $\hat{A}$ is a constant as the sum of all amplitudes, while $Y_{\min}$ means the fitness value of the best individual among the N individuals. The meaning of function $f(x_i)$ and parameter ε are the same as aforementioned in Eq. (2.2).

2.3.1.3 Displacement Operation

Displacement operation is to make displacement on each dimension of a firework and can be defined as

$$x_i^k = x_i^k + U(-A_i, A_i), \tag{2.4}$$

where $U(-A_i, A_i)$ denotes the uniform random number within the intervals of the amplitude A_i.

Algorithm 2.1 is the pseudo code of the explosion operator described in Eqs. (2.1)–(2.4).

Algorithm 2.1 Generate sparks

1: Initialization, calculate the fitness value $f(x_i)$ for each firework.
2: Calculate the number of sparks S_i.
3: Calculate the amplitude of sparks A_i.
4: $z = rand(1, dimension)$ //randomly choose z dimensions
5: **for** $k = 1 \rightarrow dimension$ **do**
6:   **if** $k \in z$ **then**
7:     $x_i^k = x_i^k + U(-A_i, A_i)$
8:   **end if**
9: **end for**

2.3.2 Mutation Operator

Suppose the position of current individual is stated as x_i^k, where i varies from 1 to N and k denotes the current dimension. The sparks of Gaussian explosion are calculated by

$$x_i^k = x_i^k * g, \tag{2.5}$$

where g is a random number in Gaussian distribution with mean 1 and variance 1 such as

$$g = \mathcal{N}(1, 1). \tag{2.6}$$

Algorithm 2.2 shows the pseudo code for Gaussian mutation.

Algorithm 2.2 Gaussian Mutation

1: Calculate the fitness value $f(x_i)$ for each firework.
2: Calculate the coefficient $g = \mathcal{N}(1, 1)$.
3: $z = rand(1, dimension)$ //randomly select z dimensions
4: **for** $k = 1 \rightarrow dimension$ **do**
5:    **if** $k \in z$ **then**
6:        $x_i^k = x_i^k * g$
7:    **end if**
8: **end for**

2.3.3 Mapping Rule

The mapping rule ensures all the individuals stay in the feasible space. If there are some outlying sparks from the boundary, they will be mapped back to their allowable scopes.

The mapping rule utilizes a modular operation and is stated as follows:

$$x_i^k = X_{LB,k} + x_i^k \% (X_{UB,k} - X_{LB,k}), \tag{2.7}$$

where x_i^k represents the positions of any sparks that lie out of bounds, while $X_{UB,k}$ and $X_{LB,k}$ stand for the maximum and minimum boundaries of a spark position. The sign % represents modular arithmetic.

2.3.4 Selection Strategy

In selection strategy, the measurement of Euclidean distance is used, where $d\left(x_i, x_j\right)$ denotes the Euclidean distance between any two individuals x_i and x_j.

$$R(x_i) = \sum_{j=1}^{K} d(x_i, x_j) = \sum_{j=1}^{K} \|x_i - x_j\|, \tag{2.8}$$

where $R(x_i)$ represents the sum of distances between individual x_i and all the other individuals. $j \in K$ means the position j belongs to set K, where K is the set of

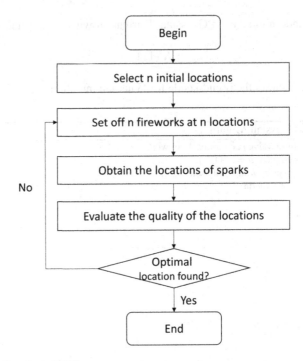

Fig. 2.3 The flowchart of FWA

combining both the sparks generated by explosion operator and mutation operator.
The roulette way is used to choose individuals for next generation, as the possibility
for choosing the individual x_i should be $p(x_i)$, which is given by

$$p(x_i) = \frac{R(x_i)}{\sum_{j \in K} R(x_j)}. \tag{2.9}$$

From Eq. (2.9), it can be seen that individuals with larger distances will have
more chances to be selected for next generation. In such a way, the diversity of the
population can be guaranteed.

The flowchart of FWA is depicted in Fig. 2.3.

The pseudo code of FWA is shown in Algorithm 2.3.

Algorithm 2.3 Pseudo code of FWA

1: Randomly select N locations for fireworks
2: **while** terminal condition is not met **do**
3:   Set off N fireworks, respectively, at the N locations:
4:   **for** all fireworks x_i **do**
5:     Calculate the number of sparks as S_i
6:     Calculate the amplitude of sparks as A_i
7:   **end for**
     //$\hat{m}$ is the number of sparks generated by Gaussian mutation
8:   **for** $k = 1 \rightarrow \hat{m}$ **do**
9:     Randomly select a firework x_i and generate a spark
10:   **end for**
11:   select the best spark and the other sparks according to selection strategy
12: **end while**

2.4 The Characteristics of FWA

FWA contains the following characteristics: explosion, instantaneity, simplicity, locality, emergent property, distribute parallelism, diversity and extendibility. The details are given below.

2.4.1 Explosion

After the first iteration of FWA, the fireworks explode within the amplitude and produce a shower of sparks. At the end of the iteration, the algorithm selects N sparks for the next generation. The N selected sparks are treated like new fireworks, preparing for explosion in the next iteration. In each iteration, the fireworks will explode, indicating the explosive characteristic of FWA.

2.4.2 Instantaneity

In each iteration, FWA calculates the number of sparks and the explosion amplitude, depending on the fitness values of the fireworks. Then, the sparks are produced by the explosion and mutation operators. Finally, the best spark is preserved at first and then the other $(N - 1)$ sparks are selected based on a selection strategy. The selected N sparks are treated as the fireworks for the next generation, while the rest of the sparks are no longer reserved. Sparks or fireworks are not kept, indicating the instantaneous characteristic of FWA.

2.4.3 Simplicity

Like other swarm intelligence algorithms, each firework in FWA only percepts its own information itself and its surrounding information, following simple rules to complete their missions. Overall, FWA is not complex, composed of simple individuals. Therefore, FWA is characteristic of simplicity.

2.4.4 Locality

In FWA, all the fireworks generate sparks within their amplitudes. Unless beyond the feasible region, sparks are confined within a certain range. Localized features of FWA reflect the powerful local search capabilities, as the algorithm can be used for local search in the latter of the search process. Therefore, FWA contains locality.

2.4.5 Emergent Property

Fireworks are in competition and collaboration with each other and the group showed high degree of intelligence which a simple individual cannot achieve. Interaction between fireworks are more complicated than a single individual's behavior. Therefore, firework algorithm has the characteristic of emergent.

2.4.6 Distributed Parallelism

In each iteration of FWA, each firework explodes and searches within different space, i.e., each firework conducts a search in different dimensions. Finally, the sparks and fireworks are combined together to choose N fireworks for the next generation. In each iteration, FWA searches the space in parallel, showing the characteristic of distribution parallelism.

2.4.7 Diversity

Population diversity is vital to the performance of any swarm intelligence algorithm. By maintaining the population diversity, the algorithm can jump out of local optima, which makes the algorithm converge to the global optimal point, which a generic optimization can hardly achieve. Therefore, swarm optimization algorithms are different from any generic optimization algorithm. The better the population diversity is, the wider the individuals are distributed. The optimal value might be easier to be found if a population is strongly diverse, as the convergence of the algorithm will not be affected significantly. Thus, population diversity is an important part of the FWA. The diversities of FWA can be concluded as follows.

2.4.7.1 The Diversity of the Number of Sparks and Explosion Amplitude

According to the explosion operator and the fitness values, each firework generates a different number of sparks within a different magnitude. The fireworks with higher fitness values produce more sparks within smaller ranges, while the fireworks with lower fitness values produce fewer sparks within larger ranges. In such a way, the diversity of the population would be guaranteed.

2.4.7.2 The Diversity of Displacement and Gaussian Mutation

FWA has two operators, explosion operator and mutation operator. In the explosion operator, a displacement is calculated according to an amplitude. The displacement is added to a position in a dimension of a selected firework. In the Gaussian mutation operator, the selected fireworks need to multiply a Gaussian random number in the position of a dimension. The explosion operator is relative to the fitness values of fireworks while the mutation operator is relative to the position of the fireworks. The two operators are different from each other, but both of them guarantee the explosion to be diverse.

2.4.7.3 The Diversity of Fireworks

Through a certain selection mechanism, the coordinates of the retained fireworks are different. As a result, these phenomena ensure the diversity of the population. In addition, in the selection strategy, sparks with a greater distance from the other sparks are more likely to be selected, which in turn, ensures the diversity of population.

2.4.8 Extendibility

In FWA, the number of sparks are uncertain and able to be determined based on the complexity of the problem in hand. The number of fireworks and sparks can be more or less, as both increase and decrease of the individuals can effectively solve the problem. Therefore, FWA has extendibility.

2.4.9 Adaptability

When solving problems using FWA, it is unnecessary for the problem to be of an explicit expression. The problem can be solved by calculating the fitness values only. Meanwhile, FWA can also solve the problems with explicit expressions, indicating its capability. Therefore, FWA is of adaptability and can be regarded as an adaptive algorithm.

2.5 Impact of Operators in FWA on Performance

2.5.1 Explosion Operator

When a spark explodes, the area around the spark is searched. If the fitness value of a spark is higher, the amplitude of the spark is larger and the number of small bursts of the spark is fewer. In this case, the sparks with better fitness values will search more carefully in smaller areas, while the sparks with worse fitness values will search in wider areas. Therefore, FWA has greater chances to find the global optimal in limited function evaluations.

The explosion operator has two parameters. The first parameter m is used to limit the total number of sparks and the second parameter N is the number of fireworks.

Function Generalized Rosenbrock is used to illustrate the impact of the explosion operator on the performance of FWA.

Figure 2.4 gives the impact of the total number of sparks m on the performance of FWA on function Generalized Rosenbrock, under the circumstance that the other parameters remain unchanged.

Experimental results show that better performances are obtained when the total number of sparks m is set between 10 and 50, while the other parameters remained unchanged.

Fig. 2.4 The impact of different number of sparks on the performance of FWA for Generalized Rosenbrock function. The *vertical axis* represents the accuracy of the experimental results

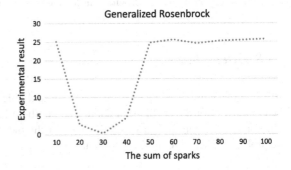

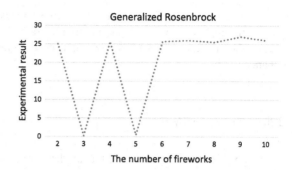

Fig. 2.5 The impact of the number of fireworks on the performance of FWA for Generalized Rosenbrock function. The *vertical axis* represents the accuracy of the experimental results

Figure 2.5 gives the impact of the total number of fireworks N on the performance of FWA on Generalized Rosenbrock function.

It can be seen from Fig. 2.5 that better performances are obtained when the total number of fireworks N is set as 3 or 5, while the other parameters remain unchanged.

Obviously, for different optimization problems, setting different values of parameters can have a certain impact on the performance of the algorithm. Better performances on function Generalized Rosenbrock are obtained when the total number of sparks is set between 20 and 40, and 3 or 5 fireworks if other parameters remain unchanged.

2.5.2 Gaussian Mutation

Gaussian mutation operator can increase the diversity of the algorithm because the sparks generated by Gaussian mutation are not limited to the area around the fireworks. In addition, since the sparks are generated between the current location and the origin by Gaussian mutation ($x_i^k <= x_i^k * g$), the performance is pretty good for functions having the optimal at the origin. For example, FWA with Gaussian mutation can easily find the location of the optimal value on function Sphere because the optimal value of function Sphere is at the origin.

Table 2.1 gives the experimental results of FWA on two functions. The functions are 30-dimensional and the algorithm iterates 300,000 times.

Table 2.1 FWA with and without Gaussian mutation on function Sphere and Generalized Rosenbrock

Status	Sphere	Generalized Rosenbrock
FWA with Gaussian mutation	0	25.209447
FWA without Gaussian mutation	1.095037	706.936069

It can be seen from Table 2.1 that the performances of FWA on both functions are greatly improved by Gaussian mutation operator. This is due to the diversity of the population increased by Gaussian mutation. The global search ability of FWA is enhanced.

2.5.3 Mapping Rule

The mapping rule is used to ensure that all the sparks are generated within the scope of the feasible space. When a firework is near the border and the amplitude of the explosion is large, the generated sparks might be out of the boundaries of the feasible space. Therefore, the sparks that are out of the boundaries will be mapped into feasible space for avoiding unnecessary computation. However, the mapping rule has its own disadvantages. For instance, it can easily pull a spark to the locations near the origin, benefiting the functions with optima near the origin.

The mapping rule adopts a modular arithmetic to ensure that the out-of-range sparks are pulled back into the feasible space.

As shown in Fig. 2.6, the feasible space is set from -100 to 100 and the limitation of explosion amplitude is set as 40. Therefore, the points beyond the boundaries will fall into the area of $-140\sim-100$ and $100\sim140$ shown as shadow areas in Fig. 2.6. According to the mapping rule, points located in these two areas are mapped into the range from 0 to 40, which is near the origin.

2.5.4 Selection Strategy

The selection strategy is to choose individuals for the next generation. The best individual is always kept for the next generation, while the remaining $(N - 1)$ individuals are selected based on Euclidean distance and the sparks further away from other sparks can be selected with larger possibilities. Thus, the diversity of FWA is guaranteed in such a way.

Fig. 2.6 The operation of mapping rule

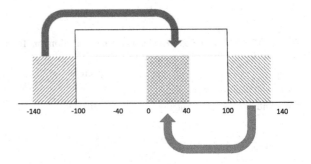

2.6 Comparison of FWA with Three Other SI Algorithms

2.6.1 Ideas Comparison Between FWA and GA

The idea of immune density is introduced into the selection strategy of FWA, so it is unnecessary to design a new selection operator. Selection strategy in FWA sounds like the selection operator in genetic algorithm, but they are different. Based on the idea of immune density, each spark in FWA is treat as an antibody in immune system. A spark (antibody) which has more similar sparks (antibodies) is chosen with a lower probability. On the contrary, a spark (antibody) which has less similar sparks (antibodies) is selected with a higher chance. Hence, the sparks (antibodies) with lower fitness values have the chance to be selected such that the diversity of the sparks (antibodies) is ensured. Compared with the genetic algorithm, the selection operator determines which individual is to be selected by the fitness values of the individuals. The selection is based on the roulette and the diversity of the population is not guaranteed.

Genetic algorithm was originally proposed by Prof. Holland of the University of Michigan [2]. At that time, Holland recognized that biological hereditary and natural evolution phenomena are similar to artificial adaptive systems. The idea of genetic and evolutionary in nature could be used to study the generation of natural, artificial adaptive systems and their relationship with the environment. He suggested to use the mechanism of genetic in study and design of artificial adaptive systems, as the swarms could be used for adaptive search and the crossover and mutation operations are vital.

The common aspects of both algorithms are as follows:

(1) Randomly initialize the initial population;
(2) Calculate the fitness value of each individual;
(3) A series of operations to be performed according to fitness values, such as selection operator, crossover operator, and mutation operator in genetic algorithm, and explosion operator and mutation operator in FWA;
(4) Select the individuals for the next generation according to the fitness values;
(5) Stop when the termination conditions are met. Otherwise, go to step 2.

From the above steps, we can see that the FWA and the genetic algorithms have a lot in common. Both randomly initialize a population, evaluate the individuals according to their fitness values and perform a certain random search. In addition, both algorithms are not guaranteed to find the optimal values.

There is no crossover operator in the FWA and the mutation operator in the FWA is totally different from that in the genetic algorithm.

Compared with the genetic algorithm, information sharing mechanisms in FWA is quite different. In the genetic algorithm, chromosomes share information with each other so the entire population moves relatively homogeneously in the feasible space. However, in the FWA, a distributed information sharing mechanism is used, while the number of sparks and the explosion amplitudes are determined by the fitness values of fireworks which are located in different areas. In addition, the fireworks are always selected from different areas and can hardly stay together due to the immune-based selection. Yet, the FWA has more mechanisms to avoid premature compared to genetic algorithm.

2.6.2 Ideas Comparison Between FWA and Two Versions of PSO

Two particle swarm optimization algorithms are introduced at first, i.e., clonal particle swarm optimization (CPSO) [3] and standard particle swarm optimization (SPSO) [4].

In the biological immune system, when the antigen gets into a living body, the immune system in the body can identify and eliminate the antigen. This process is mainly by cloning that activates antibodies, increasing their number and clearing the antigens [5]. Based on realizing the importance of the immune response, Tan and Xiao [3] made improvements to the standard PSO algorithm by proposing a cloning operator.

The process of the three algorithms is similar as explained below.

(1) Initialization. FWA initializes the fireworks, while the two kinds of PSO initialize particles.
(2) Calculate the fitness values for the individuals in Step 1.
(3) Process necessary operations. FWA processes the explosion and mutation operations, while the two kinds of PSO processes update pbest and gbest. Also, the position and speed for each particle need to be updated in each generation.
(4) Select individuals for the next generation.
(5) If the termination condition is met, the algorithm is terminated. Otherwise, go to Step 2.

From the above steps, we can see the FWA and the two kinds of PSO have much in common. They adopt random initial populations, evaluate the functions and perform the search based on the fitness values. Also, all the algorithms are not guaranteed to find the optimal solution.

However, there is no mutation operation in SPSO, while there is Gaussian mutation in CPSO and both displacement operation and Gaussian mutation in FWA. Furthermore, the Gaussian mutation in FWA plays a role on differential dimensions with the same displacement, making connections between the different dimensions. The displacement in CPSO differs in each dimension. Besides, Gaussian mutation takes place in each iteration in FWA, but runs once in several iterations in CPSO.

Compared with the two kinds of PSO, information sharing mechanism in FWA is different. In the two kinds of PSO, only gbest gives information to other particles, which is one way of information delivery, as the search process follows the information about the best particle. FWA, on the other hand, uses a distributed information sharing mechanism, so as to determine the number of sparks and explosion amplitude by the fitness values of each spark in different regions. It also needs to maintain the best firework throughout the iterative process.

Besides, FWA utilizes the idea of immune concentration to keep the diversity of the population, whereas the idea is not contained in SPSO.

2.7 Experimental Results and Analysis

2.7.1 Benchmark Functions

FWA [1] and SPSO [4] and CPSO [3] are compared on six test functions. The details of the six test functions can be seen in Appendix A.

2.7.2 Parameters Setting

- The population size is set as 5 and the spark of Gaussian mutation is set as 5.
- The total number of sparks is set as 50. Parameter a and b are set as 0.8 and 0.04.
- The constant $\hat{A}$ is set as 40. There is no lower boundary for the amplitude of explosion.
- Function is 30 dimensions, running 20 times and the number of function evaluations is limited to 400,000.

2.7.3 Experimental Results

The experimental results are shown in Table 2.2 when the functions are evaluated for 400,000 times (the accuracy is 10^{-6}).

The convergence curves are also shown in Fig. 2.7.

Table 2.2 Comparison of FWA with CPSO and SPSO (lower than 10^{-6} is treated as zero)

	FWA		CPSO		SPSO	
	Mean	Std	Mean	Std	Mean	Std
Sphere	0	0	0	0	367.1166	186.7949
Rosenbrock	12.16293	12.82113	66.58722	204.2907	5692076	4087432
Griewank	0	0	0.003693	0.011792	1.088648	0.042218
Rastrigin	0	0	6.769299	7.701368	676.1549	197.9695
Rotated Griewank	0	0	0.043401	0.042286	0.920613	0.088088
Rotated Rastrigin	0	0	23.92579	13.6093	339.2073	62.38145

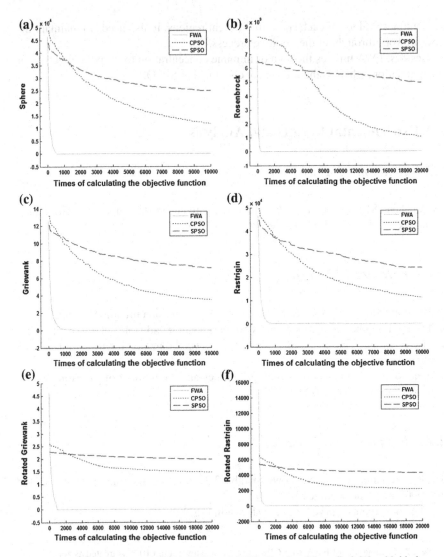

Fig. 2.7 Convergence curves of FWA, CPSO and SPSO. Reprinted from Ref. [1], with kind permission from Springer Science+Business Media

2.7.4 Analysis

It can be seen from the experimental results that FWA works significantly better than SPSO and CPSO [3] in both convergence speed and accuracy on the six test functions. Hence, FWA has good convergence and result accuracy and can be successfully applied to function optimization problems. In the comparison experiments with SPSO and CPSO, FWA reveals good advantage, which means FWA is very successful and has good prospects. In addition, since a lot of practical engineering problems can be transformed into the function optimization problem and FWA can solve function optimization problems effectively, FWA has good prospects.

However, studies on FWA is still in its infancy, while current FWA still have some problems. Further research can be concluded as follows.

(1) The main operators of FWA are realized, but they are still not perfect. For instance, Gaussian mutation achieves good performance, but it cannot solve all the problems effectively. New mechanism needs to be added to improve FWA.
(2) Parameters of the algorithm are set according to empirical experiments on simple functions. The experiments are not enough and there is no theoretical analysis for the settings. The future research is to find reasonable parameters for each problem.
(3) The function optimization experiments are conducted on a few benchmark functions. Hence, the experimental results are not convincing in some aspects. More optimization functions are needed to make the test systematic and comprehensive.
(4) FWA can be applied to many areas. For example, neural networks, fuzzy systems control and fuzzy rule learning. Yet, FWA can be applied to solving discrete issues.

2.8 Summary

This chapter described FWA in detail, including explosion operator, mutation operator, mapping rule and selection strategy. The flowchart and pseudo code of FWA were also given while the characteristics of FWA and the impact of various factors on the performance of FWA were analyzed deeply. Besides, GA, PSO, and FWA were compared. Experimental results show that, FWA had better performance compared to GA and PSO.

FWA and its applications prove that it can solve many complex optimization problems effectively. Furthermore, FWA can be parallelized and thus is suitable for dealing with big data problems. Whether for theoretical or applied researches, FWA is worth researching and might bring great scientific and economic benefits for us.

References

1. Y. Tan, Y. Zhu, Fireworks algorithm for optimization, in *Advances in Swarm Intelligence* (Springer, Berlin, 2010), pp. 355–364
2. J.H. Holland. *Adaptation in Natural and Artificial Systems: An Introductory Analysis with Applications to Biology, Control, and Artificial Intelligence*. (U Michigan Press, 1975)
3. Y. Tan, Z.M. Xiao. Clonal particle swarm optimization and its applications, in *IEEE Congress on Evolutionary Computation, CEC 2007*, (2007), pp. 2303–2309
4. J. Kennedy, R. Eberhart et al., Particle swarm optimization, in *Proceedings of IEEE International Conference on Neural Networks*, vol. 4(2) (Perth, Australia, 1995), pp. 1942–1948
5. R. Liu, H. Du, L. Jiao, Immunity poly-clonal strategy. J. Comput. Res. Dev. **41**(4), 571–576 (2004)

Chapter 3
Modeling and Theoretical Analysis of FWA

In order to describe the convergence analysis of FWA, a Markov stochastic process modeling Fireworks Algorithm has been defined and established, in the first part of this chapter, then to prove the global convergence of FWA and analyze the time complexity of FWA based on an absorbing Markov stochastic process of FWA. After that, the computation of the approximation region of expected convergence time of Fireworks Algorithm has also been given through a detailed derivation procedure. In the second part of this chapter, we will present 13 commonly used random number generators (RNGs) and also try to discuss the impact of the RNGs on the performance of FWA.

3.1 A Stochastic Process Model for FWA

Assume that the FWA [1] search is undertaken for the essential infimum, which can be defined as Eq. (3.1).

$$\psi = inf(t : \nu[n \in S | f(z) < t] > 0), \tag{3.1}$$

where $\nu[A]$ is the Lebesgue measure on the set A.

Equation (3.1) means that there must be more than one point in a subset of search space yielding functional values arbitrarily close to ψ, so that ψ is the infimum of the functional values from nonzero Lebesgue measurable set, and then the stochastic process of Fireworks Algorithm is established as follows:

Definition 3.1 $\{\xi(t)\}_{t=0}^{\infty}$ is named as the stochastic process of Fireworks Algorithm, where $\xi(t) = \{F(t), T(t)\}$ which $F(t) = \{F_1(t), F_2(t), \dots, F_n(t)\}$ denotes the position of n fireworks in the problem space in the step t; And $T(t) = \{A(t), S(t)\}$, $A(t) = \{A_1(t), A_2(t), \dots, A_n(t)\}$ denotes the explosion amplitude of n fireworks, $S(t) = \{s_1(t), s_2(t), \dots, s_n(t)\}$ denotes the explosion number of n fireworks.

© Springer-Verlag Berlin Heidelberg 2015
Y. Tan, *Fireworks Algorithm*, DOI 10.1007/978-3-662-46353-6_3

Now, an optimality region can be defined as

Definition 3.2 $R_\varepsilon = \{x \in S | f(x) - f(x^*) < \varepsilon, \varepsilon > 0\}$ is named as the optimal region of function $f(x)$, where x^* represents the optimal solution of function $f(x)$ in the problem space.

In terms of Definition 3.2, if the algorithm finds a point in the optimal region, an acceptable approximation to the global minimum of the function has been acquired by the algorithm. According to the above definition of infimum, ψ, the Lebesgue measure of optimal solution space must be nonzero, which means that $v(R_\varepsilon) > 0$ is held.

Definition 3.3 The optimal state of FWA is defined as $\xi^*(t) = \{F^*(t), T(t)\}$, where there exist $F_i(t) \in R_\varepsilon$ and $F_i(t) \in F^*(t), i \in 1, 2, \ldots, n$.

The Definition 3.3 means that the best firework in the optimal state $\xi^*(t)$ of FWA is in the optimal region R_ε. So here exist $F_i(t) \in R$ and $|f(F_i(t)) - f(x^*)| < \varepsilon$, $x^* \in R_\varepsilon$.

Lemma 3.1 *The stochastic process of FWA, $v\{\xi(t)\}_{t=0}^\infty$, is a Markov stochastic process.*

Proof Assume that $\{\xi(t)\}_{t=0}^\infty 0$ is a stochastic process of discrete times.

Because the state $\xi(t) = \{F(t), T(t)\}$ is decided by $\{F(t-1), T(t-1)\}$, so one have the following probability:

$P\{\xi(t+1) | \xi(1), \xi(2), \ldots, \xi(t)\} = P\{\xi(t+1) | \xi(t)\}$.

The above equation means that the probability of $(t+1)$th state occurring is only related to the probability of tth state occurring.

Therefore, the $\{\xi(t)\}_{t=0}^\infty$ is a Markov stochastic process.   □

Definition 3.4 (*optimal state space*) Given Y is represented as the state space of FWA's state $\xi(t)$ and $Y^* \subset Y$. Y^* is named as the optimal state space if there exists a solution $s^* \in F^*$ such that $s^* \in R_\varepsilon$ for any state $\xi(t)^* = \{F^*, T\} \in Y$.

In terms of the above definition, it means that $|f(s^*) - f(x^*)| < \varepsilon$ for any $x^* \in F^*$. If the state of the Fireworks Algorithm can arrive at the optimal state, there exists a firework in the fireworks which stays in the optimal region R_ε and the optimal solution of problem has been acquired by the FWA. After this time, the optimal solution must be in the optimal region forever.

Definition 3.5 Given a Markov stochastic process $\{\xi(t)\}_{t=0}^\infty$ and optimal state space $Y^* \subset Y$, if $\{\xi(t)\}_{t=0}^\infty$ s.t. $P\{\xi(t+1) \notin Y^* | \xi(t) \in Y^*\} = 0$, $\{\xi(t)\}_{t=0}^\infty$ is called as an absorbing Markov process.

Lemma 3.2 *The stochastic process of FWA, $\{\xi(t)\}_{t=0}^{\infty}$, is an absorbing Markov stochastic process.*

Proof First of all, according to Lemma 3.2, the stochastic process of FWA, $\{\xi(t)\}_{t=0}^{\infty}$, is a Markov stochastic process.

Then, if $F_1(t) \in F(t)$ stays in the optimal solution space R_ε, the state $\xi(t) = \{F(t), T(t)\}$ must belong to the optimal state space Y^*.

Because $F_1(t)$ is the best location in all fireworks of FWA, $f(F_1(t+1))$ is not worse than $f(F_1(t))$ in terms of the step 6 of FWA. So, the state $\xi(t+1)$ must belong to the optimal state space Y^*.

Therefore, $P\{\xi(t+1) \notin Y^* | \xi(t) \in Y^*\} = 0$, the stochastic process of FWA, $\{\xi(t)\}_{t=0}^{\infty}$, is an absorbing Markov process. $\square$

3.2 Global Convergence Theorems

In this subsection, the definition of convergence is given, which is used to analyze the convergence of Fireworks Algorithm based on the above model and its analysis.

Definition 3.6 (*convergence*) Given an absorbing Markov process $\{\xi(t)\}_{t=0}^{\infty} = \{F(t), T(t)\}$ and an optimal state space $Y^* \subset Y$, $\lambda(t) = P\{\xi(t) \in Y^*\}$ denotes the probability that the stochastic state arrives at the optimal state in step t, if $lim_{t\to\infty}\lambda(t) = 1$, $\{\xi(t)\}_{t=0}^{\infty}$ is convergence.

In terms of the above definition, the convergence of Markov stochastic process depends on the probability of $P\{\xi(t) \in Y^*\}$. If it converges to 1 with the time t, the Markov process $\{\xi(t)\}_{t=0}^{\infty}$ will be convergence. So we have the following theorem:

Theorem 3.1 *Given an absorbing Markov process $\{\xi(t)\}_{t=0}^{\infty}$ of FWA and optimal state space $Y^* \subset Y$. If $P\{\xi(t) \in Y^* | \xi(t-1) \notin Y^*\} \geq d \geq 0$ for any t and $P\{\xi(t) \in Y^* | \xi(t-1) \in Y^*\} = 1$, then $P\{\xi(t) \in Y^*\} \geq 1 - (1-d)^t$.*

Proof Let $t = 1$,

$$
\begin{aligned}
P\{\xi(1) \in Y^*\} &= P\{\xi(1) \in Y^* | \xi(0) \in Y^*\} \cdot P\{\xi(0) \in Y*\} \\
&\quad + P\{\xi(1) \in Y^* | \xi(0) \notin Y^*\} \cdot P\{\xi(0) \notin Y^*\} \\
&\geq P\{\xi(0) \in Y^*\} + d \cdot P\{\xi(0) \notin Y^*\} \\
&= P\{\xi(0) \in Y^*\} + d \cdot (1 - P\{\xi(0) \in Y^*\}) \\
&= d + (1-d) \cdot P\{\xi(0) \in Y^*\}.
\end{aligned}
$$

Because $(1 - d) \geq 0$, so $d + (1 - d) \cdot P\{\xi(0) \in Y^*\} \geq d$, then $P\{\xi(1) \in Y^*\} \geq d = 1 - (1 - d)^1$; Now, it is assumed that the $P\{\xi(t) \in Y^*\} \geq 1 - (1 - d)^t$ is held for any $t < k - 1$, and then for $t = k$, one has

$$
\begin{aligned}
P\{\xi(k) \in Y^*\} &= P\{\xi(k) \in Y^* | \xi(k - 1) \in Y^*\} \\
&\quad \cdot P\{\xi(k - 1) \in Y^*\} + P\{\xi(k) \in Y * | \xi(k - 1) \notin Y^*\} \\
&\quad \cdot P\{\xi(k - 1) \notin Y^*\} \\
&= P\{\xi(k - 1) \in Y^*\} + P\{\xi(k) \in Y^* | \xi(k - 1) \notin Y^*\} \\
&\quad \cdot P\{\xi(k - 1) \notin Y^*\} \\
&\geq P\{\xi(k - 1) \in Y^*\} + d \cdot (1 - P\{\xi(k - 1) \in Y^*\}) \\
&= d + (1 - d) \cdot P\{\xi(k - 1) \in Y*\} \\
&\geq d + (1 - d) \cdot (1 - (1 - d)^{k-1}) = 1 - (1 - d)^k.
\end{aligned}
$$

Consequently, for any $t \geq 1$, the following equation is true:

$$
P\{\xi(t) \in Y^*\} \geq 1 - (1 - d)^t. \tag{3.2}
$$

Q.E.D

FWA has the operation of mutation. For simplicity, it is assumed that the operation of mutation in FWA is only the stochastic mutation.

Theorem 3.2 *Given FWA being an absorb state Markov process $\{\xi(t)\}_{t=0}^{\infty}$ and optimal state space $Y^* \in Y$, then $\lim_{t \to \infty} \lambda(t) = 1$ which means that $\xi(t)_{t=0}^{\infty}$ will converge to the optimal state Y^*.*

Proof FWA can provide the mutation operator, so the probability that the firework of FWA arriving the optimal region R_ε from nonoptimal region via mutation operator is denoted as $P_{mu}(t)$. It is able to be expressed as follows:

$$
P_{mu} = \frac{\nu(R_\varepsilon) \cdot n}{\nu(S)}, \tag{3.3}
$$

where $\nu(S)$ is the Lebegue measure value of the problem space S with $\nu(R_\varepsilon) > 0$, $P_{mu} > 0$, n is the number of fireworks.

In terms of the stochastic Markov process $\{\xi(t)\}_{t=0}^{\infty}$ of FWA, it holds that

$$
\lambda(t) = P\{\xi(t) \in Y^* | \xi(t - 1) \notin Y^*\} = P_{mu}(t) + P_{ex}(t), \tag{3.4}
$$

where $P_{ex}(t)$ denotes the probability of the fireworks of FWA arriving at the optimal region R_ε by the firework's explosion in FWA.

So, one has

$$P\{\xi(t) \in Y^* | \xi(t-1) \in Y^*\} \geq P(mu) > 0. \tag{3.5}$$

Since the Markov process $\{\xi(t)\}_{t=0}^{\infty}$ of FWA is an absorbing Markov process that the condition of the theorem 1 is held here, the following equation can be obtained:

$$P\{\xi(t) \in Y^*\} = 1 - (1 - P_{mu}(t))^t. \tag{3.6}$$

So, we have

$$lim_{t \to \infty} P\{\xi(t) \in Y^*\} = 1. \tag{3.7}$$

As a result, the Markov process $\{\xi(t)\}_{t=0}^{\infty}$ of FWA will converge to the optimal state.   Q.E.D.

3.3 Time Complexity of FWA

3.3.1 Basic Theory of Time Complexity

The analysis of evolutionary computation methods based on Markov process model has been done by Huang Han and Hao Zhifeng, for the Evolutionary Programming [2] and Ant Conoly Optimization [3], respectively. In this section, the definitions of some important concepts and theorems on Fireworks Algorithm are presented below, which can also be referred to the literatures [2, 3] for details.

Definition 3.7 (*Expected convergence time*) Given FWA being an absorbing state Markov process $\{\xi(t)\}_{t=0}^{\infty}$ and optimal state space $Y^* \subset Y$, if γ is a stochastic nonnegative value such that: if $t \geq \gamma$, $P\{\xi(t+1) \in Y^*\} = 1$; if $0 \leq t \leq \gamma$, $P\{\xi(t+1) \notin Y^*\} < 1$, then the γ is named as the convergence time of FWA. The expected value $E\gamma$ is called as the expected convergence time of FWA.

The expected convergence time describes the expected time of arriving the global optimal solution in probability 1 at the first time. The smaller the expected value $E\gamma$ is, the faster the convergence of FWA is and the more effective FWA is. However, it can also use the Expected First Hitting Time (EFHT) as a index of convergence time which is given as follows.

Definition 3.8 (*Expected First Hitting Time*) Given FWA being an absorbing state Markov process $\{\xi(t)\}_{t=0}^{\infty}$ and optimal state space $Y^* \subset Y$; μ is a stochastic value such that: if $t = \mu$, $\xi(t) \notin Y^*$; if $0 \leq t \leq \mu$, $\xi(t) \notin Y^*$. The expected value $E\mu$ is named as Expected First Hitting Time.

The following theorem gives an approach to compute $E\lambda$:

Theorem 3.3 *Given FWA being an absorbing state Markov process $\xi(t)_{t=0}^{\infty}$ and optimal state space $Y^* \subset Y$. If $\lambda(t) = P\{\xi(t) \in Y^*\}$ and $\lim_{t \to \infty} \lambda(t) = 1$, the Expected Convergence Time is $E\gamma = \Sigma_{t=0}^{\infty}(1 - \lambda(t))$.*

Proof Since

$$\lambda(t) = P\{\xi(t) \in Y^*\} = P\{\mu \leq t\}$$
$$\Rightarrow \lambda(t) - \lambda(t-1) = P\{\mu \leq t\} - P\{\mu \leq t-1\}$$
$$\Rightarrow P\{\mu = t\} = \lambda(t) - \lambda(t-1),$$

then

$$E\mu = 0 \cdot P\{\mu = 0\} + \Sigma_{t=0}^{\infty} t \cdot P\{\mu = t\},$$
$$E\mu = \Sigma_{t=0}^{\infty} t \cdot (\lambda(t) - \lambda(t-1))$$
$$= (\lambda(1) - \lambda(0)) + 2 \cdot (\lambda(2) - \lambda(1)) + \ldots$$
$$+ t \cdot (\lambda(t) - \lambda(t-1)) + \ldots$$
$$= \Sigma_{i=1}^{\infty}(\lambda(t) - \lambda(t-1)) + \Sigma_{i=2}^{\infty}(\lambda(t) - \lambda(t-1))$$
$$+ \ldots + \Sigma_{i=t}^{\infty}(\lambda(t) - \lambda(t-1)) + \ldots$$
$$= (\lim_{t \to \infty} \lambda(t) - \lambda(1)) + (\lim_{t \to \infty} \lambda(t) - \lambda(2))$$
$$+ \ldots + (\lim_{t \to \infty} \lambda(t) - \lambda(t-1)) + \ldots$$
$$= \Sigma_{i=1}^{\infty}(\lim_{t \to \infty} \lambda(t) - \lambda(t-1)) = \Sigma_{i=1}^{\infty}(1 - \lambda(t-1))$$
$$= \Sigma_{i=1}^{\infty}(1 - \lambda(t))$$

So,

$$E\gamma = E\mu = \Sigma_{i=1}^{\infty}(1 - \lambda(t)). \qquad \text{Q.E.D.}$$

Because it is hard to obtain the value of $\lambda(t)$, so it is difficult to compute the expected convergence time $E\gamma$ exactly. So, its estimation is able to be given as follows:

The proof of Lemma 3.3, Theorems 3.4 and 3.5, and Corollary 3.1 can also been referred to [3].

Lemma 3.3 *Given two stochastic nonnegative variables, u and v, and $D_u(.)$ and $D_v(.)$ denote the distribution functions of u and v. The expected value of u and v can hold $Eu < Ev$ if $D_u(.) \geq D_v(.)$ for $t = 0, 1, 2, \ldots$*

Proof Since $D_u(t) = P\{u \le t\}$ and $D_v(t) = P\{u \le t\}(\forall t = 0, 1, 2, \dots)$,

$$Eu = 0 \cdot D_u(0) + \sum_{t=1}^{+\infty} t[D_u(t) - D_u(t-1)] = \sum_{i=1}^{+\infty}\sum_{t=1}^{+\infty}[D_u(t) - D_u(t-1)] = \sum_{i=0}^{+\infty}[1 - D_u(i)].$$

(3.8)

As a result,

$$Eu - Ev = \sum_{i=0}^{+\infty}[1 - D_u(i)] - \sum_{i=0}^{+\infty}[1 - D_v(i)] = \sum_{i=0}^{+\infty}[D_v(i) - D_u(i)] \le 0 \Rightarrow Eu \le Ev.$$

(3.9)

Q.E.D.

Theorem 3.4 *Given FWA being an absorbing state Markov process $\{\xi(t)\}_{t=0}^{\infty}$ and optimal state space $Y^* \subset Y$. If $\lambda(t) = P\{\xi(t) \in Y^*\}$ such that $0 \le D_l(t) \le \lambda(t) \le D_h(t) \le 1 (\forall t = 0, 1, 2, \dots)$ and $\lim_{t \to \infty} \lambda(t) = 1$, then:*

$$\Sigma_{i=1}^{\infty}(1 - D_l(t)) \le E\gamma \le \Sigma_{i=1}^{\infty}(1 - D_t(t)).$$

(3.10)

Proof Construct two discrete random nonnegative variables h and l, their distribution functions are $D_h(t)$ and $D_l(t)$, respectively.

Apparently, μ in Definition 3.8 is also a discrete random nonnegative variable. Its distribution function is expressed as

$$\lambda(t) = P\{\xi(t) \in Y^*\} = P\{\mu \le t\}.$$

(3.11)

Because $0 \le D_l(t) \le \lambda(t) \le D_h(t) \le 1$, $Eh \le E\mu \le El \Leftrightarrow \sum_{t=0}^{+\infty}(1 - D_h(t)) \le E\gamma = E\mu \le \sum_{t=0}^{+\infty}(1 - D_l(t))$.

Q.E.D.

Theorem 3.5 *Given FWA being an absorbing state Markov process $\{\xi(t)\}_{t=0}^{\infty}$ and optimal state space $Y^* \subset Y$; if $\lambda(t) = P\{\xi(t) \in Y^*\}$ and $0 \le a(t) \le \lambda(t) \le b(t)$, $\Sigma_{l=1}^{\infty}[(1 - \lambda(0))\Pi_{l=0}^{\infty}(1 - a(t))] \le E\gamma \le \Sigma_{t=0}^{\infty}[(1 - \lambda(0))\Pi_{i=1}^{\infty}(1 - a(t))]$.*

Proof Since
$$\lambda(t) = [1 - \lambda(t-1)]P\{\xi(t) \in Y^* | \xi(t-1) \notin Y^*\} + \lambda(t-1)P\{\xi(t) \in Y^* | \xi(t-1) \in Y^*\}(\forall t = 0, 1, 2, \dots), 1 - \lambda(t) \le [1 - a(t)][1 - \lambda(t-1)] = [1 - \lambda(0)]\prod_{i=1}^{t}[1 - a(i)].$$

According to Definition 3.3,

$$E\gamma = \sum_{i=0}^{+\infty}(1 - \lambda(i)) \leq \sum_{t=0}^{+\infty}\left[(1 - \lambda(0))\prod_{i=1}^{t}(1 - a(t))\right]. \qquad (3.12)$$

Likewise

$$E\gamma = \sum_{i=0}^{+\infty}(1 - \lambda(i)) \geq \sum_{t=0}^{+\infty}\left[(1 - \lambda(0))\prod_{i=0}^{t}(1 - b(t))\right]. \qquad (3.13)$$
$$\text{Q.E.D.}$$

Corollary 3.1 *Given FWA being an absorbing state Markov process $\{\xi(t)\}_{t=0}^{\infty}$ and optimal state space $Y^* \subset Y$; and $\lambda(t) = P\{\xi(t) \in Y^*\}$. If $a \leq P\{\xi(t+1) \in Y^*|\xi(t+1) \notin Y^*\} \leq b(a, b > 0)$ and $\lim_{t\to\infty}\lambda(t) = 1$, then the expected convergence time $E\gamma$ of FWA is such that*

$$b^{-1}[1 - \lambda(0)] \leq E\gamma \leq a^{-1}[1 - \lambda(0)]. \qquad (3.14)$$

Proof According to Definition 3.5,

$$E\gamma \leq [1 - \lambda(0)]\left[a + \sum_{t=2}^{+\infty}ta\prod_{i=0}^{t-2}(1 - a)\right]$$

$$\Rightarrow E\gamma \leq [1 - \lambda(0)]\left[a + \sum_{t=2}^{+\infty}ta(1 - a)^{t-1}\right]$$

$$\Rightarrow E\gamma \leq a[1 - \lambda(0)]\left[\sum_{t=0}^{+\infty}t(1 - a)^{t} + \sum_{t=0}^{+\infty}(1 - a)^{t}\right]$$

$$\Rightarrow E\gamma \leq a[1 - \lambda(0)]\left[\frac{1-a}{a^2} + \frac{1}{a}\right) = \frac{1}{a}[1 - \lambda(0)].$$

Likewise,

$$E\gamma \geq b^{-1}[1 - \lambda(0)], \qquad (3.15)$$

So,

$$b^{-1}[1 - \lambda(0)] \leq E\gamma \leq a^{-1}[1 - \lambda(0)]. \qquad (3.16)$$
$$\text{Q.E.D.}$$

The corollary and theorems of above indicate that the formula $P\{\xi(t) \in Y^*|\xi(t-1) \notin Y^*\}$ can give a description of the probability arriving at the optimal state from the nonoptimal state. The estimation of value range of $E\lambda$ is able to be computed by the range of value of $P\{\xi(t) \in Y^*|\xi(t-1) \notin Y^*\}$.

3.4 Deep Analysis of Time Complexity

Time complexity of FWA is to compute the expected convergence time $E\lambda$. In terms of Corollary 3.1 in the previous subsection, it is mainly related to the probability of FWA state arriving optimal region R_ε from nonoptimal region which is $P\{\xi(t+1) \in Y^*|\xi(t-1) \notin Y^*\}$. In this subsection, we will further analyze the formula in details to obtain the time complexity of FWA. Generally speaking, FWA includes three operations including explosion, mutation and selection, but the operations which directly make the Markov state of FWA get into the optimal region, are explosion and mutation, so the following theorem is able to give out.

Theorem 3.6 *Given FWA being an absorbing state Markov process $\{\xi(t)\}_{t=0}^{\infty}$ and optimal state space $Y^* \subset Y$; then FWA is such that*

$$\frac{\nu(R_\varepsilon) \times n}{\nu(S)} \leq P\{((\xi(t+1)) \in Y^*|\xi(t) \notin Y^*\}$$

$$\leq \nu(R_\varepsilon)\left(\frac{n}{\nu(S)} + \sum_{i=1}^{n} \frac{m_i}{\nu(A_i)}\right), \qquad (3.17)$$

where $\nu(R_\varepsilon)$ is the Lebegue measure value of the optimal region R_ε. $\nu(S)$ is the Lebegue measure value of the problem search region S and $v(A_i)$ is the Lebegue measure value of the explosion region A_i of ith firework.

Proof Because FWA has two operations generating the sparks or fireworks: mutation and explosion.

For mutation, it is assumed that the operation is run with randomness. The probability of one firework being mutated to the optimal region R_ε is $\frac{\nu(R_\varepsilon)}{\nu(S)}$.

So, the probability of n fireworks to be randomly mutated to the optimal region R_ε is equal to $\frac{\nu(R_\varepsilon) \times n}{\nu(S)}$.

In term of the steps of FWA, the following equation is reached:

$$P(\xi(t+1) \in Y^*|\xi(t) \notin Y^*) = \frac{\nu(R_\varepsilon) \times n}{\nu(S)} + P(exp), \qquad (3.18)$$

where $P(exp)$ is the probability which n fireworks explode and make some sparks stay in optimal region R_ε.

So, the $P(exp)$ is given by

$$P(exp) = \sum_{i=1}^{n} \frac{\nu(A_i \cap R_\varepsilon) \times m_i}{\nu(A_i)}, \qquad (3.19)$$

where A_i denotes the search space in which the ith firework explodes, m_i is the number of spark which the ith firework generates.

Because $0 \le \nu(A_i \cap R_\varepsilon) \le \nu(R_\varepsilon)$,

$$0 \le P(exp) = \sum_{i=1}^{n} \frac{\nu(A_i \cap R_\varepsilon) \times m_i}{\nu(A_i)}$$

$$\le \sum_{i=1}^{n} \frac{\nu(R_\varepsilon) \times m_i}{\nu(A_i)} = \nu(R_\varepsilon) \sum_{i=1}^{n} \frac{m_i}{\nu(A_i)}.$$

And then,

$$\frac{\nu(R_\varepsilon) \times n}{\nu(S)} \le P(\xi(t+1) \in Y^* | \xi(t) \notin Y^*)$$

$$\le \frac{\nu(R_\varepsilon) \times n}{\nu(S)} + \nu(R_\varepsilon) \sum_{i=1}^{n} \frac{m_i}{\nu(A_i)}$$

$$= \nu(R_\varepsilon) \left(\frac{n}{\nu(S)} + \sum_{i=1}^{n} \frac{m_i}{\nu(A_i)} \right)$$

It is obtained that

$$\frac{\nu(R_\varepsilon) \times n}{\nu(S)} \le P(\xi(t+1) \in Y^* | \xi(t) \notin Y^*) \le \nu(R_\varepsilon) \left(\frac{n}{\nu(S)} + \sum_{i=1}^{n} \frac{m_i}{\nu(A_i)} \right). \quad (3.20)$$

$$\text{Q.E.D.}$$

The above theorem gives the rude result because the right formula of Eq. (3.17) is very difficult to be confirmed and be computed. It is much complex for FWA to run and hard to compute the probability about it. In order to realize the probability exactly, Eq. (3.18) needs to be further investigated which is given as follows:

$$P(exp) = \sum_{i=1}^{n} \frac{\nu(S_i \cap R_\varepsilon) \times m_i}{\nu(S_i)}. \quad (3.21)$$

As it can be known, the formulae $\nu(S_i \cap R_\varepsilon)$ and m_i in the above equation play key roles to $P(exp)$ because the two formulae are dynamically changed with the running of the algorithm.

The formula $\nu(S_i \cap R_\varepsilon)$ is related to the firework location Fi. The distance between two of the fireworks selected as the next generation is as far as possible, so it can be assumed that just one firework can stay in the optimal region R_ε at the same time. In other hand, it is further assumed that there is the highest probability for the best firework to get into the optimal region R_ε.

According to the above idea of Fireworks Algorithm, $\nu(A_i) \ge \nu(A_{best})$ and $m_i \le m_{best}, i \in (1, 2, \ldots, n)$, where A_{best} and m_{best} are the exploding regions and generating sparks number of the firework whose fitness is best in all the fireworks, respectively. So, it can be derived as follows:

$$\frac{\nu(A_i \cap R_\varepsilon) \times m_i}{\nu(A_i)} < \frac{\nu(A_{best} \cap R_\varepsilon) \times m_{best}}{\nu(A_{best})}. \tag{3.22}$$

It can be considered that $(A_i \cap R_\varepsilon) \cap (A_{best} \cap R_\varepsilon) = \phi$ for $i \in (1, 2, \ldots, n)$ and $i \neq best$, especially in the early running time, so the following equation can be derived out:

$$P(exp) = \sum_{i=1}^{n} \frac{\nu(S_i \cap R_\varepsilon) \times m_i}{\nu(S_i)} < \frac{\nu(S_{best} \cap R_\varepsilon) \times m_{best}}{\nu(S_{best})} < \frac{\nu(R_\varepsilon) \times m_{best}}{\nu(S_{best})}. \tag{3.23}$$

So, Eq. (3.17) can be expressed as follows:

$$\frac{\nu(R_\varepsilon) \times n}{\nu(S)} \leq P(\xi(t+1) \in Y^* | \xi(t) \notin Y^*)$$

$$\leq \nu(R_\varepsilon) \left(\frac{n}{\nu(S)} + \frac{m_{best}}{\nu(S_{best})} \right). \tag{3.24}$$

The above Eq. (3.24) is more meaningful than Eq. (3.17), which tells that the best firework is more important than others. From Eq. (3.24) and Corollary 3.1, let $a = \frac{\nu(R_\varepsilon) \times n}{\nu(S)}$ and $b = \nu(R_\varepsilon) \left(\frac{n}{\nu(S)} + \frac{m_{best}}{\nu(S_{best})} \right)$, so it leads to the following equation:

$$\frac{\nu(S) \times \nu(S_{best})}{\nu(R_\varepsilon) \times (n \times \nu(S_{best}) + m_{best} \times \nu(S))} \times (1 - \lambda(0)) \leq E\gamma \leq \frac{\nu(S)}{\nu(R_\varepsilon) \times n} \times (1 - \lambda(0)), \tag{3.25}$$

where $\lambda(t) = P\{\xi(t) \in Y^*\}$.

According to FWA, the initialization of n fireworks is penetrated at random. The following results can be acquired:

Since $\lambda(0) = P\{\xi(0) \in Y^*\} \ll 1, 1 - \lambda(0) = 1$, thus

$$\frac{\nu(S) \times \nu(S_{best})}{\nu(R_\varepsilon) \times (n \times \nu(S_{best}) + m_{best} \times \nu(S))} \leq E\gamma \leq \frac{\nu(S)}{\nu(R_\varepsilon) \times n}. \tag{3.26}$$

Corollary 3.2 *FWA's expected convergence time $E\lambda$ is such that:*

$$\frac{\nu(S) \times \nu(S_{best})}{\nu(R_\varepsilon) \times (n \times \nu(S_{best}) + m_{best} \times \nu(S))} \leq E\gamma \leq \frac{\nu(S)}{\nu(R_\varepsilon) \times n}. \tag{3.27}$$

From Eq. (3.24), the more lager value of R_ε is, the smaller value of $\nu(S)$ is beneficial to the efficiency of FWA, but the two values are directly related to the search problems. Equation (3.23) indicates that $\nu(S_{best})$ and m_{best} are very important for estimating FWA's expected convergence time. But the above result is just obtained under the condition of some assumptions. The more exact analysis might be further done through considering the details of some equations of FWA.

In summary, like other swarm intelligence algorithms, FWA is able to be expressed as a Markov process. After some concepts of Markov stochastic process of Fireworks Algorithm, FWA's global convergence has been proved, and then the approximate region of expected convergence time of FWA has also been computed accordingly. Moreover, the time complexity of FWA is also to be analyzed in terms of an absorbing Markov process. Although the results given here are preliminary and incomplete, some theorems could provide a guide for studying the theory of Fireworks Algorithm for future.

3.5 Influence of Random Number Generators on FWA

Random numbers are widely used in intelligent optimization algorithms such as Genetic Algorithm (GA), Ant Colony Optimization (ACO) and Particle Swarm Optimization (PSO), Fireworks Algorithm (FWA) [1], just to name a few. Random numbers are usually generated by deterministic algorithms called Random Number Generators (RNGs) and play a key role in driving the search process. The performance of RNGs can be analyzed theoretically using criteria such as period and lattice structure [4, 5], or by systematic statistical test [6]. However, none of these analyses are relevant directly to RNGs' impact on optimization algorithms like PSO, FWA.

It is interesting to ask how RNGs can effect FWA as well as other stochastic methods. Clerc [7] replaced the conventional RNGs with a short length list of numbers (i.e., a RNG with a very short period) and empirically studied the performance of PSO. The experiments show that, at least for the moment, there is no sure way to build a "good" list for high performance. Thus, RNGs with certain degree of randomness are necessary for the success of stochastic search.

Bastos-Filho et al. [8, 9] studied the impact of the quality of CPU- and GPU-based RNGs on the performance of PSO. The experiments show that PSO needs RNGs with minimum quality and no significant improvements were achieved when comparing high-quality RNGs to medium-quality RNGs. Only Linear Congruential Generator (LCG) [4] and Xorshift algorithms [10] were compared for CPUs, and only one method for generating random numbers in an ad hoc manner on GPUs was adopted for comparing GPUs.

In general, RNGs shipped with math libraries of programming languages or other specific random libraries are used when implementing intelligent optimization algorithms. These RNGs generate random numbers of very diverse qualities with different efficiency. A comparative study on the impact of these popular RNGs will be helpful when implementing intelligent algorithms for solving optimization problems.

In this section, we selected 13 widely used, highly optimized, uniformly distributed RNGs and applied them to FWA for empirically comparing their impact on the optimization performance of FWA. Nine well-known benchmark functions were implemented for comparisons. All the experiments were conducted on the GPU for fast execution. Two novel strategies, league scoring strategy, and lose-rank strategy were introduced to conduct a systematic comparison on these RNGs' performance. Though the work is limited to the impact on FWA, other intelligent algorithms can also be studied in similar framework.

3.5.1 Random Number Generators

According to the source of randomness, random number generators fall into three categories [11]: true random number generators (TRNGs), quasirandom number generators (QRNGs), and pseudorandom number generators (PRNGs).

TRNGs utilize physical sources of randomness to provide truly unpredictable numbers. These generators are usually slow and unrepeatable, and usually need the support of specific hardware [5]. So TRNGs are hardly used in the field of stochastic optimization. QRNGs are designed to evenly fill an n-dimensional space with points. Though quite useful, they are not widely used in the domain of optimization. PRNGs are used to generate pseudorandom sequences of numbers that satisfy most of the statistical properties of a truly random sequence but are generated by a deterministic algorithm. PRNGs are the most common RNGs of the three groups, and provided by almost all programming languages. There also exist many well-optimized PRNGs for open access. As we only discuss PRNGs in this chapter, we will use random numbers and pseudorandom numbers alternatively henceforth.

Random numbers can subject to various distributions, such as uniform, normal, and cauchy distributions. Of all the distributions, uniform distribution is the simplest and the most important one. Not only uniform random numbers are widely used in many different domains, but also they are used as the base generators for generating random numbers subject to other distributions. Many methods, like transformation methods and rejection methods, can be used to convert uniformly distributed numbers to ones with specific nonuniform distributions [5, 12].

As here we only study uniform distribution, the remainder of the section merely introduces RNGs with uniform distribution. RNGs for generating uniform distribution random numbers can be classified into two groups, according to the basic arithmetic operations utilized: RNGs based on modulo arithmetic and RNGs based on binary arithmetic.

3.5.2 Modular Arithmetic Based RNGs

RNGs of this type yield sequences of random numbers by means of linear recurrence modulo m, where m is a large integer.

Linear Congruential Generator (LCG): LCG is one of the best-known random number generators. LCG is defined by the following recurrence relation:

$$x_i = a \cdot x_{i-1} + c \mod m, \tag{3.28}$$

where x is the sequence of the generated random numbers and $m > 0, 0 < a < m$, and $0 \leq c, x_0 < m$. If uniform distribution on $[0, 1)$ is need, then use $u = \frac{x}{m}$ as the output sequence.

For LCG, a, c and m should be carefully chosen to make sure that maximum period can be achieved [4]. LCG can be easily implemented on computer hardware which can provide modulo arithmetic by storage-bit truncation. RNG using LCG is shipped with C library (rand()) as well as many other languages such as Java (java.lang.Random). LCG has a relatively short period (at most 2^{32} for 32-bit integer) compared to other more complicated ones.

A special condition of LCG is when $c = 0$, which presents a class of multiplicative congruential generators (MCG) [13]. Multiple carefully selected MCGs can be combined into more complicated algorithms such as Whichmann–Hill generator [14].

Multiple Recursive Generator (MRG): MRG is a derivative of LCG and can achieve much longer period. An MRG of order k is defined as follows:

$$x_i = (a_1 \cdot x_{i-1} + a_2 \cdot x_{i-2} + \cdots + a_k \cdot x_{i-k}) \mod m. \tag{3.29}$$

The recurrence has maximal period length $m^k - 1$, if tuple $(a_1, \ldots, a_k)$ has certain properties [4].

Though these generators can provide both good statistical quality and long periods, a necessary condition for a good figure of merit with respect to the spectral test is that $\sum_{i=1}^{k} a_i^2$ be large [15]. However, the relatively prime moduli require complex algorithms using 32-bit multiplications and divisions, so they are not suitable for current GPUs.

Combined Multiple Recursive Generator (CMR): CMR combines multiple MRGs and can obtain better statistical properties and longer periods compared with a single MRG. A well-known implementation of CMR, CMR32k3a [15], combines two MRGs:

$$\begin{aligned} x_i &= a_{11} \cdot x_{i-1} + a_{12} \cdot x_{i-2} + x_{13} \cdot x_{i-3} \mod m_1 \\ y_i &= a_{21} \cdot y_{i-1} + a_{22} \cdot y_{i-2} + x_{23} \cdot y_{i-3} \mod m_2 \\ z_i &= x_i - y_i \mod m_1 \end{aligned} \tag{3.30}$$

where z forms the required sequence.

Note that combining the two multiple recursive generators (MRG) will lead to sequences with better statistical properties in high dimensions and longer periods compared with those generated from a single MRG. The combined generator described above has a period length of approximately 2^{191}.

3.5.3 Binary Arithmetic Based RNGs

RNGs of this type are defined directly in terms of bit strings and sequences. As computers are fast for binary arithmetic operations, binary arithmetic-based RNGs can be more efficient than modulo arithmetic-based ones.

Xorshift: Xorshift [10] produces random numbers by means of repeated use of bitwise exclusive-or (xor, $\oplus$) and shift ($\ll$ for left and $\gg$ for right) operations.

A xorshift with four seeds (x, y, z, w) can be implemented as follows:

$$
\begin{aligned}
t &= (x \oplus (x_i \ll a)) \\
x &= y \\
y &= z \\
z &= w \\
w &= (w \oplus (w \gg b)) \oplus (t \oplus (t \gg c))
\end{aligned}
\tag{3.31}
$$

where w forms the required sequence.

With a carefully selected tuple (a, b, c), the generated sequence can have a period as long as $2^{128} - 1$.

Mersenne Twister (MT): MT [16] is one of the most widely respected RNGs, it is a twisted Generalized Feedback Shift Register (GFSR). The underlying algorithm of MT is as follows:

- Set r w-bit numbers $(x_i, i = 1, 2, \ldots, r)$ randomly as initial values.
- Let

$$
A = \begin{pmatrix} 0 & I_{w-1} \\ a_w & a_{w-1} \cdots a_1 \end{pmatrix},
\tag{3.32}
$$

where I_{w-1} is the $(w-1) \times (w-1)$ identity matrix and $a_i, i = 1, \ldots, w$ take values of either 0 or 1. Define

$$
x_{i+r} = \left(x_{i+s} \oplus \left(x_i^{(w:(l+1))} | x_{i+1}^{(l:1)} \right) A \right),
\tag{3.33}
$$

where $x_i^{(w:(l+1))} | x_{i+1}^{(l:1)}$ indicates the concatenation of the most significant (upper) $w - l$ bits of x_i and the least significant l bits of x_{i+1}.

- Perform the following operations sequentially:

$$
\begin{aligned}
z &= x_{i+r} \oplus (x_{i+r} \gg t_1) \\
z &= z \oplus ((z \ll t_2) \,\&\, m_1) \\
z &= z \oplus ((z \ll t_3) \,\&\, m_2) \\
z &= z \oplus (x \gg t_4) \\
u_{i+r} &= z/(2^w - 1)
\end{aligned}
\tag{3.34}
$$

where t_1, t_2, t_3 and t_4 are integers and m_1 and m_2 are bit masks and '&' is a bitwise AND operation.

$u_{i+r}, i = 1, 2, \ldots$ form the required sequence on interval $(0, 1]$.

With proper parameter values, MT can generate sequence with a period as long as $2^{19,937}$ and extremely good statistical properties [16]. Strategies for selecting good initial values are studied in [17], while Saito et al. [18] proposed efficient implementation for fast execution on GPUs.

3.5.4 Experimental Setup

In this section, we describe the experimental environment and parameter settings in detail.

3.5.4.1 Testbed

We conducted our experiments on a PC running 64-bit Windows 7 Professional with 8G DDR3 Memory and Intel core I5-2310 (@2.9 GHz 3.1 GHz). The GPU used for implementing FWA in the experiments is NVIDIA GeForce GTX 560 Ti with 384 CUDA cores. The program was implemented with C and compiled with Visual Studio 2010 and CUDA 5.5.

3.5.4.2 RNGs Used for Comparison

Besides functions provided by programming languages, many libraries with well-implemented RNGs are available, such as AMD's ACML [21] and Boost Random Number Library [22] targeted at CPUs and specific implementations [11, 17, 23] for GPU platform.

Among all these candidates, Math Kernel Library (MKL) [24] (for CPU) and CURAND [25] (for GPU) were selected for the experiments considering the following reasons: (1) RNGs provided by the two libraries cover the most popular RNG algorithms, and (2) both MKL and CURAND are well optimized for our hardware platform (I5 CPU and GeForce 560 Ti GPU), so a fair comparison of efficiency can be expected. So experiments with these two libraries are broadly covered in terms of types of RNGs and present a fair comparison in terms of time efficiency.

As LCG is widely shipped by standard library of various programming language, we added a RNG with LCG (C's rand()). The RNGs used in the experiments are listed by Table 3.1.

3.5.4.3 Benchmark Functions

For an extensive experiment, GPU-based test suit, cuROB [26] was used. cuROB has as many as 37 test functions including unimodal, multimodal, hybrid and combination functions, and it supports any dimension.

Table 3.1 Random number generators tested

No.	Algorithm	Description	Note
1	xorshift	Implemented using the xorshift algorithm [10], created with generator type CURAND_RNG_PSEUDO_XORWOW	CURAND with CUDA Toolkit 5.5
2	xorshift	Same algorithm as 1, faster but probably statistically weaker, set ordering to CURAND_ORDERING_PSEUDO_SEEDED	
3	Combined Multiple Recursive	Implemented using the Combined Multiple Recursive algorithm [15], created with generator type CURAND_RNG_PSEUDO_MRG32K3A	
4	Mersenne Twister	Implemented using the Mersenne Twister algorithm with parameters customized for operation on the GPU [18], created with generator type CURAND_RNG_PSEUDO_MTGP32	
5	Multiplicative Congruential	Implemented using the 31-bit Multiplicative Congruential algorithm [13], create with parameter VSL_BRNG_MCG31	MKL 11.1
6	Generalized Feedback Shift Register	Implemented using the 32-bit generalized feedback shift register algorithms, create with parameter VSL_BRNG_R250 [19]	
7	Combined Multiple Recursive	Implemented using Combined Multiple Recursive algorithm [15], create with parameter VSL_BRNG_MRG32K3A	
8	Multiplicative Congruential	Implemented using the 59-bit Multiplicative Congruential algorithm from NAG Numerical Libraries [14], create with parameter VSL_BRNG_MCG59	
9	Wichmann-Hill	Implemented using the Wichmann-Hill algorithm from NAG Numerical Libraries [14], create with parameter VSL_BRNG_WH	
10	Mersenne Twister	Implemented using the Mersenne Twister algorithm MT19937 [16], create with parameter VSL_BRNG_MT19937	
11	Mersenne Twister	Implemented using the Mersenne Twister algorithms MT2203 [20] with a set of 6024 configurations. Parameters of the generators provide mutual independence of the corresponding sequences, create with parameter VSL_BRNG_MT2203	
12	Mersenne Twister	Implemented using the SIMD-oriented Fast Mersenne Twister algorithm SFMT19937 [17], create with parameter VSL_BRNG_SFMT19937	
13	Linear Congruential	Implementing using Linear Congruential algorithm with $a = 1103515245$, $c = 12345$, $m = 2^{32}$, only high 16 bits are used as output	MS Visual Studio C library rand()

3.5.5 Experimental Results and Analysis

This section presents the experimental results. Both efficiency of RNGs and solution quality of FWA using each RNG are described and analyzed in detail.

3.5.5.1 Solution Quality

In all experiments, 51 independent trials were performed for each function, where 10,000 iterations were executed for each trial. 51 integer numbers were randomly generated from uniform distribution as seeds for each trail, and all RNGs shared the same 51 seeds. All fireworks were initialized randomly within the whole feasible search space and the initialization was shared by all RNGs (to be exactly, RNG No. 10 was used for the purpose of initialization).

For a particular function, the solution quality can be compared between any two RNGs with statistical test. But there is no direct way to compare groups of RNGs (13 in our experiments).

Lose-Rank Strategy: Lose rank can be calculated as follows: For a certain RNG, say R1, set its lose rank to 0. R1 compares its solutions for a function with those of all other RNGs one after another. If R1 is statistically worse than some RNGs, then add 1 to its lose rank. In this way, we can calculate all RNGs' lose ranks for all functions.

The idea underlying lose rank is that if some RNG performs significantly worse in terms of solution quality, then it has a relative large lose rank.

Following this strategy, we can find that the lose ranks of almost all functions are zeros (for this reason, the lose rank values are omitted here). Based on the observation, there exist no significant bad RNGs, and there is no outright good ones. There is no strong reason to prefer any RNG to others as far as its impact on solution quality is concerned.

3.5.5.2 Efficiency of RNGs

We ran each RNG program to generate random numbers in batch of different sizes and test the speed. The results are presented in Table 3.2 (Note that C's rand() does not support batch generating of random numbers, so the average speed was measured instead).

In general, RNGs based on both CPUs and GPUs achieve better performance by generating batches of random numbers, and GPUs need larger batch size to get peak performance than CPUs. In the condition of large batch size, CURAND (No.1–No.4) can be several to tens fold faster than MKL for the same algorithms.

Modulo arithmetic-based RNGs are less efficient than binary arithmetic ones, just as aforementioned. Combined Multiple Recursive algorithm (No. 7) and Wichmann–Hill algorithm (No. 9) present the slowest RNGs, followed by Multiplicative

Table 3.2 RNG efficiency comparison under different batch size (# of random numbers per nanosecond)

Batch Size	GPU CURAND				CPU MKL								C
	1	2	3	4	5	6	7	8	9	10	11	12	13
1	0.3	0.3	0.3	0.3	17.5	17.3	14.8	15.9	7.8	15.2	14.7	15.2	36.5
10	2.7	2.6	1.9	2.6	95.2	116.3	54.6	109.9	56.8	93.5	87.7	129.9	
20	4.8	4.9	3.2	5.1	144.9	166.7	67.1	192.3	79.4	181.8	161.3	227.3	
50	10.9	11.0	7.2	12.8	294.1	227.3	112.4	285.7	120.5	344.8	294.1	454.5	
100	21.1	21.6	14.2	25.8	400.0	263.2	138.9	357.1	142.9	434.8	384.6	714.3	
200	41.7	43.7	28.9	49.8	555.6	285.7	156.3	416.7	158.7	500.0	555.6	1111.1	
500	104.2	114.9	76.9	125.0	625.0	416.7	178.6	454.5	166.7	625.0	666.7	1428.6	
1000	200.0	243.9	163.9	232.6	666.7	555.6	185.2	454.5	172.4	666.7	769.2	1111.1	
2000	312.5	476.2	344.8	434.8	666.7	666.7	188.7	454.5	172.4	714.3	769.2	1250.0	
5000	500.0	1250.0	1000.0	909.1	769.2	769.2	192.3	476.2	172.4	833.3	833.3	1428.6	
10000	588.2	2500.0	1428.6	1250.0	714.3	833.3	192.3	476.2	175.4	833.3	833.3	1428.6	
20000	666.7	3333.3	1666.7	1666.7	714.3	833.3	192.3	476.2	175.4	769.2	769.2	1428.6	
50000	769.2	10000.0	2500.0	3333.3	769.2	833.3	192.3	476.2	175.4	769.2	769.2	1428.6	
100000	714.3	10000.0	2500.0	5000.0	714.3	833.3	192.3	476.2	175.4	769.2	769.2	1428.6	
200000	714.3	10000.0	2500.0	10000.0	714.3	714.3	192.3	476.2	175.4	769.2	769.2	1666.7	

Congruential (No. 8). As a comparison, Mersenne Twister algorithm presents the fastest RNGs. CPU-based SFMT19937 (No. 12) can be one order of magnitude faster than CPU-based CMR32K3A (No. 7), while the GPU version (No. 4) can be 5-fold faster than the CPU implementation. Considering the good statistical property of MT [16, 17, 20], it makes the best RNG of all the RNGs concerned.

3.5.5.3 Remarks

In general, both CPU- and GPU-based RNGs can achieve best performance when generating blocks of random numbers that are as large as possible. Fewer calls to generate many random numbers is more efficient than many calls generating only a few random numbers. GPU-based RNGs can be several fold even one order of magnitude faster than their CPU-based counterparts, moreover Mersenne Twister algorithm presents the most efficient RNG.

Finally, despite only FWA using uniformly distributed RNGs was discussed here, however, the proposed strategies could be also extended to compare any other intelligent algorithm on real-world optimization problems or benchmark test suites. RNGs for nonuniform distributions could also be exploited and studied in the framework of the proposed strategies.

3.6 Summary

In this chapter, first of all, the modeling, global convergence, and time complexity of FWA were presented in detail, then the influence of random number generators to the performance of FWA was investigated.

For theoretical analysis of FWA, this chapter defined some concepts of Markov stochastic process of FWA and proved that FWA is global convergence, and then computed the approximate region of expected convergence time of FWA. Although the results given here are preliminary and incomplete, some theorems could provide a guide for studying the theory of Fireworks Algorithm in the future.

For random generator influence on FWA, though different RNGs have various statistical strengths, no significant disparity was observed in FWA by experiments. Even the most common linear congruential generator performs very well despite the fact that random number sequences generated by LCG are of lower quality in terms of randomness compared to other more complicated RNGs. As a result, it is reasonable to utilize the most efficient algorithms for random number generation used by an intelligent optimization like FWA.

References

1. Y. Tan, Y. Zhu, Fireworks algorithm for optimization, in *Advances in Swarm Intelligence* (Springer, Berlin, 2010), pp. 355–364
2. H. Huang, Z. Hao, Y. Qin, Time complexity of evolutionary programming. J. Comput. Res. Dev. **45**(11), 1850–1857 (2008)
3. H. Huang, Z. Hao, W. Chunguo, Y. Qin, Time convergence speed of ant colony optimization. J. Comput. **30**(8), 1344–1353 (2007)
4. D.E. Knuth, *The Art of Computer Programming, Volume 2: Seminumerical Algorithms* (Amsterdam, London, 1969)
5. P. L'Ecuyer, Random number generation, in *Handbook of Computational Statistics*. Springer Handbooks of Computational Statistics, ed. by J.E. Gentle, W.K. Hardle, Y. Mori (Springer, Berlin, 2012), pp. 35–71. doi:10.1007/978-3-642-21551-3_3. ISBN: 978-3-642-21550-6
6. P. L'Ecuyer, R. Simard, TestU01: A C library for empirical testing of random number generators. ACM Trans. Math. Softw. **33**(4), (2007). doi:10.1145/1268776.1268777. ISSN: 0098-3500
7. M. Clerc, List-based optimisers: experiments and open questions. Int. J. Swarm Intell. Res. (IJSIR) **4**(4), 23–38 (2013)
8. C.J.A. Bastos-Filho, J.D. Andrade, M.R.S. Pita, A.D. Ramos, Impact of the quality of random numbers generators on the performance of particle swarm optimization, in *IEEE International Conference on Systems, Man and Cybernetics, SMC 2009* (2009), pp. 4988–4993. doi:10.1109/ICSMC.2009.5346366
9. C.J.A. Bastos-Filho, M.A.C. Oliveira, D.N.O. Nascimento, A.D. Ramos, Impact of the random number generator quality on particle swarm optimization algorithm running on graphic processor units, in *2010 10th International Conference on Hybrid Intelligent Systems (HIS)* (2010), pp. 85–90
10. G. Marsaglia, Xorshift RNGs. J. Stat. Softw. **8**(14), 1–6 (2003)
11. L. Howes, D. Thomas, Efficient random number generation and application using CUDA, *GPU Gems 3* (Addison-Wesley Professional, Boston, 2007), pp. 805–830
12. J.A. Rice, *Mathematical Statistics and Data Analysis* (Thomson Higher Education, Belmont, 2007)
13. P. L'Ecuyer, Tables of linear congruential generators of different sizes and good lattice structure. Math. Comput. **68**(225), 249–260 (1999). doi:10.1090/S0025-5718-99-00996-5. ISSN: 0025-5718
14. The Numerical Algorithms Group Ltd, NAG Library Manual. Mark **23**, (2011)
15. P. L'Ecuyer, Good parameters and implementations for combined multiple recursive random number generators. Oper. Res. **47**(1), 159–164 (1999)
16. M. Matsumoto, T. Nishimura, Mersenne twister: a 623-dimensionally equidistributed uniform pseudo-random number generator. ACM Trans. Model. Comput. Simul. (TOMACS) **8**(1), 3–30 (1998)
17. M. Saito, M. Matsumoto, SIMD-oriented fast mersenne twister: a 128-bit pseudorandom number generator, in *Monte Carlo and Quasi-Monte Carlo Methods 2006*, ed. by A. Keller, S. Heinrich, H. Niederreiter (Springer, Berlin, 2008), pp. 607–622. doi:10.1007/978-3-540-74496-2_36. ISBN: 978-3-540-74495-5
18. M. Saito, M. Matsumoto, Variants of mersenne twister suitable for graphic processors. ACM Trans. Math. Softw. **39**(2), 12:1–12:20 (2013). doi:10.1007/978-3-540-74496-2_36. ISBN: 978-3-540-74495-5
19. S. Kirkpatrick, E.P. Stoll, A very fast shift-register sequence random number generator. J. Comput. Phys. **40**(2), 517–526 (1981)
20. M. Matsumoto, T. Nishimura, Dynamic creation of pseudorandom number generators, in *Monte Carlo and Quasi-Monte Carlo Methods 1998*, ed. by H. Niederreiter, J. Spanier (Springer, Berlin, 2000), pp. 56–69
21. AMD Inc. Core Math Library (ACML)
22. The Boost Random Number Library

23. W.B. Langdon, A fast high quality pseudo random number generator for NVIDIA CUDA, in *Proceedings of the 11th Annual Conference Companion on Genetic and Evolutionary Computation Conference: Late Breaking Papers, GECCO'09* (ACM, New York, 2009), pp. 2511–2514. doi:10.1145/1570256.1570353. ISBN: 978-1-60558-505-5
24. Intel Corp. The Math Kernel Library
25. NVIDIA Corp. CURAND Library Programming Guide v5.5 (2013)
26. K. Ding, Y. Tan, cuROB: A GPU-based Test Suit for Real-Parameter Optimization, in *Advances in Swarm Intelligence*. Lecture Notes in Computer Science, vol. 8794, ed. by Y. Tan, Y.H. Shi, C.A. Coello Coello (Springer, Berlin, 2014), pp. 66–78

Part II
FWA Variants

Since the invention of the Fireworks Algorithm (FWA), it has attracted a lot of attention in the community of swarm intelligence (SI). Recently, many improvement works on FWA have been proposed and studied in-depth. These improvement works can be classified into two categories, one category is based on the analysis of limitations of FWA and then make the related improvements, and in the other category, FWA is hybridized with other SI algorithms.

In the following chapters, we present several important improvement works till date. In Chap. 4, we introduce the fireworks algorithm based on function approximation approaches. In Chap. 5, we introduce another improvement work which proposes new calculation methods for explosion amplitude and explosion sparks number in FWA. In Chap. 6, the enhanced fireworks algorithm (EFWA) is presented, which gives the comprehensive analysis of FWA and points out five limitations in FWA which leads to the related improvement work on the limitations.

In Chaps. 7 and 8, two adaptive algorithms, namely the dynamic search in FWA (dynFWA) and adaptive fireworks algorithm (AFWA), are proposed based on automatically adjusting the explosion amplitude in FWA. The performances of dynFWA and AFWA are greatly enhanced over the previous work on FWA. By incorporating a probabilistically oriented explosion mechanism (POEM) into the conventional FWA, a novel cooperative fireworks algorithm (CoFWA) is proposed in Chap. 9 to enhance the interactions among the individual fireworks in the swarm. In Chap. 10, hybrid works between FWA and other SI algorithms are given.

Chapter 4
FWA Based on Function Approximation Approaches

In this chapter, we present an empirical study on the influence of approximation approaches on accelerating the fireworks algorithm search by elite strategy [1]. We use three sampling data methods to approximate fitness landscape, i.e., the best fitness sampling method, the sampling distance near the best fitness individual sampling method, and the random sampling method. For each approximation method; we conduct a series of combinative evaluations with different sampling methods and sampling numbers for accelerating fireworks algorithm. The experimental evaluations on benchmark functions show that this elite strategy can enhance the fireworks algorithm search capability effectively. We also analyze and discuss the related issues on the influence of approximation model, sampling method, and sampling number on the fireworks algorithm acceleration performance.

4.1 Introduction

Evolutionary computation (EC) acceleration is a promising study direction in recent EC community [2]. On the one hand, the acceleration approach study is practical fundamental to establish the EC convergence theory, on the other hand, accelerated EC algorithms (ECs) benefit their practical applications in industrial society. Approximating ECs fitness landscape and conducting an elite strategy is one of the effective approaches to accelerate ECs, Reference [3] proposed an approximation approach in original n-D dimensional search space to accelerate EC and Refs. [4, 5] extended its work to approximate fitness landscape in projected one-dimensional search space.

Fireworks algorithm is a new ECs and a population-based optimization technique [6]. Compared with the other ECs, fireworks algorithm presents a different search manner, which is to search the nearby space by simulating the explosion phenomenon. Since it was proposed, fireworks algorithm has shown its significance and superiority in dealing with the optimization problems of non-negative matrix factorization [7, 8]. Although the fireworks algorithm has shown its advantage than the other EC algorithms, it still has the further improvement possibility by elite strategy. To obtain

© Springer-Verlag Berlin Heidelberg 2015
Y. Tan, *Fireworks Algorithm*, DOI 10.1007/978-3-662-46353-6_4

elite from approximated fitness landscape, in this paper, we used some variations of sampling methods for approximating fireworks algorithm fitness landscape. It includes the best fitness sampling method, the sampling distance near the best fitness individual sampling method and the random sampling method. We apply an elite strategy to enhance fireworks algorithm search capability with different sampling methods and different sampling data number.

The originalities of the proposed method are: (1) this study discovers that elite strategy is an efficient method to accelerate the fireworks algorithm search; (2) the influence of sampling methods on acceleration performance of fireworks algorithm is investigated; and (3) the influence of sampling number on acceleration performance of the fireworks algorithm in each sampling method is also studied.

4.2 Fireworks Algorithm

The fireworks algorithm generates offspring candidates around their parent individual in search space as if the fireworks explosion generates sparks around its explosion point. Given a single objective function $f : \Omega \subseteq R^D \rightarrow R$, it is supposed that fireworks algorithm is to find a point $x \in \Omega$, which has the minimal fitness. It is assumed that the population size of fireworks is N and the population size of explosion spark is m and Gaussian mutation sparks is $\hat{m}$. Each fireworks $i (i = 1, 2, ..., N)$ in a population has the following properties: a current position X_i, a current explosion amplitude A_i and the number of generated spark s_i. The fireworks generates a number of sparks within a fixed explosion amplitude. AS to the optimization problem f, a point with better fitness is considered as a potential solution, which the optima locate nearby with high chance, vice versa.

4.3 Fireworks Algorithm Acceleration by Elite Strategy

4.3.1 Motivation

It is a promising study topic on approximating fitness landscape to accelerate ECs search, Ref. [9] investigated the fitness landscape approximation approaches and recent related topic on surrogate-assisted EC was reported in Ref. [10]. A novel fitness landscape approximation method using Fourier Transform was introduced in [11, 12]. References [3–5] conducted some related works on fitness landscape approximation for accelerating ECs in D and one-dimensional search space, respectively. There are several issues that we are still not clear regarding the influence of the different sampling method with the different number to the ECs approximation model accuracy and acceleration performance. However, previous works show little concern on the influence of the different sampling methods with the different number to the ECs approximation model accuracy and acceleration performance. Some related research topics need to be investigated furthermore. First, there is no related report on the different sampling method influence to the approximation model accuracy and fireworks algorithm acceleration. In this study, we use three sampling data, i.e., best sampling, distance near the best fitness individual sampling method, and random sampling to investigate the sampling method influence to the model accuracy and fireworks algorithm acceleration. Second, we also want to investigate the influence of different sampling number on the approximation model accuracy and fireworks algorithm acceleration. In this work, we set the different sampling number (3, 5, 10) in each sampling method to conduct a set of comparative evaluations to study on those issues.

4.3.2 Sampling Methods

1. Best Sampling Method, selects the best K individuals as sampling data.
2. Distance Near the Best Fitness Individual Sampling Method, selects the nearest K individuals to the best individual using Euclidean distance as sampling data.
3. Random Sampling Method, selects K individuals randomly as sampling data.

4.3.3 Fireworks Algorithm with an Elite Strategy

Our elite strategy for approximating fitness landscape uses only one of the D parameter axes at a time instead of all D parameter axes, and projects individuals onto each 1 regression space. Each of the D-dimensional regression spaces has M projected individuals, which come from the different sampling methods with a different sampling number. We approximate the landscape of each one-dimensional regression space using the projected M individuals and select the elite from the D approximated one-dimensional landscape shapes.

In our evaluations, the sampling number is set as 3, 5, and 10; i.e., $K = 3, 5$, and 10 to check the sampling number's influence to the acceleration performance. We test least squares approximation for approximating the one-dimensional regression search spaces with one and two degree polynomial functions. The elite are generated from the resulting approximated shapes, see Fig. 4.1.

The actual least square regression functions used is polynomial curve fitting, given by the Eq. (4.1).

$$\begin{pmatrix} x_{11}^t & x_{12}^{t-1} & \cdots & x_{1K}^0 \\ x_{21}^t & x_{21}^{t-1} & \cdots & x_{2K}^0 \\ \vdots & \vdots & & \vdots \\ x_{D1}^t & x_{D2}^{t-1} & \cdots & x_{DK}^0 \end{pmatrix} \begin{pmatrix} a_1 \\ a_2 \\ \vdots \\ a_K \end{pmatrix} = \begin{pmatrix} y_1 \\ y_2 \\ \vdots \\ y_K \end{pmatrix} \tag{4.1}$$

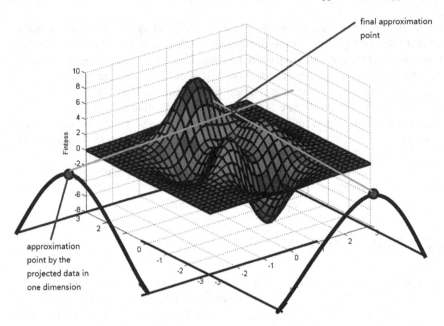

Fig. 4.1 Original D dimensional space and one-dimensional spaces obtained by reducing the dimensions of the original one

where x_{ij}, $(i = 1, 2, \ldots, D)$ and $(j = 1, 2, \ldots, K)$ are the projected individual of point set x_i, $(i = 1, 2, \ldots, K)$ and their fitness values among (x_{ij}, y_j) in the ith one-dimensional regression space for $(i = 1, 2, \ldots, D)$ and $(j = 1, 2, \ldots, K)$, a_0, $a_1, \ldots, a_k$ are the parameters obtained by least squares method, t is the power of polynomial function.

Least square approximation by two degree polynomial function (LS2) simplifies a regression space with a nonlinear curve, and it is easy to obtain its inflection point from its gradient, using the inflection point as the elite. Linear least square approximation (LS1) uses a linear function to approximate the regression space. Its gradient is either descent or ascent. A safer approach, taking into account both descent and ascent, is to select the average point of the linear approximation line as the elite.

The proposed methods replace the worst individual in each generation with the selected elite. Although we cannot deny the small possibility that the global optimum is located near the worst individual, the possibility that the worst individual will become a parent in the next generation is also low; removing the worst individual therefore presents the least risk and is a reasonable choice.

The whole flowing of fireworks algorithm with an elite strategy shows in Algorithm 4.1.

Algorithm 4.1 Pseudocode of Fireworks Algorithm with an Elite Strategy

---

1: initialization.
2: randomly select N fireworks at N locations
3: **for** $t = 1$ to $maxIter$ **do**
4:    evaluate the N fireworks using objective function
5:    calculate the number of spark and explosion amplitude of each fireworks
6:    generate explosion sparks by algorithm 2.1
7:    generate Gaussian mutation sparks by 2.2
8:    obtain the optimal candidate in the **Set** which includes generating explosion sparks, $\hat{m}$ Gaussian
      mutation sparks and N fireworks
9:    randomly select the other $N - 1$ fireworks
10:   obtain N fireworks for the next iteration
11:   approximate fitness landscape in each projected one-dimensional search space
12:   obtain a spark from approximated curves by Elite Strategy
13:   **if** *Elite Fitness > Worst Individual Fitness* **then**
14:      replace the worst individual
15:   **end if**
16: **end for**
17: return the optima

4.4 Experimental Evaluations

4.4.1 Experimental Design

To investigate the influence of sampling methods and sampling number on the acceleration performance, we used the fireworks algorithm as the optimization method to study our research proposals. Ten benchmark functions are selected as a test suite [13], Table A.4 shows their range, optima, and characteristic. In our experiments, we compare our proposed acceleration approaches by least squares approximation with a normal fireworks algorithm, conduct t-test for each proposed acceleration methods, and of course, we checked normality of data distribution before applying t-test.

Here, we abbreviate the fireworks algorithm where the search space is regressed by a two degree least square approximation as FWA-LS2, where it is regressed by a line power function least squares approximation as FWA-LS1. The best distance near the best fitness individual and random sampling method are abbreviated as BST, DIS, and RAN, followed by the sampling number. These abbreviations are also used in Tables 4.1, 4.2, and 4.3.

In this method, the number of fireworks (N) and Gaussian mutation firework ($\hat{m}$) are both set as 8 and other parameters are set as in Ref. [6]. Experimental evaluations run 50 trails of 1000 generations on each ten benchmark functions independently. All the benchmark functions are set as 10 dimensions. The experimental platform for all evaluations is MATLAB®2011b, based on an Intel(R) Core(TM) i7-2600 CPU, 4G RAM machine, and runs under Windows®7.

Table 4.1 Mean and variance of 10 benchmark functions (the variance is parenthesis) used in experimental evaluations, the bold font numbers show the better final results

Methods	F1	F2	F3	F4	F5
N	−4.2E+02 (1.5E+03)	6.6E+02 (4.8E+05)	4.7E+06 (1.5E+13)	2.4E+03 (2.5E+06)	2.8E+03 (5.4E+06)
LS1-BST3	−4.2E+02 (4.7E+03)	7.4E+02 (1.0E+06)	4.0E+06 (8.4E+12)	2.7E+03 (2.2E+06)	3.5E+03 (8.6E+06)
LS1-BST5	−4.1E+02 (3.1E+04)	4.4E+02 (6.1E+05)	4.5E+06 (8.8E+12)	2.1E+03 (1.8E+06)	2.9E+03 (8.7E+06)
LS1-BST10	−4.4E+02 (4.5E+02)	3.5E+02 (4.6E+05)	5.2E+06 (2.1E+13)	2.2E+03 (2.7E+06)	3.4E+03 (7.2E+06)
LS2-BST3	−4.5E+02 (1.2E−01)	−8.0E+01 (6.5E+04)	1.8E+06 (3.0E+12)	4.2E+02 (2.3E+05)	1.9E+03 (2.4E+06)
LS2-BST5	−4.5E+02 (1.3E−01)	−1.0E+02 (8.5E+04)	1.6E+06 (2.4E+12)	2.1E+02 (1.7E+05)	2.2E+03 (4.6E+06)
LS2-BST10	−4.5E+02 (8.8E−02)	−6.2E+01 (9.1E+04)	1.4E+06 (1.8E+12)	−3.1E−01 (5.9E+04)	1.4E+03 (2.2E+06)
LS1-DIS3	−4.3E+02 (5.0E+02)	4.2E+02 (5.6E+05)	3.9E+06 (1.0E+13)	2.3E+03 (2.3E+06)	3.3E+03 (6.5E+06)
LS1-DIS5	−4.1E+02 (8.8E+03)	6.3E+02 (6.7E+05)	4.9E+06 (1.8E+13)	2.1E+03 (2.5E+06)	3.3E+03 (5.8E+06)
LS1-DIS10	−4.0E+02 (4.3E+04)	4.6E+02 (4.8E+05)	5.3E+06 (3.1E+13)	2.3E+03 (3.2E+06)	3.3E+03 (7.8E+06)
LS2-DIS3	−4.4E+02 (4.0E+02)	1.2E+02 (1.6E+05)	2.0E+06 (2.0E+12)	7.5E+02 (4.0E+05)	2.3E+03 (4.9E+06)
LS2-DIS5	−4.4E+02 (2.6E+02)	−3.0E+00 (8.6E+04)	1.5E+06 (1.4E+12)	5.7E+02 (2.7E+05)	2.4E+03 (3.0E+06)
LS2-DIS10	−4.4E+02 (2.6E+02)	−5.1E+00 (7.6E+04)	1.4E+06 (1.9E+12)	3.2E+02 (2.1E+05)	2.0E+03 (3.7E+06)
LS1-RAN3	−3.8E+02 (8.7E+04)	7.8E+02 (9.0E+05)	4.9E+06 (2.8E+13)	2.2E+03 (2.1E+06)	2.9E+03 (5.7E+06)
LS1-RAN5	−3.7E+02 (1.4E+05)	7.0E+02 (8.3E+05)	5.0E+06 (2.8E+13)	2.4E+03 (2.5E+06)	2.4E+03 (6.1E+06)
LS1-RAN10	−3.8E+02 (1.7E+05)	5.7E+02 (7.4E+05)	4.5E+06 (1.6E+13)	2.5E+03 (2.0E+06)	3.2E+03 (6.5E+06)
LS2-RAN3	**−4.5E+02 (4.5E−02)**	−3.1E+02 (9.5E+03)	9.1E+05 (6.6E+11)	−1.5E+02 (3.9E+04)	1.8E+03 (1.7E+06)
LS2-RAN5	−4.5E+02 (9.5E−02)	**−3.5E+02 (5.2E+03)**	9.5E+05 (6.0E+11)	**−2.5E+02 (1.1E+04)**	1.2E+03 (1.1E+06)
LS2-RAN10	−4.5E+02 (5.1E+00)	−3.3E+02 (3.5E+03)	**8.4E+05 (2.9E−11)**	−2.4E+02 (1.3E+04)	**−1.3E+02 (2.1E+05)**

(continued)

Table 4.1 (continued)

Methods	F6	F7	F8	F9	F10
N	4.6E+04 (1.9E+10)	−1.8E+02 (2.4E+01)	−1.2E+02 (6.3E−03)	−3.1E+02 (1.2E+02)	−2.8E+02 (3.1E+02)
LS1-BST3	1.5E+05 (1.0E+12)	−1.8E+02 (7.7E+01)	−1.2E+02 (9.6E−03)	−3.1E+02 (1.1E+02)	−2.8E+02 (3.6E+02)
LS1-BST5	8.9E+03 (5.2E+08)	−1.8E+02 (6.5E+00)	−1.2E+02 (8.4E−03)	−3.1E+02 (1.2E+02)	−2.8E+02 (3.7E+02)
LS1-BST10	2.1E+04 (6.9E+09)	−1.8E+02 (1.2E+01)	−1.2E+02 (8.7E−03)	−3.1E+02 (1.4E+02)	−2.8E+02 (3.8E+02)
LS2-BST3	**1.6E+03(3.5E+06)**	−1.8E+02 (1.1E+00)	−1.2E+02 (6.5E−03)	−3.2E+02 (3.8E+01)	−2.8E+02 (2.2E+02)
LS2-BST5	1.6E+03 (3.8E+06)	−1.8E+02 (1.0E+00)	−1.2E+02 (1.1E−02)	−3.2E+02 (3.8E+01)	−2.8E+02 (2.9E+02)
LS2-BST10	2.0E+03 (6.2E+06)	−1.8E+02 (9.7E−01)	**−1.2E+02(4.8E−03)**	**−3.2E+02(2.2E+01)**	−2.8E+02 (3.3E+02)
LS1-DIS3	3.6E+03 (1.4E+07)	−1.8E+02 (2.4E+01)	−1.2E+02 (1.0E−02)	−3.1E+02 (1.1E+02)	−2.8E+02 (2.9E+02)
LS1-DIS5	1.1E+04 (1.3E+09)	−1.8E+02 (1.4E+01)	−1.2E+02 (8.7E−03)	−3.1E+02 (1.1E+02)	−2.8E+02 (2.5E+02)
LS1-DIS10	1.3E+04 (1.4E+09)	−1.8E+02 (1.7E+01)	−1.2E+02 (1.4E−02)	−3.1E+02 (1.5E+02)	−2.8E+02 (3.4E+02)
LS2-DIS3	3.5E+03 (1.4E+07)	−1.8E+02 (1.6E+00)	−1.2E+02 (6.2E−03)	−3.1E+02 (9.2E+01)	−2.9E+02 (1.9E+02)
LS2-DIS5	1.5E+04 (5.9E+09)	−1.8E+02 (5.4E−01)	−1.2E+02 (7.6E−03)	−3.2E+02 (4.4E+01)	−2.9E+02 (1.9E+02)
LS2-DIS10	4.2E+03 (2.8E+07)	−1.8E+02 (4.9E−01)	−1.2E+02 (6.9E−03)	−3.1E+02 (5.7E+01)	**−2.9E+02 (1.7E+02)**
LS1-RAN3	1.9E+04 (4.6E+09)	−1.8E+02 (1.4E+01)	−1.2E+02 (9.7E−03)	−3.1E+02 (1.3E+02)	−2.8E+02 (3.0E+02)
LS1-RAN5	7.1E+03 (1.7E+08)	−1.8E+02 (3.1E+01)	−1.2E+02 (8.7E−03)	−3.1E+02 (9.8E+01)	−2.8E+02 (3.1E+02)
LS1-RAN10	5.1E+03 (1.3E+08)	−1.8E+02 (9.0E+00)	−1.2E+02 (6.2E−03)	−3.1E+02 (1.4E+02)	−2.8E+02 (3.6E+02)
LS2-RAN3	1.6E+03 (4.1E+06)	**−1.8E+02(6.3E−01)**	−1.2E+02 (9.6E−03)	−3.2E+02 (5.1E+01)	−2.8E+02 (2.6E+02)
LS2-RAN5	1.6E+03 (4.3E+06)	−1.8E+02 (4.3E−01)	−1.2E+02 (6.0E−03)	−3.2E+02 (7.0E+01)	−2.9E+02 (2.3E+02)
LS2-RAN10	2.7E+03 (8.2E+06)	−1.8E+02 (4.4E−01)	−1.2E+02 (6.8E−03)	−3.2E+02 (7.5E+01)	−2.9E+02 (3.6E+02)

Table 4.2 T-Test results of 10 benchmark functions used in experimental evaluations, the bold font means the significance of the proposed acceleration methods in the significant level of 0.05

Methods	F1	F2	F3	F4	F5	F6	F7	F8	F9	F10	Average
LS1-BST3	0.30733215	0.66264958	0.29076284	0.293890	0.19412898	0.48179234	0.87950749	0.35946016	0.91104486	0.50163682	0.48822053
LS1-BST5	0.52975586	0.12835392	0.66625593	0.43016709	0.77425095	0.06589619	0.20030846	0.71035100	0.22041523	0.41588405	0.41416387
LS1-BST10	0.08056825	0.02470911	0.60396748	0.68870988	0.20200305	0.26278522	0.49376467	0.30779018	0.74328076	0.21277440	0.36203530
LS2-BST3	**0.00018873**	**1.7289E−09**	**6.3823E−09**	**1.5510E−11**	**0.03135001**	**0.02731704**	**0.00341214**	0.57552981	**3.1363E−07**	0.66158560	**0.12993900**
LS2-BST5	**0.00019108**	**1.0157E−09**	**9.0085E−07**	**4.6657E−13**	0.2130529	**0.02736166**	**0.00199084**	0.58023683	**1.6390E−07**	0.70201959	**0.15248540**
LS2-BST10	**0.00018142**	**4.6875E−09**	**2.9097E−07**	**1.3887E−14**	**0.00064191**	**0.02850594**	**0.00105950**	**0.00710928**	**5.25E−10**	0.26088098	**0.02983793**
LS1-DIS3	0.3024598	0.09248131	0.21053517	0.73353761	0.29185093	**0.03449843**	0.95869120	0.1610230	0.56215176	0.28643497	0.3633664
LS1-DIS5	0.30193380	0.82493130	0.83781266	0.43509969	0.26738590	0.0834036	0.8580748	0.66146725	0.3738602	0.90766802	0.55516379
LS1-DIS10	0.35673754	0.14957042	0.59834329	0.85930014	0.26509955	0.10248507	0.25256032	0.13459419	0.83652087	0.15412697	0.37093384
LS2-DIS3	**0.04020126**	**1.023E−05**	**9.0772E−06**	**5.572E−09**	0.25845547	**0.03421305**	**0.01218165**	0.50861468	0.21696933	0.06585411	**0.1136509**
LS2-DIS5	**0.02593316**	**4.1032E−08**	**4.1816E−07**	**1.8205E−10**	0.40383607	0.16633869	**0.00131148**	0.56170367	**0.00011524**	0.09590750	**0.12551463**
LS2-DIS10	**0.02499335**	**3.3676E−08**	**2.9331E−08**	**2.7265E−12**	0.08826582	**0.03718129**	**0.000463**	0.96882973	**0.04129824**	**0.00233455**	**0.1163366**
LS1-RAN3	0.29955990	0.47553783	0.92415835	0.63494715	0.7833978	0.22088026	0.5596284	0.1089092	0.59620154	0.13525320	0.47384736
LS1-RAN5	0.25031038	0.82796216	0.8422587	0.95198690	0.41726152	0.05208138	0.5428202	0.57430298	0.93416252	0.81746090	0.62106077
LS1-RAN10	0.41702221	0.5579850	0.7661850	0.77668166	0.32993593	**0.04174701**	0.10787766	0.61451596	0.24905846	0.63211117	0.44931201
LS2-RAN3	**0.00016607**	**3.1592E−13**	**6.1902E−09**	**2.4789E−15**	**0.01630269**	**0.02736504**	**0.00033605**	0.19381760	**3.4057E−05**	0.47844205	**0.07164636**
LS2-RAN5	**0.00018380**	**8.4637E−14**	**7.9995E−09**	**6.7861E−16**	**3.2460E−05**	**0.02735219**	**0.00038332**	0.96308009	**0.00044558**	**0.00543778**	**0.09969152**
LS2-RAN10	**0.00084498**	**1.4595E−13**	**3.6374E−09**	**7.6255E−16**	**1.1666E−11**	**0.03114742**	**0.00105663**	0.54632155	**0.02897558**	**0.01357420**	**0.06219204**

Table 4.3 Confidence interval of p-value of approximation methods, sampling method, and sampling number method, the sampling number for each method is 60, i.e., it follows a normal distribution according to the central limit theorem, and the confident probability is 0.95, i.e., u = 1.96

Methods	Mean	Variance	Confident interval
LS1	0.45534488	0.07738241	[0.43576443, 0.47492533]
LS2	**0.1001438**	**0.04668558**	**[0.08833074, 0.11195690]**
BST	**0.26278034**	**0.07998594**	**[0.24254110, 0.28301957]**
DIS	0.27416103	0.09252915	[0.25074792, 0.29757413]
RAN	0.29629168	0.11027739	[0.26838765, 0.32419570]
Sampling # 3	0.27344509	0.08367126	[0.25227334, 0.29461684]
Sampling # 5	0.32801333	0.11433117	[0.29908355, 0.35694311]
Sampling # 10	**0.23177462**	**0.08064302**	**[0.21136912, 0.25218012]**

4.4.2 Experimental Results

Table 4.1 shows the mean and variance of ten benchmark functions, and Table 4.2 shows the t-test results of each acceleration method compared to the normal fireworks algorithm.

From those experimental results, we can conclude that

(1) Elite strategy is an effective approach to enhance fireworks algorithm search capability.
(2) The random sampling method has the better acceleration performance than the others.
(3) For some benchmark functions, best sampling method has the same acceleration performance as the random sampling method.
(4) The distance sampling method is not more effective than the other sampling methods, i.e., best and random sampling method.
(5) In the distance sampling method, the approximation model by the more sampling number have the better performance than the other approaches, such as F9 and F10.
(6) For all the benchmark functions, LS2 methods are better LS1 methods.

4.5 Discussion

Table 4.3 shows the confidence interval of the p value for each approximation approach. We analyze the influence of approximation approaches from the experimental results of Tables 4.1, 4.2, and 4.3.

4.5.1 Fireworks Algorithm Acceleration Performance

Fireworks algorithm is a new EC algorithm, which is inspired by the nature phenomenon of firework explosion. The crucial mechanism of fireworks algorithm is the multichange strategy, i.e., one firework makes several sparks for implementing the multipath searches in the optimization process, which is the principal difference from evolution strategy. This mechanism presents the fireworks algorithm optimization capability and originality. This kind of multichange strategy can be applied to other ECs for obtaining the improved optimization performance.

It is one of the main proposals to obtain the fitness landscape to enhance fireworks algorithm performance, we use polynomial regression of one and two degree function as the approximation model. From the evaluation result, we can conclude that our proposed elite strategy by obtaining the fitness landscape is an effective method to enhance fireworks algorithm search capability. Most of our used sampling methods for approximating the fitness landscape in projected one-dimensional space with different sampling number can accelerate the search of fireworks algorithm.

4.5.2 Approximation Methods

We use one and two degree polynomial function as the approximation model to approximate fitness landscape into linear and nonlinear search space. Experimental results in Table 4.2 suggests that two degree model is more effective to accelerate the fireworks algorithm search than the one degree model in most of the benchmark functions. It is because that two degree model, i.e., nonlinear model can extend the search space range than the linear model, where the fireworks algorithm conducts a search within the limited search space.

Nonlinear models (LS2) are better approximating the fitness landscape for the ten benchmark functions than linear models (LS1) from Table 4.3. From our comparative evaluations, nonlinear model shows its strong capability to extend the potential global optima region in the search space. For most of the real-world EC application, nonlinear models are the beneficial regression model to approximate the fitness landscape both in approximation model accuracy and acceleration performance.

4.5.3 Sampling Methods

In the experimental evaluations, we investigate three sampling methods, i.e., best sampling method, distance near the best fitness individual sampling method, and random sampling methods. For the most of the benchmark functions, the acceleration performance with the best sampling method and random sampling are better than the distance sampling.

On the one hand, the spark with the better fitness locate in the global optima region, and the approximation model fitted by those data can obtain the accurate model for accelerating. On the other hand, the spark distance near the spark with the best fitness does not mean they are also close to the global optima region, and the approximation accuracy and acceleration performance by distance sampling are worse than that by the other sampling methods. From our comparative evaluation, we found those two points.

Sampling methods cannot take effect in isolation, which have a better acceleration performance only with the proper approximation methods. In our study, the proper approximation methods are the two degree polynomial model, i.e., nonlinear model.

Best sampling method can enhance fireworks algorithm search effectively for all the benchmark functions except F10.

Distance near the best fitness individual sampling method is effective in most of the cases, except F5, F8, and F10. However, it needs more computational cost in computing the distance, in the practical points of view, it is a useless sampling method.

The random sampling method can accelerate all the benchmark functions with nonlinear approximation model form Table 4.2. The elite obtained by this method can increase the population diversity. This is the new discovery from our study work. Because of the random sampling method is with the less computational cost in the sampling process than best and distance sampling methods, which need more search and sort operations, it is the beneficial sampling method to the application of the proposed method in the practical points of view.

If the sampling data belong to a certain distribution, may be the acceleration performance can be improved with the concrete fitness landscape, the future research will involve this direction.

4.5.4 Sampling Data Number

The sampling data number is crucial for obtaining an accurate approximation model to accelerate the fireworks algorithm search. Theoretically, more number of sampling data means obtaining more accuracy approximation model and better acceleration performance.

From Table 4.1, the acceleration performance by the best sampling with LS2 and five and ten sampling have better performance than that with three sampling data.

From Tables 4.2 and 4.3, distance sampling method with 10 sampling data are effective in F1, F2, F3, F4, F6, F7, F9, and F10, it can show the relationship between sampling data number and acceleration performance, i.e., the more sampling number means more better approximation and acceleration performance.

Sampling data number decides the computational cost in the approximation regression computing, so we should select the proper sampling number to obtain approximation model by balancing the computational cost and acceleration performance for a concrete landscape or a real-world application.

4.6 Summary

In the chapter, we investigated a series of methods to enhance the fireworks algorithm by elite strategy. There are three sampling methods, which we used in our methods to approximate fireworks algorithm fitness landscape. We conduct a set of comparative evaluations to study on the sampling method and number influence on the approximation accuracy and acceleration performance. From our experimental evaluations, there are five discoveries below.

(1) Our proposed elite strategy is an efficient method to enhance fireworks algorithm search capability significantly.
(2) Sampling method cannot take effect in isolation, it must be with a proper approximation model to accelerate the FWA search for a certain benchmark function, i.e., fitness landscape.
(3) For some cases, the best sampling method and the random sampling method have the same acceleration performance.
(4) It can be obtained by a fitness landscape approximation accuracy and acceleration performance with the more sampling data number and a proper approximation model. In this study, the better approximation model is the nonlinear model (LS2) based on the experiments.
(5) In the practical points of view, the random sampling method is a better tool to obtain the higher acceleration performance in both computational time and final solution quality.

References

1. Y. Pei, S. Zheng, Y. Tan, H. Takagi, An empirical study on influence of approximation approaches on enhancing fireworks algorithm, in *Proceedings of the 2012 IEEE Congress on System, Man and Cybernetics*, (IEEE, 2012), pp. 1322–1327
2. Y. Pei, H. Takagi, A survey on accelerating evolutionary computation approaches. in *2011 International Conference of Soft Computing and Pattern Recognition (SoCPaR)* (IEEE, 2011), pp. 201–206
3. T. Ingu, H. Takagi, Accelerating a GA convergence by fitting a single-peak function, in *1999 IEEE International Fuzzy Systems Conference Proceedings, FUZZ-IEEE'99*, vol. 3 (IEEE, 1999) pp. 1415–1420
4. Y. Pei, H. Takagi, Accelerating evolutionary computation with elite obtained in projected one-dimensional spaces. in *2011 Fifth International Conference on Genetic and Evolutionary Computing (ICGEC)*, (IEEE, 2011), pp. 89–92
5. Y. Pei, H. Takagi, Comparative evaluations of evolutionary computation with elite obtained in reduced dimensional spaces, in *2011 Third International Conference on Intelligent Networking and Collaborative Systems (INCoS)* (IEEE, 2011), pp. 35–40
6. Y. Tan, Y. Zhu, Fireworks algorithm for optimization, in *Advances in Swarm Intelligence* (Springer, Berlin, 2010), pp. 355–364
7. A.N. Langville, C.D. Meyer, R. Albright, J. Cox, and D. Duling, Utilizing nonnegative matrix factorization for e-mail classification problems. In *Survey of Text Mining III: Application and Theory* (Wiley New York, 2010) pp. 57–80

8. A. Janecek, Y. Tan, Iterative improvement of the multiplicative update NMF algorithm using nature-inspired optimization in *2011 Seventh International Conference on Natural Computation (ICNC)*, vol. 3 (IEEE, 2011), vol. 3, pp .1668–1672
9. Y. Jin, A comprehensive survey of fitness approximation in evolutionary computation. Soft Comput. **9**(1), 3–12 (2005)
10. Y. Jin, Surrogate-assisted evolutionary computation: recent advances and future challenges. Swarm Evol. Comput. **1**(2), 61–70 (2011)
11. Y. Pei, H. Takagi, Fourier analysis of the fitness landscape for evolutionary search acceleration. in *2012 IEEE Congress on Evolutionary Computation (CEC)* (IEEE, 2012), pp. 1–7
12. Y. Pei, H.Takagi, Comparative study on fitness landscape approximation with fourier transform. in *2012 Sixth International Conference on Genetic and Evolutionary Computing (ICGEC)* (IEEE, 2012), pp. 400–403
13. P.N. Suganthan, N. Hansen, J.J. Liang, K. Deb, Y.-P. Chen, A. Auger, et al., Problem definitions and evaluation criteria for the CEC 2005 special session on realparameter optimization. KanGAL Report 2005005 (2005)

Chapter 5
FWA with Controlling Exploration and Exploitation

Since FWA was proposed in 2010 [1], it has shown its significance and superiority in dealing with the optimization problems. However, calculation of the number of explosion sparks and the amplitude of firework explosion of FWA should dynamically control the exploration and exploitation of searching space with iteration. The mutation operator of FWA needs to generate the search diversity. This chapter provides a new method to calculate the number of explosion sparks and the amplitude of firework explosions. By designing a transfer function, the rank number of a firework is mapped to the scale of the calculation of scope and spark number of a firework explosion. A parameter is used to dynamically control the exploration and exploitation of FWA with iteration going on. In addition, this chapter uses a new random mutation operator to control the diversity of FWA search.

5.1 Some Improvements on Operations in FWA

5.1.1 The Amplitude and Number of Sparks

In FWA, the fitness of firework is needed when to compute the scope and sparks number of firework explosion. In order to improve the calculation of the number of firework sparks and the amplitude of firework explosion, we use the sequence number of fireworks to compute the two quantities. Therefore, a transfer function must be designed to map the sequence number of fireworks to function value that is used to better calculate the amplitude and spark number of a firework explosion.

The transfer function is cited from the sigmoid function as the Eq. (5.1):

$$f(x) = \frac{1}{1 + e^x}.$$

(5.1)

© Springer-Verlag Berlin Heidelberg 2015
Y. Tan, *Fireworks Algorithm*, DOI 10.1007/978-3-662-46353-6_5

The function is further improved and is added a parameter a as the following Eq. (5.2):

$$f(x) = \frac{1}{1 + e^{(x-1)/a}},\tag{5.2}$$

where a is a control parameter to change the shape of the above transfer function. Equation (5.2) can transfer the sequence number of firework rank of fitness to different values of function, which is used to calculate the spark number and amplitude of firework explosion. The function of Eq. (5.2) is named as transfer function. When the parameter $a = 1, 5, 9, 13$ and 21, the figures of the transfer function with different parameter values are plotted as Fig. 5.1.

From Fig. 5.1, it can be found that the function fitness of different sequence numbers are more and more mean as the parameter a is increasing. So, the calculating number of explosion sparks is designed as

$$S_n = m\frac{f(n)}{\sum_{n=1}^{N} f(n)},\tag{5.3}$$

where m is the total of number of spark, n is the sequence number of a firework, and S_n denotes the spark number of the nth firework explosion.

What is more, the calculating the amplitude of firework is designed as

$$A_n = A\frac{f(N - n + 1)}{\sum_{n=1}^{N} f(N - n + 1)},\tag{5.4}$$

where A is the maximum amplitude of firework explosion, n is the sequence number of a firework, and A_n denotes the amplitude of the nth firework explosion.

In Eqs. (5.3) and (5.4), N is the total number of firework in FWA. The function $f(x)$ is the Eq. (5.2), whose parameter a is varied with the iteration from 20 to 1. With variable parameter a, the explosion number of spark and explosion amplitude of fireworks are dynamically changed as the iteration goes on.

Figures 5.2 and 5.3 plot the number of sparks and amplitude of explosion of eight fireworks with iterations which are calculated by using Eqs. (5.3) and (5.4).

Figure 5.2 shows that the number of spark generated by the firework of FWA with the modified equation is very regular. For the fireworks with better fitness values, the number of sparks is more and more with the iteration, while for the worse fireworks, the number of sparks is less and less. Figure 5.3 shows that the amplitude of the better firework explosion is smaller and smaller as the iteration goes on, while the amplitude of bad firework explosion is bigger and bigger with iteration. The policy can embody the idea of algorithm that a good firework has more number of generating sparks with less amplitude of explosion, while a bad firework generates less number of sparks with large amplitudes of explosion.

There might be a global optimal solution near the good firework. Hence, the explosion of a good firework undertakes the local searching, while a bad firework explosion undertakes the global exploration in a wider solution space. In the proposed

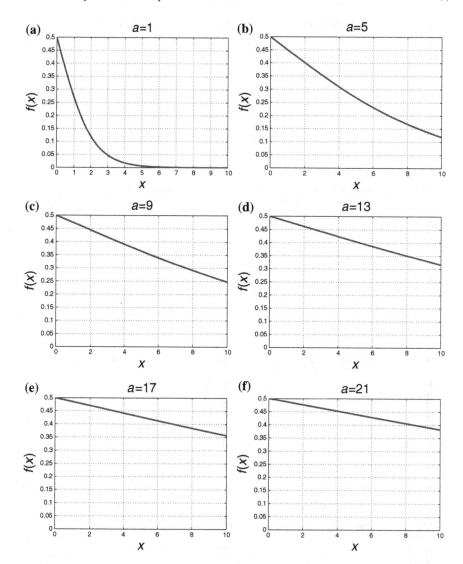

Fig. 5.1 The curves of transfer function with different parameters. Reprinted from Ref. [2], with kind permission from Springer Science+Business Media

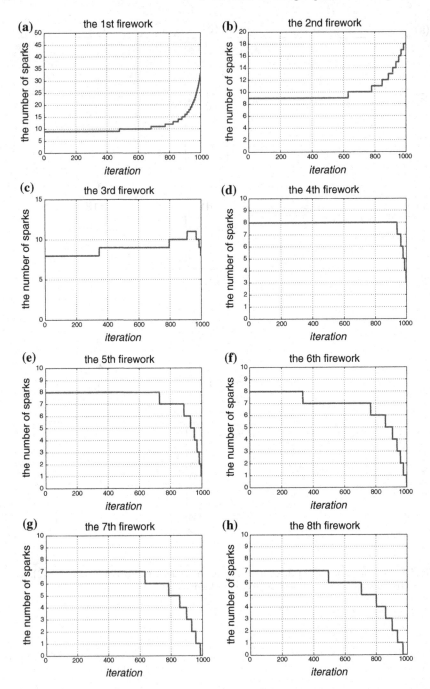

Fig. 5.2 The number of sparks for every firework in modifying FWA. Reprinted from Ref. [2], with kind permission from Springer Science+Business Media

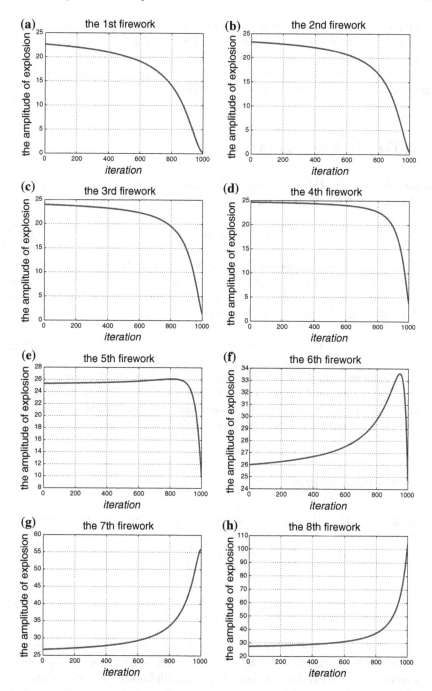

Fig. 5.3 The amplitude of explosion for every firework in modifying FWA. Reprinted from Ref. [2], with kind permission from Springer Science+Business Media

method, the dynamic change of number of sparks and the amplitude of explosion with iteration can embody that the global exploration is done in the early time of algorithm running, while the local exploitation is enhanced during the later time of algorithm's iteration. So, the new calculation of equations can better control the exploitation and exploration of FWA with iteration.

5.1.2 The Mutation Improvement

To keep the diversity, FWA employed a Gaussian mutation to generate sparks. The kth dimensions of the ith firework, x_i^k, mutates as x_i^k by the following equation:

$$x_i^k = x_i^k * Gaussian(1, 1).$$ (5.5)

However, the above mutation operation makes original FWA easily converged to zero point of the search space, and it is difficult for FWA to generate the diversity. In order to increase the diversity of FWA, the random mutation is employed to make the firework mutated. The mutation formula is

$$x_i^k = x_i^k + rand() * (X_{UB,k} - X_{LB,k}),$$ (5.6)

where x_i^k denotes that the position of the kth dimensions of ith firework, $X_{LB,k}$ denotes that the minimal bound of the kth dimensions of the ith firework, $X_{UB,k}$ denotes that maximal bound of the kth dimensions of the ith firework. The function $rand()$ gains the sampling value in the interval [0, 1] with the uniform distribution.

5.1.3 Selection Strategy

Original FWA selects N locations for the next generation fireworks by the Eqs. (5.7) and (5.8):

$$R(x_i) = \sum_{j \in K} d(x_i, x_j) = \sum_{j \in K} ||x_i - x_j||,$$ (5.7)

$$p(x_i) = \frac{R(x_i)}{\sum_{j \in K} R(x_i)},$$ (5.8)

where x_i is the location of ith spark or firework, $d(x_i, x_j)$ is the distance between two sparks or fireworks. K is the set of sparks and fireworks generated in current generation. The $p(x_i)$ is the probability which the ith firework or spark is selected as firework of next generation.

Equations (5.7) and (5.8) do not consider the fitness of sparks of the fireworks for the selection of next generation fireworks' location. This is not in consistence with the idea of Eqs. (5.3) and (5.4). Equations (5.3) and (5.4) use the sequence number to sort the fireworks' fitness and to calculate the sparks number and the amplitude of fireworks explosion, but Eqs. (5.7) and (5.8) of the original FWA do not consider the fitness to select the fireworks' location. Therefore, there are two methods to be provided to modify the selection operator.

1. Fitness selection using a roulette
   Like the original FWA, the best of the set will be selected first. Then the others are selected based on fitness proportion using a roulette. So, the selection probability of every spark or firework must be calculated by

$$p(x_i) = \frac{y_{max} - f(x_i)}{\sum_{i \in k} (y_{max} - f(x_i))},$$ (5.9)

where y_{max} is the maximum value of the objective function among the set k which consists of the fireworks and sparks in the current generation. The other fireworks will be selected using the roulette according to the probability gained by Eq. (5.9).

2. Best fitness selection
   In Zheng et al. [3], used a random selection operator to replace the previous time consuming one. It is as the following, when the algorithm determined the number of fireworks of every generation, all the sparks and fireworks of the current generation are sorted according to their fitness and then the best sparks or fireworks with the best fitness are selected as the location of next generation. The method is very simple and in accordance with the new calculation of explosion number and explosion amplitude of fireworks in Eqs. (5.3) and (5.4).

5.2 Experiment and Analysis

5.2.1 Experimental Design

In order to evaluate the performance of the improved FWA, 14 benchmark functions provided by CEC 2005 are employed [4]. These benchmark functions include five unimodal functions and nine multimodal functions. The optimal fitness of these functions is not zero and is added bias. These functions are shifted and the optimal locations are shifted to different locations from zero point in the solution space. More details on the benchmark functions can be seen in [4].

In order to test the performance of improved FWA, the improved FWA with best fitness selection and random mutation (IFWABS), the improved FWA with the fitness selection using the roulette and random mutation (IFWAFS), the original FWA, and the global PSO are compared with each other. The global PSO is employed the decreasing weight w from 0.9 to 0.4 proposed in [5] and the neighbor particles of

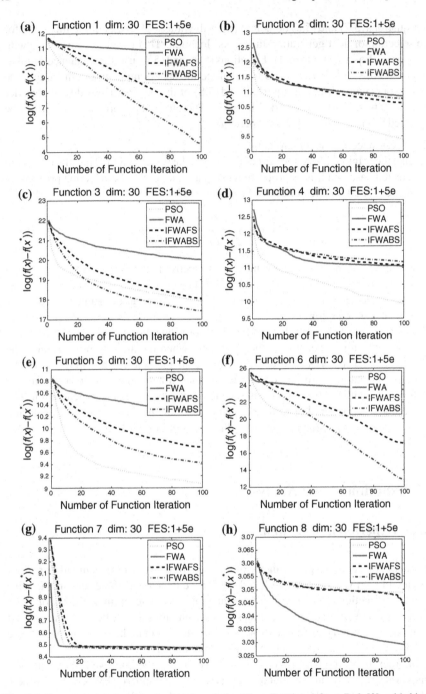

Fig. 5.4 Convergence curves on the benchmark functions. Reprinted from Ref. [2], with kind permission from Springer Science+Business Media

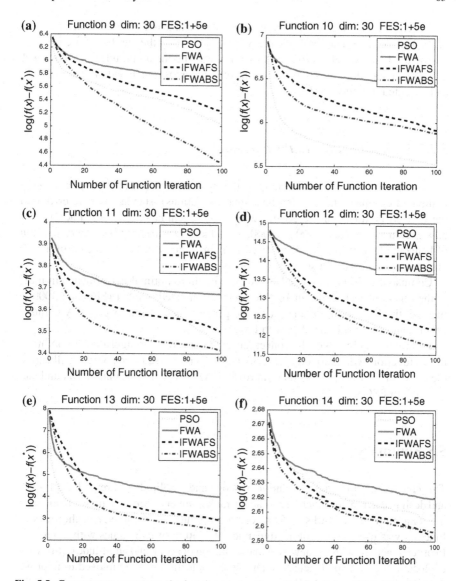

Fig. 5.5 Convergence curves on the benchmark functions. Reprinted from Ref. [2], with kind permission from Springer Science+Business Media

each particle is all particles. The particle population size is 100. The factor $c1$ and $c2$ of PSO are set as 2. The FWA and the improved FWA set the number of fireworks as 10, the total number of explosion sparks S as 80, and the amplitude of explosion A as the range length of problem space. The experiment is conducted in Matlab 2012b and executed in windows 7.

5.2.2 Experimental Results and Analysis

The experiment is conducted to compute the mean error fitness ($f(x) - f(x^*)$ $f(x^*)$ is the real optimal fitness of the benchmark functions), standard square error and the best error fitness in the 25 run on the 14 benchmark functions. Each run of all algorithm is evaluated 1000, 10000, 10000 and $D * 10000$ (D is the dimension of benchmark function), respectively. Each algorithm will be conducted in the 10 dimensions and 30 dimensions, respectively.

Compared to PSO, improved FWA exhibits more optimal performance in most of the functions, especially in 100000 Fitness Evaluated number (FEs) and 300000 FEs. As the FEs are more and more, the performance of improved FWA is much better than PSO that it matches with PSO.

Figures 5.4 and 5.5 plot the convergence process for four algorithms to optimize the 14 functions with 300000 FEs in 30 dimensions. These figures visually illustrate the effect of four algorithms that improved FWA is excel to original FWA and can match with PSO.

5.3 Summary

This chapter modified the calculation of scope and amplitude of firework explosion, and designed a transfer function to map the rank number of firework fitness to allocate the total number of sparks and explosion amplitude. A parameter in the transfer function was used to control the dynamic calculation of two values with iterations. In order to accord with the new idea of calculation of scope and amplitude of firework explosion, the best spark selection and fitness selection were employed to improve the selection operator of FWA at last.

References

1. Y. Tan, Y. Zhu, Fireworks algorithm for optimization, in *Advances in Swarm Intelligence* (Springer, Berlin, 2010), pp. 355–364
2. J. Liu, S. Zheng, Y. Tan, The improvement on controlling exploration and exploitation of firework algorithm. Advance in Swarm Intelligence, vol. 7928 (Springer 2013), pp.11–23

3. S. Zheng, A. Janecek, Y. Tan, Enhanced fireworks algorithm, in *2013 IEEE Congress on Evolutionary Computation (CEC)* (IEEE, 2013), pp. 2069–2077

4. P.N. Suganthan, N. Hansen, J.J. Liang, K. Deb, Y.-P. Chen, A. Auger, et al., Problem definitions and evaluation criteria for the CEC 2005, special session on real-parameter optimization KanGAL Report 2005005 (2005)

5. Y. Shi, R. Eberhart, A modified particle swarm optimizer, in *The 1988 IEEE International Conference on Evolutionary Computation Proceedings, 1998. IEEE World Congress on Computational Intelligence* (IEEE 1988), pp. 69–73

Chapter 6
Enhanced Fireworks Algorithm

In this chapter, we present an improved version of the recently developed Fireworks Algorithm (FWA) based on several modifications, according to our previous work "Enhanced Fireworks Algorithm" [1]. A comprehensive study on the operators of conventional FWA revealed that the algorithm works surprisingly well on benchmark functions which have their optimum at the origin of the search space. However, when being applied on shifted functions, the quality of the results of conventional FWA deteriorates severely and worsens with increasing shift values, i.e., with increasing distance between function optimum and origin of the search space. Moreover, compared to other meta-heuristic optimization algorithms, FWA has high computational cost per iteration. In order to tackle these limitations, we present five major improvements on FWA: (i) a new minimal explosion amplitude check, (ii) a new operator for generating explosion sparks, (iii) a new mapping strategy for sparks which are out of the search space, (iv) a new operator for generating Gaussian sparks, and (v) a new operator for selecting the population for the next iteration.

The resulting algorithm is called *Enhanced Fireworks Algorithm* (*EFWA*). Experimental evaluation on 12 benchmark functions with different shift values shows that EFWA outperforms conventional FWA in terms of convergence capabilities, while reducing the runtime significantly.

6.1 Properties of Conventional FWA

As already mentioned in the original FWA [2], FWA outperformed SPSO and CPSO significantly and converged in most cases toward the function optimum already after a few iterations. However, when applying FWA on shifted functions the results worsen progressively with increasing distance between function optimum and origin of the search space. By investigating the operators of FWA, we found that some of them create [map] sparks at [to] locations which are close to the origin of the search space, independent of the function optimum. This behavior is mostly caused by the mapping operator and the Gaussian sparks operator. In terms of runtime we found that the

© Springer-Verlag Berlin Heidelberg 2015
Y. Tan, *Fireworks Algorithm*, DOI 10.1007/978-3-662-46353-6_6

cost per iteration of FWA is significantly higher than that of most other optimization algorithms. In Sect. 6.2 we analyze all operators of FWA in detail and point out which operators are responsible for its actual behavior and its high computational cost. In summary, conventional FWA has the following drawbacks:

(i) For functions which have their optimum at the origin, FWA will find the optimal solution very fast. However, not due to the intelligence of the algorithm but due to the specific mapping and Gaussian mutation operators which map/create sparks close to the origin.

(ii) For functions which have their optimum far away from the origin, FWA has to face the two drawbacks that the mapping operator rebounds most solutions which are out of the search space to locations which are far away from the function optimum, and that the mutation operator creates many sparks at locations close to the origin (i.e., again far away from the optimum).

(iii) FWA has a high computational cost per iteration.

6.2 The Proposed EFWA

In this section, we present the new operators of EFWA in detail after indicating the limitations of each operator in conventional FWA.

6.2.1 A New Minimal Explosion Amplitude Check (MEAC)

Equation (2.3) shows how the explosion amplitude for each firework is calculated in conventional FWA. A firework with better fitness will have a smaller explosion amplitude while a firework with lower fitness has a larger explosion amplitude. Although this idea seems reasonable, the explosion amplitude of the fireworks having the best (or a very good) fitness will usually be very small, i.e., close to 0 according to Eq. (2.3). If the explosion amplitude is [close to] zero, the explosion sparks will be located at [almost] the same location as the firework itself. As a result, it may happen that the location of the best firework cannot be improved until another firework has found a better location. In order to avoid this problem, we introduce a lower bound $A_{\min}$ of the explosion amplitude, which is based on the progress of the algorithm. During the early phase of the search, $A_{\min}$ is set to a higher value in order to facilitate exploration, with increasing number of evaluations, $A_{\min}$ is decreased in order to allow for better exploitation capabilities around good locations. For each dimension k, the explosion amplitude A_i^k is bound as follows:

$$A_{ik} = \begin{cases} A_{\min,k} & \text{if } A_{i,k} < A_{\min,k}, \\ A_{ik} & \text{otherwise,} \end{cases} \tag{6.1}$$

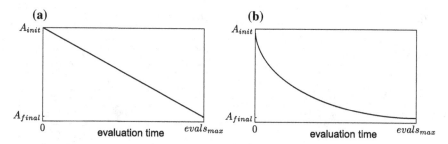

Fig. 6.1 Linearly and nonlinearly decreasing minimal explosion amplitude. **a** Linear decrease. **b** Nonlinear decrease

A new value for A_{min} is calculated in each iteration. In this work, we use two different modes to calculate A_{min}. The first approach is based on a linearly decreasing function (cf. Eq. (6.2)), and the other approach is based on a nonlinearly decreasing function (cf. Eq. (6.3)).

$$A_{min,k}(t) = A_{init} - \frac{A_{init} - A_{final}}{evals_{max}} * t, \tag{6.2}$$

$$A_{min,k}(t) = A_{init} - \frac{A_{init} - A_{final}}{evals_{max}} \sqrt{(2 * evals_{max} - t)t}, \tag{6.3}$$

In both equations, t refers the number of function evaluation at the beginning of the current iteration, and $evals_{max}$ is the maximum number of evaluations. A_{init} and A_{final} are the initial and final minimum explosion amplitude, respectively. Compared to the linear decrease of A_{min}, the nonlinear decrease enhances exploitation already at an earlier stage of the algorithm (i.e., after fewer iterations). Figure 6.1 shows a graphical representation of Eqs. (6.2) and (6.3).

6.2.2 A New Operator for Generating Explosion Sparks

In FWA, the offset displacement for explosion sparks of each firework is only calculated once and the same value is added to the location of selected dimensions. Obviously, adding the same value in each dimension leads to a bad local search ability. To avoid this problem, we calculate a different offset displacement for selected dimensions (those dimensions, where z^k equals 1). Algorithm 6.1 shows the proposed process of generating explosion sparks in EFWA.

Figure 6.2 shows the difference between the generation of explosion sparks in FWA and EFWA. As discussed before, the offset displacement in FWA is similar for all selected dimensions, in EFWA a different offset displacement is calculated in each dimension.

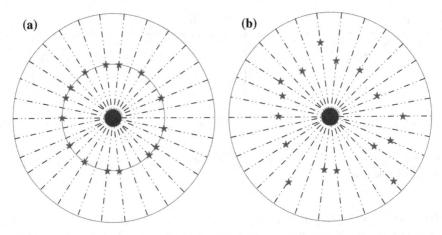

Fig. 6.2 Generation of explosion sparks in FWA and EFWA. **a** FWA. **b** EFWA

Algorithm 6.1 Generating "explosion sparks" in EFWA.

1: Initialize the location of the "explosion sparks": $x_i = X_i$,
2: Set $z^k = round(U(0, 1))$, $k = 1, 2, ..., D.$,
3: **for** each dimension of x_{ik}, where $z^k == 1$ **do**
4:      Calculate offset displacement: $\Delta X_{ik} = A_i \times U(-1, 1)$,
5:      $x_{ik} = x_{ik} + \Delta X_{ik}$,
6:      **if** x_{ik} out of bounds **then**
7:          map x_{ik} to the potential space (see next subsection),
8:      **end if**
9: **end for**

6.2.3 A New Mapping Operator

In conventional FWA, when the location of a new spark exceeds the search range in dimension k, the new spark will be mapped to another location according to the following equation:

$$x_{ik} = X_{LB,k} + |x_{ik}| \% (X_{UB,k} - X_{LB,k}),   \qquad (6.4)$$

where $X_{UB,k}$ $X_{LB,k}$ denotes the upper search bound and lower search bound in the feasible range. In many cases, a spark will exceed the allowed search space only by a rather small value. Moreover, as the search space is often equally distributed ($X_{LB,k} \equiv -X_{UB,k}$), the adjusted position x_{ik} will be very close to the origin in many cases. The following example is used to explain this comment: Consider an optimization problem within the search space $[-20, 20]$. If, in dimension k, a new spark is created at the point $x_{ik} = 21$, it will be mapped to the location $x_{ik} = -20 + |21| \% (40)$. Since the result of the modulo operation $21 \% (40) = 21$, x_{ik} will be mapped to the location $x_{ik} = 1$, which is already very close to the origin.

In cases where $X_{LB,k} \equiv -X_{UB,k}$, this mapping operator is partly responsible for drawbacks (i) and (ii) as mentioned in Sect. 6.1. In order to avoid the problems caused by the conventional mapping operator we replace this method with a uniform random mapping operator, which maps the sparks to any location in the search space with uniform distribution.

6.2.4 A New Operator for Generating Gaussian Sparks

Together with the mapping operator, the Gaussian mutation operator is the main reason why conventional FWA works significantly better than other optimization algorithms for functions which have their optimum at the origin (cf. the results in [2]). Figure 6.3a, b show the location of the Gaussian sparks for a two-dimensional Ackley function with the optimum at [0, 0] and [−70, −55], respectively. In each iteration, the location of (only) the Gaussian sparks is plotted and not deleted. The location of the Gaussian sparks in Fig. 6.3a indicates that most sparks are located at the origin, i.e., close to the optimum. Moreover, it can be seen that the areas close to the coordinate axes are also more crowded than other parts of the search space. Figure 6.3b reveals an interesting fact about the location of the Gaussian sparks for the shifted Ackley function. Even though the optimum is now far away from the origin, many sparks are located close to the center. Obviously, many sparks in Fig. 6.3a are not located close to the center because of the intelligence of the algorithm, but rather because many Gaussian sparks are created near to the origin of the search space, independent of the location of the function optimum. We point out that Fig. 6.3a, b were created using the mapping operator of conventional FWA.

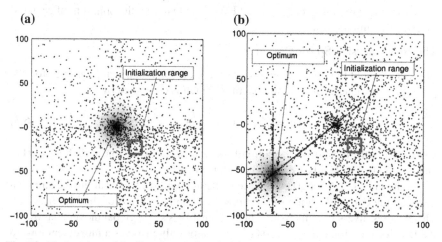

Fig. 6.3 The locations of the Gaussian sparks using the conventional FWA (Ackely function using 100 000 function evaluations). **a** No shift—optimum at origin. **b** Shift—optimum at (−70, −55)

The reason for this behavior is the calculation of the Gaussian sparks as shown in the following equation:

$$x_{ik} = X_{ik} * g, \tag{6.5}$$

Here, g is set to a random value from a normal distribution with expected value and variance both set to 1. In cases where g is close to 0, x_{ik} will be close to 0 as well. As a result, many Gaussian sparks will be located close to the origin of the search space in dimension k. Moreover, for large g, many Gaussian sparks are created at locations which are out of bounds. In this case, the mapping operator of conventional FWA will map the newly created spark to a location which is in many cases close to the origin. Another problem of the conventional Gaussian sparks operator is the fact that fireworks which are already located close to the origin of the search space cannot escape from this location; if firework X_{ik} is close to zero, the location of spark x_{ik} be close to zero as well, since $x_{ik} = X_{ik} * g$.

Initialization

Since conventional FWA is able to converge toward the optimum already after very few function evaluations (we again refer to the result in [2]), we also analyzed the behavior of FWA during the first iteration. Figure 6.4a–d shows the distribution of the explosion and Gaussian sparks, respectively, directly after the initialization with different initialization ranges. We repeatedly initialized the fireworks, created the two types of sparks, and plotted their locations until we reached 5 000 function evaluations (around 100 repetitions). The distribution of the sparks is (expectedly) independent of the function optimum, and shows a similar behavior for different initialization ranges, which were set to dim 1: [15, 30]; dim 2: [15, 30] for Fig. 6.4a, b and dim 1: [60, 75]; dim 2: [30, 45] for Fig. 6.4c, d. Obviously, some Gaussian sparks are located very close to the origin of the function, independent of the initialization range. This is another indication why conventional FWA is able to find the optimum of centered functions within a few iterations.

The New Gaussian Spark Operator

In order to avoid the problems of the conventional Gaussian mutation operator, we propose a new Gaussian mutation operator which is computed by

$$x_{ik} = X_{ik} + (X_{Bk} - X_{ik}) \times e, \tag{6.6}$$

where, X_B is the location of the currently best firework/explosion spark found so far, and e $= \mathcal{N}(0, 1)$. Details are given in Algorithm 6.2.

As shown in Fig. 6.5, the new mutation operator will stretch out along the direction between the current location of the firework and the location of the best firework. This ensures diversity of the search but also involves some global movement toward the best location found so far. This new operator only involves a movement toward the origin of the search space if the currently best firework is located at the origin.

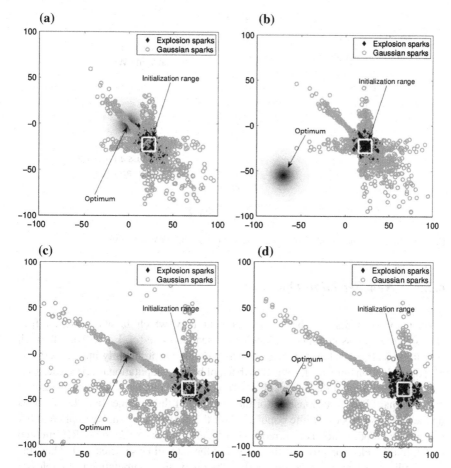

Fig. 6.4 Sparks after initialization in conventional FWA. **a** No shift—optimum at origin. **b** Shift—optimum at $(-70, -55)$. **c** No shift—optimum at origin. **d** Shift—optimum at $(-70, -55)$

Algorithm 6.2 Generating "Gaussian sparks" in EFWA

1: Initialize the location of the "Gaussian sparks": $x_i = X_i$
2: Set $z^k = round(U(0, 1))$, $k = 1, 2, ..., D$
3: Calculate offset displacement: $e = \mathcal{N}(0, 1)$
4: **for** each dimension x_{ik}, where $z^k == 1$ **do**
5:     $x_{ik} = x_{ik} + (X_{Bk} - x_{ik}) * e$, where X_B is the position of the best firework found so far.
6:     **if** x_{ik} out of bounds **then**
7:        mapping the position according to Eq. (6.4)
8:     **end if**
9: **end for**

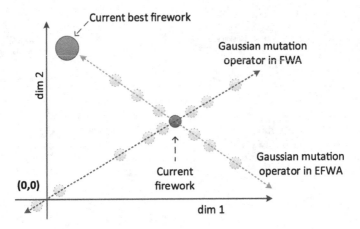

Fig. 6.5 Difference between the Gaussian sparks operator in FWA and EFWA

6.2.5 A New Selection Operator

FWA involves a distance-based selection strategy which favors to select fire-works/sparks in less crowded regions of the search space (cf. Eq. (2.9)). Although selecting locations in low crowded regions with higher probability increases diversity, this selection operator has the drawback of being computational very expensive. A runtime profiling of the original FWA code revealed that the selection operator of conventional FWA is responsible for the majority of the runtime. In order to speed up the selection process of the population for the next generation, we apply another selection method, which is referred to as *Elitism-Random Selection* (ERP) method [1, 3]. In this selection process, the optima of the set will be selected first. Then, the other individuals are selected randomly. Obviously, the computational complexity of ERP is only linear with respect to the number of fireworks, and therefore reduces the runtime of EFWA significantly.

6.3 Experiments

In order to investigate the performances of the new operators, we compare conventional FWA not only to the newly proposed Enhanced Fireworks Algorithm (EFWA), but also three variants of EFWA (denoted as eFWA-X) which can be regarded as a hybridization of FWA and EFWA. While EFWA uses all newly proposed operators, the eFWA-X variants use only some of them. Table 6.1 summarizes which of the new operators are used by the different algorithms. The abbreviations of the operators refer to EXP: new operator for generating explosion sparks; MAP: new mapping operator; GAU: new operator for generating Gaussian sparks; AMP 1: minimal explosion am-

Table 6.1 Algorithms and new operators

	EXP	MAP	GAU	AMP 1	AMP 2	SEL
FWA	○	○	○	○	○	○
eFWA-I	✓	✓	✓	○	○	○
eFWA-II	✓	✓	✓	✓	○	○
eFWA-III	✓	✓	✓	○	✓	○
EFWA	✓	✓	✓	○	✓	✓

plitude check linear decrease; AMP 2: minimal explosion amplitude check nonlinear decrease; SEL: new selection operator. E.g., eFWA-I uses the new operators EXP, MAP and GAU, the new operator AMP is not included, and the selection operator is taken from conventional FWA. Besides comparing different FWA variants with each other, Standard PSO (SPSO) is used for performance comparison.

6.3.1 Experimental Setup

Twelve functions are selected as a test suite. Table 6.1 illustrates the names, numbers, search space (*Range*), optimal locations (*Opt* $\mathbf{x}$), fitness at the optimal location (*Opt* $f(\mathbf{x})$), and dimensions (*Dim.*). For each function, the initial range is set to $[X_{UB,k}/2, X_{UB,k}]$, where $X_{UB,k}$ is the upper bound of the search space in the kth dimension. In the experiments, a number of shift values are added to these basic functions in order to shift the global optimum. We used seven different shift indexes in order to analyze the influence of different shift values on the performance of the algorithms (see Table 6.2). For each shift value SI, the position of the function will be shifted (in each dimension) by the corresponding shift value SV. If SI is equal to zero, the function is not shifted. For example, for function f_1 and an SI of 6, the function will be shifted in each dimension by $0.7 * ((100 - (-100))/2) = 70$, while the search range remains unaffected.

For all experiments, A_{init} and A_{final} (Eqs. 6.2 and 6.3) are set to $(X_{UB,k} - X_{LB,k}) \times 0.02$ and $(X_{UB,k} - X_{LB,k}) \times 0.001$, respectively. All other parameters of (E)FWA are taken from [2], and SPSO parameters are taken from [4]. As experimental platform we used MATLAB 2011b, running Win 7 on an Intel Core i7-2600 CPU;

Table 6.2 Shift index (SI) and shift value (SV)

Shift index	Shift value	Shift index	Shift value	Shift index	Shift value
1	$0.05 * \frac{X_{UB,k}-X_{LB,k}}{2}$	2	$0.1 * \frac{X_{UB,k}-X_{LB,k}}{2}$	3	$0.2 * \frac{X_{UB,k}-X_{LB,k}}{2}$
4	$0.3 * \frac{X_{UB,k}-X_{LB,k}}{2}$	5	$0.5 * \frac{X_{UB,k}-X_{LB,k}}{2}$	6	$0.7 * \frac{X_{UB,k}-X_{LB,k}}{2}$

A shift index of zero indicates that the function is not shifted

3.7GHZ; 8GB RAM. Compared to conventional FWA, the only additional parameters of EFWA are A_{init} and A_{final}, which can be fixed as fractions of the search space.

6.3.2 Experimental Results

In this section, we first evaluate the influence of the newly proposed operators presented in Sect. 6.2. After that, we compare EFWA with conventional FWA and with SPSO, our baseline reference. The final results after 300 000 function evaluations are presented for all 12 benchmark functions in Table 6.3 as mean fitness and standard deviation over 30 runs.

Evaluation of EXP, MAP and GAU. Since these three[1] operators are responsible for the fact that conventional FWA performs better on function which have their optimum at the origin of the search space than on functions whose optimum is far away from the center (cf. Sect. 6.1), all eFWA variants use the new operators EXP, MAP and GAU. As expected, when SI = 0 conventional FWA reaches the optimum of all functions that have their optimum at 0.0^D (marked in red font in Table 6.3). However, as already mentioned, this good performance is not due to the intelligence of the algorithm but rather because the Gaussian explosion operator and the mapping operator of conventional FWA create/map many sparks to locations which are very close to the search space. Indeed, Table 6.3 reveals that when SI = 0, conventional FWA only fails to find the optimum of two function, f_3 and f_7—both functions have their optimum at 1.0^D, and thus not at the origin of the search space. With increasing SI, the results of FWA worsen significantly for most functions, especially so for f_1, f_2, f_5, f_7, f_{11}, f_{12}. When comparing the results of conventional FWA with eFWA-I (using the new operators EXP, MAP, GAU), it can be seen that changing these three operators alone does not improve the performance of the algorithm. Although the results are more stable with respect to different shift values, in most cases they are worse than the results of conventional FWA. In the next paragraph, we will discuss how the AMP operator is able to improve eFWA-I.

Evaluation of AMP. Table 6.3 reveals that the minimal explosion amplitude check strategy (AMP) is crucial for improving the diversity of the fireworks. The difference between the results of eFWA-I, which does not use an explosion amplitude check, and eFWA-II (linear decrease of A_{min}^k) and eFWA-III (nonlinear decrease of A_{min}^k) are obvious. Both, eFWA-II and eFWA-III, clearly outperform eFWA-I. Comparing the results of eFWA-II and III, it can be seen that eFWA-III achieves slightly better results. Hence, the nonlinear decrease of the minimum explosion amplitude A_{min}^k is preferred over the linear decrease of A_{min}^k, and will be used in EFWA.

Evaluation of SEL. Compared to eFWA-III, EFWA replaces the time consuming distance-based selection operator with the new selection operator (cf. Sect. 6.2.5). In

---

[1] We note that the questionable part of the explosion sparks operator of conventional FWA is the mapping operation, not the explosion operation itself.

Table 6.3 Mean and variance (in parenthesis) of all 12 benchmark functions used in the experiments (SI = shift index)

SI	Alg.	f_1	f_2	f_3	f_4	f_5	f_6	f_7	f_8	f_9	f_{10}	f_{11}	f_{12}
0	SPSO	0	3.712886	9.025685	18.8362	0.000387	162.4213	0	−1.03163	3.000000	0.000000	0	0
	FWA	0	0	18.06771	0	0	0	0.013432	−1.03163	3.000001	0.000000	0	0
	eFWA-I	1751.919	7097.559	4622089	5.747068	1.123267	172.9351	3961796	−1.03162	3.000035	0.005529	388.1999	20224.25
	eFWA-II	0.166553	1.065167	107.821	0.150611	0.273287	145.9337	0.006831	−1.03163	3.000000	0.000000	0.007313	1.139661
	eFWA-III	0.082755	0.313803	91.52661	0.082682	0.143627	144.4078	0.002717	−1.03163	3.000000	0.000000	0.002982	0.501991
	EFWA	0.080583	0.327939	110.6504	10.20939	0.134513	128.0998	0.00354	−1.03163	3.000000	0.000000	0.003003	0.477865
1	SPSO	0	3.875658	10.29035	19.73264	0.000777	168.5133	0	−1.03163	3.000000	0.000000	0	0
	FWA	0.235929	62.19819	1.368985	0.157826	0.10509	4.260847	0.012899	−1.03163	3.000007	0.001946	0.017149	2.086486
	eFWA-I	1683.079	7163.763	4675799	5.595889	1.148833	174.0394	3888660	−1.03162	3.000032	0.005156	363.8366	20400.76
	eFWA-II	0.166322	0.934341	123.2805	0.125657	0.26553	149.7567	0.006147	−1.03163	3.000000	0.000000	0.006905	1.120488
	eFWA-III	0.082681	0.301201	108.301	0.127529	0.13763	146.8661	0.00256	−1.03163	3.000000	0.000000	0.003115	0.486991
	EFWA	0.07894	0.303336	129.4147	14.78846	0.131399	133.1501	0.002475	−1.03163	3.000000	0.000000	0.002899	0.47585
2	SPSO	0	5.589772	11.93113	19.80728	0.000483	172.6362	0	−1.03163	3.000000	0.000000	0	0
	FWA	0.661069	227.831	88.48228	0.854293	0.210448	6.658164	0.068989	−1.03163	3.000005	0.004376	0.046928	5.556603
	eFWA-I	1712.366	7540.234	4624418	5.77933	1.206808	178.2714	3171013	−1.03162	3.000043	0.005719	361.0415	19108.3
	eFWA-II	0.168955	0.933509	112.6466	0.123641	0.263951	147.7272	0.006349	−1.03163	3.000000	0.000000	0.006618	1.146244
	eFWA-III	0.083509	0.303672	124.3955	0.140883	0.135765	163.6453	0.002603	−1.03163	3.000000	0.000000	0.002971	0.53305
	EFWA	0.07936	0.319263	137.1596	17.50123	0.13835	133.8353	0.003588	−1.03163	3.000000	0.000000	0.003048	0.487051
3	SPSO	0	9.897378	10.72043	19.91163	0.000918	164.4082	0	−1.03163	3.000000	0.000000	0	0
	FWA	1.495644	643.9931	139.0133	1.417975	0.300615	4.434904	0.213188	−1.03163	3.000014	0.006803	0.144981	14.3453
	eFWA-I	1700.841	7237.548	5578362	6.128872	1.304753	183.6675	3144835	−1.03162	3.000055	0.004777	371.1394	20822.16
	eFWA-II	0.16799	0.962124	104.893	1.141007	0.269353	144.2484	0.006098	−1.03163	3.000000	0.000000	0.00725	1.147999
	eFWA-III	0.082844	0.317519	107.7358	2.909734	0.140745	153.2287	0.002806	−1.03163	3.000000	0.000000	0.002972	0.507572
	EFWA	0.079119	0.32688	124.8834	19.53764	0.135506	121.61	0.00263	−1.03163	3.000000	0.000000	0.00286	0.479914

(continued)

Table 6.3 (continued)

SI	Alg.	f_1	f_2	f_3	f_4	f_5	f_6	f_7	f_8	f_9	f_{10}	f_{11}	f_{12}
4	SPSO	0	17.95454	11.89914	19.95621	0.001015	153.9394	0	−1.03163	3.000000	0	0	0
	FWA	2.353416	1722.845	267.3387	1.784102	0.354713	8.558857	0.353507	−1.03163	3.000015	0.005517	0.235881	26.67778
	eFWA-I	1740.594	6539.821	5085283	8.49947	1.403223	186.7392	3024996	−1.03162	3.000059	0.004584	391.184	19512.98
	eFWA-II	0.176647	0.941755	130.0864	13.86739	0.260521	139.4024	0.006635	−1.03163	3.000000	0.000000	0.006542	1.051908
	eFWA-III	0.084896	0.316917	128.5722	12.17068	0.137188	149.618	0.003242	−1.03163	3.000000	0.000000	0.003134	0.491694
	EFWA	0.077153	0.313821	109.0857	19.97647	0.135921	125.1156	0.002854	−1.03163	3.000000	0.000000	0.003035	0.492438
5	SPSO	0	40.33175	13.87386	19.98846	0.00116	119.374	0.000215	−1.03163	3.000000	0.000000	0	0
	FWA	3.337647	6204.057	429.1847	2.187983	0.375887	8.765354	0.582387	−1.03163	3.000033	0.005833	0.413908	42.90086
	eFWA-I	1637.047	6123.058	6141640	14.33566	1.452845	212.6149	3064550	−1.03162	3.000151	0.005151	435.9372	18802.31
	eFWA-II	0.168983	1.024643	145.3226	20.01804	0.263803	165.1913	0.006224	−1.03163	3.00000	0.000000	0.00708	1.090932
	eFWA-III	0.085402	0.299152	124.9419	19.93473	0.142257	175.968	0.002433	−1.03163	3.000000	0.000000	0.002926	0.521624
	EFWA	0.081641	0.316406	187.6413	20.0044	0.132645	143.9208	0.002553	−1.03163	3.000000	0.000000	0.002859	0.484936
6	SPSO	0	146.8231	16.32905	19.99701	0.005713	118.2048	0	−1.03163	30.00000	0.000000	0	0
	FWA	3.445464	7729.23	941.3545	2.396496	0.348149	9.385204	0.8201	−1.03163	30.00243	0.007451	0.416452	46.33276
	eFWA-I	1710.271	5362.155	17299855	19.97842	1.455054	248.5832	3939765	−1.03162	30.00258	0.005344	637.0806	21882.77
	eFWA-II	0.177206	0.890351	182.0592	20.02658	0.259169	202.2376	0.005894	−1.03163	30.00000	0.000000	0.006879	1.130264
	eFWA-III	0.080248	0.323097	168.1323	20.01205	0.142437	201.9406	0.002692	−1.03163	30.00000	0.000000	0.003025	0.471663
	EFWA	0.077676	0.309704	133.5418	20.01094	0.133103	165.2688	0.002882	−1.03163	30.00000	0.000000	0.002854	0.486914

terms of convergence, final fitness and standard deviation, there is almost no difference between the selection operators. The results of eFWA-III and EFWA are almost
identical for all functions (cf. Table 6.3). The only exceptions are function f_4, with
advantages for the distance-based selection operator (eFWA-III), and f_6, with advantages for the elitism-random selection operator (EFWA). In terms of computation cost
the picture is different. The new selection operator decreases the runtime of EFWA
drastically, as shown in Fig. 6.6 for the function f_{10}. As can be seen, the runtime
of EFWA is much shorter than the runtimes of conventional FWA or the eFWA-X
variants, which all use the conventional distance-based selection operator. We note
that the fractions of runtimes for all other functions are very similar to Fig. 6.6.

EFWA versus conventional FWA. A comparison between all eFWA-X variants
and EFWA reveals that EFWA is the best algorithm in terms of convergence, final
result, and runtime. Hence, in the remaining parts of this section we use EFWA
for comparison with conventional FWA and with the baseline algorithm SPSO. To
show whether the improvement of EFWA over conventional FWA is significant or
not, a number of t-tests were conducted and the recorded p-values are given in
Table 6.4. The null hypothesis that EFWA achieves similar mean results than FWA
is tested against the alternative that EFWA achieves better mean results over FWA.
If $p < 0.05$, and the mean results of EFWA are lower than the results of FWA, the
null hypothesis will be rejected and the test will be reported as significant; otherwise
it will be reported as not significant. Hence, bold values in Table 6.4 indicate that the
improvement of EFWA is significant for the corresponding function / SI combination.
We draw the following conclusions from Tables 6.3 and 6.4.

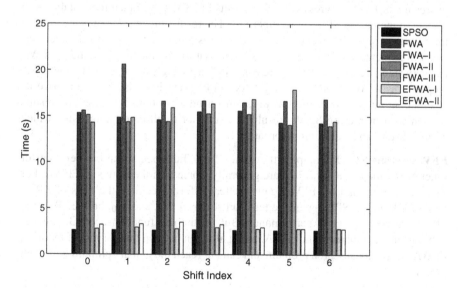

Fig. 6.6 Runtime of FWA, eFWA-X and SPSO

Table 6.4 t-test results for EFWA versus conventional FWA

EFWA	SI = 0	SI = 1	SI = 2	SI = 3	SI = 4	SI = 5	SI = 6
f_1	0	0.000000	0.000000	0.000000	0.000000	0.000000	0.000000
f_2	0	0.000000	0.000000	0.000000	0.000000	0.000000	0.000000
f_3	0.000006	0	0.005094	0.141752	0.000003	0.000119	0.000000
f_4	0	0	0	0	0	0	0
f_5	0	0.000175	0.000000	0.000000	0.000000	0.000000	0.000000
f_6	0	0	0	0	0	0	0
f_7	0.007443	0.000000	0.000000	0.000000	0.000000	0.000000	0.000000
f_8	0.322126	0.322126	0.083202	0.047784	0.000530	0.000003	0.000005
f_9	0.026484	0.000003	0.006833	0.000003	0.000003	0.000001	0.000000
f_{10}	NaN	0.002984	0.000003	0.000000	0.000000	0.000000	0.000000
f_{11}	0	0.000000	0.000000	0.000000	0.000000	0.000000	0.000000
f_{12}	0	0.000000	0.000000	0.000000	0.000000	0.000000	0.000000

Functions are *not* shifted (i.e, SI = 0): FWA achieves better results than EFWA for all functions which have their optimum at the origin, and for function f_3 which has the optimum at 1.0^D. For the other functions which have their optimum at a different location than the origin (f_7, f_8, and f_9), we can see that EFWA can improve the results of FWA for functions f_7 and f_9. For function f_8, both algorithms achieved similar results, however, with smaller variance for EFWA.

Functions are shifted (i.e., SI ! = 0): With increasing SI, EFWA is able to improve the results of FWA progressively. The results of EFWA are much more stable with respect to increasing SI for most functions. The results in Table 6.4 indicate that the improvement of EFWA over conventional FWA is significant for almost all functions, when SI is increased. However, we also note that for the functions f_4 and f_6, FWA has very stable results even for increasing SI. Surprisingly, for these functions FWA outperforms EFWA. For function f_4. As can be seen, FWA is able to improve its results continuously over the full duration of the algorithm. Although the results worsen with increasing SI, the results are superior to all other algorithms, which often get stuck in local minima for larger SI.

EFWA versus SPSO. Compare the results from Table 6.3, it can be seen that for function f_2, the results of SPSO are generally worse than the results of EFWA. For f_4 and f_6, for small SI, EFWA gains better performance than the ones of SPSO while for large SI, SPSO gains better results. For f_8, f_9, and f_{10}, both EFWA and SPSO achieve very good performance. But for the rest functions, SPSO achieves better results compared with EFWA, and it also can be seen that for most functions, EFWA has the local search problem and can not reach the global optimum with high precision.

6.3.3 Experimental Observations

From the results of our experimental evaluation, we conclude the following observations:

- In general, EFWA shows significant improvements over conventional FWA for most functions.
- With increasing shift values, EFWA achieves much better results than FWA.
- SPSO achieve better results than EFWA for most functions.
- The results of EFWA remain almost unaffected even if the optimum of the function is shifted toward the edges of the search space.
- The new operator AMP, which limits the lower bound of the explosion amplitude, is an important tool to balance between exploration and exploitation abilities.
- EFWA reduces the runtime of FWA by a factor of six.

6.3.4 Influences of Dimension Selection Methods on FWA and EFWA

In [1], we presented the Enhanced Fireworks Algorithm in details. The EFWA is proposed based on the FWA, which makes five improvements. In [2], in the explosion operator and Gaussian sparks operator, while generating an explosion spark or Gaussian spark, the number of dimension should be calculated at first with uniform distribution, In FWA, it first calculate the number of dimension that will be selected at first and then perform the explosion process and Gaussian mutation process. The number of dimension selected is $z = ceil(rand * D)$, where the z is under mean distribution in $[1, D]$ where D is the total number of dimension of the optimization function f.

In [1], in the proposed EFWA, in the dimension select method, it can be seen that the dimension selection method is $z^k = round(rand(0, 1))$, $k = 1, 2, ..., D.$, for each dimension, it has the half probability to be selected independently. Thus, the total dimension number selected is under binomial distribution in $[1 - D]$.

However, the experimental results in [1] is still based on the dimension selection method in conventional FWA, i.e., the total selected dimension number is under mean distribution, which will lead that if others do experiments according to the pesudocode of EFWA and will not be able to get the results which are same as the results in [1]. Thus, here we present an discussion about it. Here, we also want to point out that those two dimension selection methods differ in performance and the previous dimension selection method used in FWA will makes the performance better if we keep previous the dimension selection method in EFWA.

Moreover, in [1], we also present the results of SPSO for further comparison between PSO and FWA, in the experimental results about SPSO presented in [1], it is surprising to find that with the increasing of SI, the SPSO results sometimes fails to converge toward the optimum, at that moment, we are not clear about the reasons for this strange phenomenon, later, we conduct further experiments over SPSO and finally, find that PSO algorithm is very sensitive to the different *boundary conditions*, while in the implementation of SPSO in [1], the boundary conditions is that while the particle exceeds the search bound, the position of the particle will be set to the bound. Thus, in the SPSO presented in [1], the SPSO fails to converge sometimes while the SI increases.

For more details about the discussion, please visit http://www.cil.pku.edu.cn/ research/fwa/index.html.

6.4 Summary

In this chapter, an enhanced version of FWA was presented. Based on the detailed analysis of operators of conventional FWA, five new operators to overcome FWA's limitations were presented, which included the minimal explosion amplitude check strategy, the new mapping operator, the new operator for generating explosion sparks, the new Gaussian mutation operator and selection operator. Experimental results demonstrated that EFWA is a significant improvement version of FWA.

References

1. S. Zheng, A. Janecek, Y. Tan, Enhanced fireworks algorithm, in *2013 IEEE Congress on Evolutionary Computation (CEC)* (IEEE, 2013), pp. 2069–2077
2. Y. Tan, Y. Zhu, Fireworks algorithm for optimization, in *Advances in Swarm Intelligence* (Springer, Berlin, 2010), pp. 355–364
3. A.P. Engelbrecht *Fundamentals of Computational Swarm Intelligence* (Wiley, New York, 2006)
4. D. Bratton, J. Kennedy, Defining a standard for particle swarm optimization, in *IEEE Swarm Intelligence Symposium, SIS 2007* (IEEE, 2007), pp. 120–127

Chapter 7
Fireworks Algorithm with Dynamic Search

In this chapter, we present a dynamic version of Fireworks Algorithm, which is an improved version of the recently developed enhanced fireworks algorithm (EFWA) based on an adaptive dynamic local search mechanism. In EFWA, the explosion amplitude of each firework is computed based on the quality of the firework's current location. This explosion amplitude is limited by a lower bound which decreases with the number of iterations in order to avoid the explosion amplitude to be [close to] zero, and in order to enhance global search abilities at the beginning and local search abilities toward the later phase of the algorithm. As the explosion amplitude in EFWA depends solely on the fireworks' fitness and the current number of iterations, this procedure does not allow for an adaptive optimization process. To deal with these limitations, a dynamic search fireworks algorithm (dynFWA) is proposed, which uses a dynamic explosion amplitude for the firework at the current best position. If the fitness of the best firework could be improved, the explosion amplitude will increase to speed up convergence. On the contrary, if the current position of the best firework could not be improved, the explosion amplitude will decrease in order to narrow the search area. In addition, we show that one of the EFWA operators, i.e., Gaussian mutation operator, can be removed in dynFWA without a loss in accuracy—this makes dynFWA computationally more efficient than EFWA.

7.1 Introduction

The fireworks algorithm (FWA) [1] is a recently developed SI algorithm based on simulating the explosion process of real fireworks exploding and illuminating the night sky. In FWA, the fireworks (i.e., individuals) are let off to the potential search space and an explosion process is initiated for each firework. This stochastic *explosion* process is one of the key features of FWA. After the explosion, a shower of sparks fills

© Springer-Verlag Berlin Heidelberg 2015
Y. Tan, *Fireworks Algorithm*, DOI 10.1007/978-3-662-46353-6_7

the local space around the firework. Both fireworks as well as the newly generated sparks represent potential solutions in the search space. A principle FWA works as follows: At first, N fireworks are initialized randomly, and their quality (i.e., fitness) is evaluated in order to determine the explosion amplitude and the number of sparks for each firework. Subsequently, the fireworks explode and generate different types of sparks within their local space. Finally, N candidate fireworks are selected among the set of candidates, which includes the newly generated sparks as well as the N original fireworks. The algorithm continues the search until a termination criterion (time, maximum number of iteration or fitness evaluation, or convergence) is reached.

Later, Zheng S. et al. proposed the enhanced fireworks algorithm (EFWA) which incorporates five modifications compared to conventional FWA in order to eliminate the drawbacks of the original algorithm: (i) a new minimal explosion amplitude check, (ii) a new operator for generating explosion sparks, (iii) a new mapping strategy for sparks which are out of the search space, (iv) a new operator for generating Gaussian sparks, and (v) a new operator for selecting the population for the next iteration.

In FWA and EFWA, the explosion amplitude value is one of the key parameters which is used to balance between the local and global search capabilities of the algorithms. The fitness of the current location of each firework is used to calculate the explosion amplitude. The main idea is that a firework with *better* fitness (i.e., smaller fitness value for minimization problems[1]) can generate a *larger population* of explosion sparks within a *smaller range*, i.e., with a small explosion amplitude, while fireworks with poorer fitness (i.e., higher fitness value) can only generate a smaller population within a larger range, i.e., with bigger explosion amplitude. As a result, fireworks at good locations will perform local search in a narrow range around the current location, while fireworks with higher fitness will perform global search in a wider range.

The minimal explosion amplitude check (MEAC) strategy in EFWA [2] introduced a lower bound for this explosion amplitude in order to avoid that the explosion amplitude of the best firework found so far is set to zero. This strategy decreases the explosion amplitude solely with the current number of function evaluations, which heavily depends on the predefined number of iterations for the algorithm. Experimental results indicate that this strategy does not allow for efficient local search around the current best solution. Therefore, in this chapter, we present an appropriate strategy for varying the explosion amplitude dynamically based on the current success of the optimization process. Additionally, we show that it is possible to remove the rather time-consuming Gaussian sparks operator of EFWA in dynFWA without loss in optimization accuracy.

---

[1]In this chapter, without loss of generality, the optimization problem f is assumed to be a minimization problem.

7.2 Properties of Minimal Explosion Amplitude Check (MEAC) Strategy in EFWA

For simplicity and convinces of description later on, we use the following definitions:

Definition 7.1 Core Firework (CF): In each iteration, the firework at the currently best location is marked as core firework (CF). Thus, for minimization problems, among the set C of all fireworks the firework X_{CF} is selected as CF *iff*

$$\forall X_i \in C: f(X_{CF}) \leq f(X_i) \tag{7.1}$$

Definition 7.2 Local Minimum Space and Local Minimum Point: Given an objective function f, in a continuous space $\Psi \subseteq \Omega$, there $\exists$ only one point $\mathbf{x}$, $\exists \varepsilon$, and $f(x_i) - f(\mathbf{x}) \geqslant 0$, for $\forall x_i$, $|x_i - \mathbf{x}| \leqslant \varepsilon$, then $\mathbf{x}$ is a local minimum point. For region S, if these is only one local minimal point in it, then S is a local minimum space.

A firework with better fitness can generate a larger population of explosion sparks within a smaller range, i.e., with a small explosion amplitude. On the contrary, fireworks with poorer fitness can only generate a smaller population within a larger range, i.e., with higher explosion amplitude. This allows to balance between exploration and exploitation capabilities of the algorithm. *Exploration* refers to the ability of the algorithm to explore various regions of the search space in order to locate promising good solutions, while *exploitation* refers to the ability to conduct a thorough search within a smaller area recognized as promising in order to find the optimal solution (cf. [3]). Exploration is achieved by those fireworks which have a large explosion amplitude (i.e., poorer fitness), since they have the capability to escape from local minima. Exploitation is achieved by those fireworks which have a small explosion amplitude (i.e., better fitness), since they reinforce the local search ability in promising areas.

In EFWA, the MEAC Strategy (cf. Sect. 6.2.1) enforces the exploration capabilities at the early phase of the algorithm (larger $A_{\min}^k \Rightarrow$ global search), while at the final phase of the algorithm the exploitation capabilities are enforced (smaller $A_{\min}^k \Rightarrow$ local search). This is further enforced by the nonlinear decrease of $A_{\min}^k$ (cf. Eq. 6.3). Obviously, this procedure decreases the explosion amplitude solely with the current number of function evaluations which heavily depends on the predefined number of iterations for the algorithm. The explosion amplitude strategy should consider the optimization process information rather than solely the information about the current iteration (or evaluation) count. In order to tackle this problem we propose a dynamic explosion strategy is proposed for the CF in order to enhance the local search ability.

7.3 The dynFWA

In dynFWA, fireworks are separated into two groups. The first group consists of the CF, while the second group consists of all remaining fireworks, i.e., non-CF. The responsibility of the CF is to perform a local search around the best location found so far, while the responsibility of the second group is to maintain the global search ability. For both groups, the explosion amplitude is a key feature in order to efficiently and effectively improve the current locations of the fireworks. However, for the CF, the selection of the explosion amplitude is even more important due to its high influence on the convergence speed toward the local minimum point within the local minimum space and the property that it is always selected in the optimization process. In contrast to EFWA, the explosion amplitude in dynFWA of the CF is not calculated by Eq. (2.3). Instead, the local information of the optimization process (i.e., the information if the algorithm has improved its best location during the last iterations) is used for the calculation of the explosion amplitude for the CF. For all fireworks in the second group (non-CFs), the explosion amplitude is calculated similarly to EFWA, i.e., by using Eq. 2.3, however, without using the MEAC from Eq. (6.1).

7.3.1 Dynamic Explosion Amplitude for the First Group (CF)

The CF stores the best solution found so far. Let us define $\hat{X}_b$ as the "best" newly created explosion spark of all fireworks in the swarm, and $\Delta_f = f(\hat{X}_b) - f(X_{CF})$. Based on the value of Δ_f, for minimization problems, there are two situations:

7.3.1.1 One or Several Explosion Sparks Have Found a Better Position, i.e., $\Delta_f < 0$

It is possible that (i) an explosion spark generated by the CF has found the best position, or that (ii) an explosion spark generated by a *different* firework than the CF has found the best position. Both cases indicate that the swarm has found a new promising position and that $\hat{X}_b$ will be the CF for the next iteration.

(i) In most cases, $\hat{X}_b$ has been created by the CF. In such cases, in order to speed up the convergence of the algorithm, the explosion amplitude of the CF for the next iteration will be increased compared to the current iteration. This situation is illustrated in Fig. 7.1a, b.

(ii) In other cases—though with a lower probability—a firework different from the CF will create $\hat{X}_b$. This situation happens more frequently during the earlier phase of the optimization process than during later iterations. In such cases, $\hat{X}_b$ will become the new CF for the next iteration (recall that the best location among all individuals is always selected for the next iteration). Since the position of the

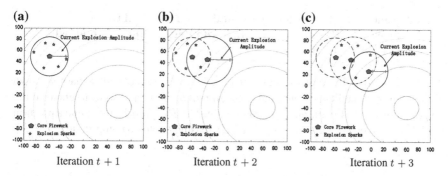

Fig. 7.1 Illustration of the amplification/reduction of the CF's explosion amplitude. In **a** the radius of the circle with *dashed red line* indicates the explosion amplitude of the CF in iteration t, while the *circle* with *solid black line* indicates the explosion amplitude in iteration $t + 1$; the increased explosion amplitude indicates that in this situation, a better position has been found by the explosion sparks. In iteration $t + 2$ (**b**), the CF is able to further improve its location, and, as a result, the explosion amplitude of the CF is further increased. **c** Shows an example when the fitness of the CF could not be improved. In this case, the CF's explosion amplitude is decreased in iteration $t + 3$ (Color figure online)

CF is changed, the current explosion amplitude which considers the optimization information of the position of the current CF will not be effective to the newly selected CF ($\hat{X}_b$). However, it is possible that $\hat{X}_b$ is located in rather close proximity to the previous CF: since the CF creates the large number of sparks among all fireworks, the random selection method may select several sparks created by the CF, which are initially located in close proximity to the CF. If so, the same consideration as in (i) applies, and the explosion amplitude of the CF will be increased. If $\hat{X}_b$ is created by a firework which is not in close proximity to the CF, the explosion amplitude can be reinitialized to the predefined value. However, since it is difficult to define "close" proximity, we do not compute the distance between $\hat{X}_b$ and X_{CF} but rely on the dynamic explosion amplitude update ability. Similarly to (i), the explosion amplitude is increased. If the new CF cannot improve its location in the next iteration, the new CF is able to adjust the explosion amplitude itself dynamically in the following iterations.

We underline the idea why an increasing explosion amplitude may accelerate the convergence speed. Assume that the current position of the CF is far away from the global/local minimum. Increasing the explosion amplitude is a direct and effective approach in order to increase the step size toward the global/local optimum in each iteration, i.e., it allows for faster movements toward the optimum. However, we note that usually the probability to find a position with better fitness decreases with increasing explosion amplitude due to the increased search space (obviously, this depends to a large extent on the optimization function).

7.3.1.2 None of the Explosion Sparks of the CF Nor of All Other Fireworks Has Found a Position with Better Fitness Compared to the CF, i.e., $\Delta_f \geq 0$

In this situation, the explosion amplitude of the CF is reduced to narrow down the search in a smaller region around the current location and so as to enhance the exploitation capability of the local search of the CF. The probability to find a position with better fitness usually increases with decreasing explosion amplitude.

7.3.2 Dynamic Explosion Amplitude Update for CF

Based on the previous discussion, the final proposed algorithm is as follows.

Algorithm 7.1 Dynamic explosion amplitude update for CF

Require: Define:
$\quad X_{CF}$ is the current location of the CF;
$\quad \hat{X}_b$ is the best location among all explosion sparks;
$\quad A_{CF}$ is the current explosion amplitude of the CF;
$\quad C_a$ is the amplification coefficient;
$\quad C_r$ is the reduction coefficient;
Ensure:
1: **if** $f(\hat{X}_b) - f(X_{CF}) < 0$ **then**
2: $\quad A_{CF} \leftarrow A_{CF} * C_a$;
3: **else**
4: $\quad A_{CF} \leftarrow A_{CF} * C_r$;
5: **end if**

Algorithm 7.1 summarizes the dynamic update strategy as discussed in Sect. 7.3.1. Figure 7.2 shows the process of the amplification/reduction during the optimization of the sphere function for 1,000 iterations (algorithm dynFWA). As can be seen from Fig. 7.2, the reduction and amplification explosion amplitudes happen in an alternating manner. Obviously, there are more reduction than amplification phases, which is partly caused by the values of C_a and C_r, which are set to 1.2 and 0.9, respectively, and by the fact that the explosion amplitude is initially set to the size of the search space, which is a rather large number in the first iteration (cf. Sect. 7.4.1).

7.3.3 Some Theoretical Analysis and Considerations

In what is following, we discuss why a reduction of the explosion amplitude increases the probability to find a better location. A Taylor series is used to represent the properties of the local region around the CF. Assume a continuously differentiable

Fig. 7.2 The reduction and amplification of CF's explosion amplitude (using function f_1 (sphere function) from [4], cf. Sect. 7.4)

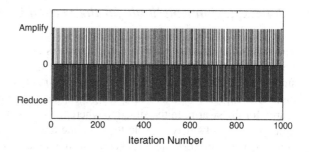

second-order optimization function g with k dimensions. If the position of the CF is *not* a local/global minimal point, and A_{CF} is the current explosion amplitude, then

$$g(\mathbf{x}) - g(X_{CF}) = \nabla g(X_{CF})^T (\mathbf{x} - X_{CF}) + \frac{1}{2}(\mathbf{x} - X_{CF})H(\mathbf{x})(\mathbf{x} - X_{CF}), \quad (7.2)$$

where

$$H(\mathbf{x}) = \left[\frac{\partial^2 g}{\partial x_i \partial x_j} \right]_{k \times k}. \quad (7.3)$$

According to the definition of "local minimal space and local minimal point" in Sect. 7.2 (i.e., X_{CF} is not a local/global minimum point), there $\exists \varepsilon$, $\forall \mathbf{x}$ in

$$S = \{\mathbf{x} | \, |\mathbf{x} - X_{CF}| \leqslant \varepsilon\}, \quad (7.4)$$

and

$$g(\mathbf{x}) - g(X_{CF}) = \nabla g(X_{CF})^T (\mathbf{x} - X_{CF}) + o(\nabla g(X_{CF})^T (\mathbf{x} - X_{CF})), \quad (7.5)$$

where $o(.)$ means low order.

From the Taylor series, if $\varepsilon \to 0$, then in region S, if there exists a point $\mathbf{x}_1$ and $\mathbf{x}_1 - X_{CF} = \Delta \mathbf{x}$, then there exists a point $\mathbf{x}_2$ and $\mathbf{x}_2 - X_{CF} = -\Delta \mathbf{x}$. Under this circumstance, the probability of generating a spark with smaller fitness than the CF is very high (i.e., $(g(\mathbf{x}_1) - g(X_{CF})) * (g(\mathbf{x}_2) - g(X_{CF})) < 0$). In case the CF does not find a better position while generating a number of sparks, it is likely that $A_{CF} \geqslant \varepsilon$. We cannot expect the property of region $T = \{\mathbf{x} | \varepsilon \leqslant |\mathbf{x} - X_{CF}| \leqslant A_{CF}\}$ that in region T whether there exists a position with better fitness compared to the CF, thus, if the CF generates sparks with uniform distribution in each dimension, the probability p' that a spark is located in S is $p' = \frac{\|S\|}{\|S\| + \|T\|}$, where $\| \, \|$ denotes the hypervolume of this region. If the CF does not find a better position, the explosion amplitude A_{CF} is reduced to increase the probability p' that the CF can generate a spark in region S thus to increase the probability finding a point with smaller fitness than CF.

7.3.4 Explosion Amplitude for the Second Group (Non-CF)

The explosion amplitudes for non-CF fireworks are calculated based on Eq. (2.3), i.e., similar to EFWA but without the minimum explosion amplitude check (MEAC) strategy. Compared to the CF, these non-CF fireworks can only create a smaller number of explosion sparks within a larger explosion amplitude to perform the global search for the swarm. In the situation where the CF gets stuck in local minima, this group of fireworks may be able to allow the algorithm to escape from premature convergence, since these fireworks continue the search in different areas of the search space.

7.3.5 Elimination of the Gaussian Sparks Operator

The motivation behind the Gaussian sparks (cf. [1]) is to further increase the diversity of the swarm. In EFWA, the Gaussian sparks are calculated by

$$X_{ik} = X_{ik} + (X_{Bk} - X_{ik}) \times e, \tag{7.6}$$

where X_{Bk} is the kth dimension value of location of the currently best firework(denoted), and $e = Gaussian(0, 1)$ (cf. Sect. 6.2.4). From Fig. 7.3, it can be seen that the newly generated sparks will be located along the direction between a selected firework i and the CF. Any newly generated Gaussian sparks will be either (1) close to the CF, (2) close to firework i, or (3) located along the direction between the CF and firework i, however, with some distance to both, the CF and firework i. In the first two cases, the operator will have similar performance as the explosion sparks generated by the CF and firework i, respectively. In the third situation, the Gaussian spark can be interpreted as an explosion spark generated by a firework with large explosion amplitude. Thus, based on the above analysis it can be stated that in many situations Gaussian sparks will not be able to effectively increase the diversity of the swarm.

Fig. 7.3 Gaussian mutation operator in EFWA

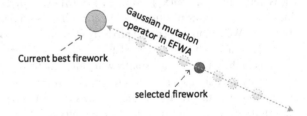

7.3.6 Framework of dynFWA

Based on the operators discussed in the previous subsections, the final algorithm called dynamic search fireworks algorithm (dynFWA) is presented in Algorithm 7.2.

Algorithm 7.2 Framework of dynFWA

1: Initialize N fireworks and evaluate the quality
2: Initialize the explosion amplitude for CF
3: **while** termination criteria are not met **do**
4: Calculate number of explosion sparks (cf. Eq. 2.3)
5: Calculate explosion amplitude for non-CF (cf. Eq. 2.3)
6: **for** each firework **do**
7:   Generate explosion sparks
8:   Map sparks at invalid locations back to search space
9:   Evaluate quality of explosion sparks
10: **end for**
11: Update explosion amplitude of CF (cf. Algorithm 7.1)
12: Select N fireworks for next iteration
13: **end while**

7.4 Experiments

To investigate the performance of the proposed dynFWA algorithm as well as the necessity of the Gaussian mutation operator (i.e., the performance of dynFWA removing this operator), we compare two EFWA variants and two dynFWA variants. Each algorithm is tested with and without Gaussian mutation operator, respectively. Besides comparing dynFWA and EFWA, the most recent version of SPSO (SPSO2011, [5]) is also used for performance comparison. In the following, we briefly describe the five algorithms used for experimental evaluation:

- EFWA—the baseline algorithm as presented in [2][2];
- EFWA-NG—in this algorithm, the Gaussian sparks operator has been removed from EFWA;
- dynFWA-G—this algorithm implements the dynFWA algorithm as described in Sect. 7.3 *including* the Gaussian mutation operator.
- dynFWA—similar as dynFWA-G but *without* Gaussian mutation operator.

---

[2]We note that experimental results in the EFWA [2] are unintentionally based on the dimension selection method of conventional FWA [1], which varies the number of adapted dimensions uniformly among all dimension. More details can be found at http://www.cil.pku.edu.cn/research/FWA/index.html. All results and pseudocodes in this chapter are based on the published pseudocode in [2].

• SPSO2011—the most recent SPSO variant. Compared to earlier versions of SPSO it features an improved velocity update by exploiting the idea of *rotational invariance* for the velocity update instead of sequential dimension-by-dimension update of older versions of SPSO (cf. [5]).[3]

7.4.1 Experimental Setup

Similar to EFWA, the number of fireworks in dynFWA is set to 5, but in dynFWA, the maximum number of explosion sparks in each iteration is set to 150. The reduction and amplification factors C_r and C_a of dynFWA are empirically set to 0.9 and 1.2, respectively, and A_{CF} is initially set to the size of the search space in order to maintain a high exploration capability at the beginning. All other parameters for dynFWA and all EFWA parameters are identical to [2], SPSO2011 parameters are listed in Table A.4 [5]. For each algorithm we performed 51 runs on each optimization function; the final mean results after 300 000 function evaluations are presented. As experimental platform we used MATLAB 2011b (Windows 7; Intel Core i7-2600 CPU @ 3.7 GHZ; 8 GB RAM). To validate the performance of the proposed algorithms we used the recent CEC 2013 benchmark suite that includes 28 different benchmark functions as listed in [4].

7.4.2 Experimental Results

In this subsection, we first evaluate the influence of removing the Gaussian sparks operator as discussed in Sect. 7.3.5. After that we evaluate the performance of dynFWA and compare its performance to EFWA and SPSO2011.

7.4.2.1 Evaluation of Gaussian Mutation Operator

To evaluate whether the results of EFWA and dynFWA improve or deteriorate after removing the Gaussian sparks operator we compare the results of EFWA and EFWA-NG, and the results of dynFWA-G and dynFWA, respectively. In order to validate the improvement between any two algorithms, the Wilcoxon signed-rank test is conducted (cf. Sect. 7.4.2). Assume that data X, Y are fitness results for a given number of runs of two different algorithms. If the mean value of X is smaller than Y and the Wilcoxon signed-rank test under 5 % significance level is true, then it is

---

[3] In addition to the results (median, maximum, minimal) in [5], detailed results of SPSO2011 which include the results of each single run were also submitted to the *CEC2013 competition* held by P.N. Suganthan. These results are available at http://goo.gl/pXB1WH. The mean fitness error results are computed based on the results in the folder "1534" of file "Results-of-22-papers.zip".

Table 7.1 Wilcoxon signed-rank test results for EFWA versus EFWA-NG and dynFWA-G versus dynFWA (bold values indicate the performance difference is significant)

F.	EFWA versus EFWA-NG			dynFWA-G versus dynFWA		
	EFWA	EFWA-NG	p-value	dynFWA-G	dynFWA	p-value
f_1	−1.3999E+3	−1.3999E+3	**2.316E−3**	−1.4000E+3	−1.4000E+3	1.000E+0
f_2	6.8926E+5	6.5258E+5	4.256E−1	7.6981E+5	8.6937E+5	1.801E−1
f_3	7.7586E+7	6.4974E+7	8.956E−1	1.2007E+8	1.2317E+8	6.393E−1
f_4	−1.0989E+3	−1.0989E+3	7.858E−1	−1.0863E+3	−1.0896E+3	**3.183E−2**
f_5	−9.9992E+2	−9.9992E+2	**4.290E−2**	−1.0000E+3	−1.0000E+3	1.463E−1
f_6	−8.5073E+2	−8.4462E+2	1.654E−1	−8.6524E+2	−8.6995E+2	9.156E−2
f_7	−6.2634E+2	−6.2991E+2	9.552E−1	−6.9946E+2	−7.0010E+2	6.663E−1
f_8	−6.7907E+2	−6.7906E+2	9.776E−1	−6.7909E+2	−6.7910E+2	4.997E−1
f_9	−5.6846E+2	−5.6889E+2	5.178E−1	−5.7435E+2	−5.7587E+2	1.711E−1
f_{10}	−4.9916E+2	−4.9918E+2	3.732E−1	−4.9994E+2	−4.9995E+2	3.591E−1
f_{11}	5.8198E+0	3.5430E+1	5.830E−2	−2.9978E+2	−2.9589E+2	6.127E−1
f_{12}	3.9944E+2	4.1107E+2	6.193E−1	−1.4993E+2	−1.4222E+2	4.762E−1
f_{13}	2.9857E+2	2.8909E+2	8.220E−1	5.4523E+1	5.3830E+1	8.513E−1
f_{14}	2.7240E+3	2.9344E+3	**4.101E−2**	2.8909E+3	2.9180E+3	8.147E−1
f_{15}	4.4595E+3	4.5515E+3	6.869E−1	3.9186E+3	4.0227E+3	4.879E−1
f_{16}	2.0063E+2	2.0056E+2	2.811E−1	2.0056E+2	2.0058E+2	7.358E−1
f_{17}	6.2461E+2	6.3152E+2	9.179E−1	4.5397E+2	4.4261E+2	1.197E−1
f_{18}	5.7361E+2	5.6953E+2	6.938E−1	5.8801E+2	5.8782E+2	8.660E−1
f_{19}	5.1022E+2	5.1012E+2	9.402E−1	5.0750E+2	5.0726E+2	6.193E−1
f_{20}	6.1466E+2	6.1457E+2	**1.559E−2**	6.1309E+2	6.1328E+2	3.632E−1
f_{21}	1.1178E+3	1.1362E+3	**6.910E−4**	9.9532E+2	1.0102E+3	6.431E−1
f_{22}	6.3181E+3	6.3674E+3	9.776E−1	4.1463E+3	4.1262E+3	9.402E−1
f_{23}	7.5809E+3	7.5707E+3	7.217E−1	5.6661E+3	5.6526E+3	9.402E−1
f_{24}	1.3452E+3	1.3611E+3	**1.079E−2**	1.2738E+3	1.2729E+3	8.586E−1
f_{25}	1.4426E+3	1.4435E+3	8.734E−1	1.3964E+3	1.3970E+3	8.882E−1
f_{26}	1.5461E+3	1.5400E+3	2.687E−1	1.4744E+3	1.4607E+3	1.337E−1
f_{27}	2.6210E+3	2.5780E+3	3.534E−1	2.2721E+3	2.2804E+3	8.147E−1
f_{28}	4.7651E+3	4.9949E+3	6.460E−1	1.7686E+3	1.6961E+3	3.555E−1

believed that the results of X are significant better than Y. A comparison between EFWA versus EFWA-NG and dynFWA-G versus dynFWA are given in Table 7.1:

- EFWA-NG performs slightly better than EFWA on 16 functions, while it performs slightly worse than EFWA on 12 functions. However, for five functions the Wilcoxon signed-rank test indicates that EFWA is significantly better than EFWA-NG, while for one function EFWA-NG is significant better than EFWA. Hence, for EFWA these results suggest that the Gaussian mutation operator should not be removed, although EFWA-NG is faster in terms of runtime (see next section).

- In general, the performance of dynFWA and dynFWA-G is very similar. Only for function f_4 dynFWA performs significantly better than dynFWA-G. This indicates that dynFWA without the Gaussian mutation operator achieves slightly better results than dynFWA-G, and is also faster in terms of runtime (see next section).

In the rest of this chapter, we use the best EFWA variant (EFWA *with* Gaussian sparks operator) and the best dynFWA variant (dynFWA *without* Gaussian mutation operator) for further comparison.

7.4.2.2 Comparison of dynFWA and EFWA

Table 7.2 shows the average fitness values over 51 runs for each function for the three algorithms SPSO2011, EFWA and dynFWA, and the corresponding rank of each algorithm. Moreover, at the bottom of Table 7.2 we present the mean fitness rank for each algorithm. Table 7.4 shows the runtime of each algorithm—the runtime of the fastest algorithm (dynFWA) is set to 1, for all other algorithms, the proportional runtimes compared to dynFWA are presented.

A comparison between the proposed dynFWA and EFWA reveals that dynFWA outperforms EFWA in terms of average fitness rank and runtime. It can be seen that dynFWA achieves better mean fitness results than EFWA on 23 functions except function f_2, f_3, f_4, f_{14}, f_{18}. The mean fitness rank results suggest that dynFWA gains great advantages over EFWA. To test whether the improvement of dynFWA over EFWA is significant or not, a number of Wilcoxon signed-rank tests were conducted and the corresponding p-values are presented in Table 7.3. The results indicate that the improvement of dynFWA is significant compared to EFWA for 22 benchmark functions. Moreover, in terms of computational complexity it can be seen that dynFWA significantly reduces the runtime of EFWA for the same number of function evaluations. This is mostly caused by the removal of the Gaussian mutation operator that is computationally rather expensive.

7.4.2.3 Comparison of dynFWA and SPSO2011

A comparison between the proposed dynFWA and the latest SPSO variant indicates dynFWA is able to achieve a better mean rank compared to SPSO2011 (cf. Table 7.2). In total, dynFWA achieves better results (smaller mean fitness) than SPSO2011 on 17 functions, while SPSO2011 is better than dynFWA on 10 functions, for one function, the results are identical. In terms of computational complexity we measured the runtime of SPSO2007, however, since the results of SPSO2011 are adopted from the [5] literature, we cannot compare the runtimes as they depend on the implementation and the infrastructure. Table 7.4 shows that compared to SPSO2007, dynFWA has

Table 7.2 Mean fitness on the benchmark functions and mean fitness rank of SPSO2011, EFWA and dynFWA

F.	SPSO2011	Rank	EFWA	Rank	dynFWA	Rank
f_1	**−1.4000E+3**	**1**	−1.3999E+3	3	**−1.4000E+3**	**1**
f_2	**3.3719E+5**	**1**	6.8926E+5	2	8.6937E+5	3
f_3	2.8841E+8	3	**7.7586E+7**	**1**	1.2317E+8	2
f_4	3.7543E+4	3	**−1.0989E+3**	**1**	−1.0896E+3	2
f_5	**−1.0000E+3**	**1**	−9.9992E+2	3	−1.0000E+3	2
f_6	−8.6210E+2	2	−8.5073E+2	3	**−8.6995E+2**	**1**
f_7	**−7.1208E+2**	**1**	−6.2634E+2	3	−7.0010E+2	2
f_8	−6.7908E+2	2	−6.7907E+2	3	**−6.7910E+2**	**1**
f_9	−5.7123E+2	2	−5.6846E+2	3	**−5.7587E+2**	**1**
f_{10}	−4.9966E+2	2	−4.9916E+2	3	**−4.9995E+2**	**1**
f_{11}	−2.9504E+2	2	5.8198E+0	3	**−2.9589E+2**	**1**
f_{12}	**−1.9604E+2**	**1**	3.9944E+2	3	−1.4222E+2	2
f_{13}	**−6.1406E+0**	**1**	2.9857E+2	3	5.3830E+1	2
f_{14}	3.8910E+3	3	**2.7240E+3**	**1**	2.9180E+3	2
f_{15}	**3.9093E+3**	**1**	4.4595E+3	3	4.0227E+3	2
f_{16}	2.0131E+2	3	2.0063E+2	2	**2.0058E+2**	**1**
f_{17}	**4.1626E+2**	**1**	6.2461E+2	3	4.4261E+2	2
f_{18}	**5.2063E+2**	**1**	5.7361E+2	2	5.8782E+2	3
f_{19}	5.0951E+2	2	5.1022E+2	3	**5.0726E+2**	**1**
f_{20}	6.1346E+2	2	6.1466E+2	3	**6.1328E+2**	**1**
f_{21}	**1.0088E+3**	**1**	1.1178E+3	3	1.0102E+3	2
f_{22}	5.0988E+3	2	6.3181E+3	3	**4.1262E+3**	**1**
f_{23}	5.7313E+3	2	7.5809E+3	3	**5.6526E+3**	**1**
f_{24}	**1.2667E+3**	**1**	1.3452E+3	3	1.2729E+3	2
f_{25}	1.3993E+3	2	1.4426E+3	3	**1.3970E+3**	**1**
f_{26}	1.4861E+3	2	1.5461E+3	3	**1.4607E+3**	**1**
f_{27}	2.3046E+3	2	2.6210E+3	3	**2.2804E+3**	**1**
f_{28}	1.8013E+3	2	4.7651E+3	3	**1.6961E+3**	**1**
Mean Rank						
	SPSO2011	1.75	EFWA	2.68	dynFWA	**1.54**

almost the same runtime. Since the SPSO2011 operators (new velocity update strategies) appear to be more complex, therefore, probably at least its time-consuming as same as the SPSO2007 (cf. [6]), we expect SPSO2011 to have a similar or slightly higher computationally complexity compared to initial version of SPSO2007.

Table 7.3 Wilcoxon signed-rank test results for dynFWA versus EFWA (bold values indicate the significant improvement)

F.	f_1	f_2	f_3	f_4	f_5	f_6	f_7
p-value	**0.00E+0**	6.94E−3	9.90E−2	**0.00E+0**	**0.00E+0**	**1.58E−3**	**0.00E+0**
F.	f_8	f_9	f_{10}	f_{11}	f_{12}	f_{13}	f_{14}
p-value	**1.73E−2**	**0.00E+0**	**0.00E+0**	**0.00E+0**	**0.00E+0**	**0.00E+0**	1.41E−1
F.	f_{15}	f_{16}	f_{17}	f_{18}	f_{19}	f_{20}	f_{21}
p-value	**5.10E−5**	3.20E−1	**0.00E+0**	6.35E−2	**1.41E−4**	**0.00E+0**	**0.00E+0**
F.	f_{22}	f_{23}	f_{24}	f_{25}	f_{26}	f_{27}	f_{28}
p-value	**0.00E+0**	**0.00E+0**	**0.00E+0**	**0.00E+0**	**0.00E+0**	**0.00E+0**	**0.00E+0**

Table 7.4 Runtime comparison

SPSO2007	1.01	EFWA	1.30	dynFWA-G	1.17
SPSO2011	/	EFWA-NG	1.03	dynFWA	**1**

7.5 Summary

In this chapter, we presented the *dynamic search Firework Algorithm (dynFWA)*, an improved version of the recently developed enhanced fireworks algorithm (EFWA). dynFWA uses a dynamic explosion amplitude for the core firework (CF), i.e., the firework at the current best position. This dynamic explosion amplitude depends on the quality of the current local search around the CF. The main task for the CF is to perform a local search, while the responsibility for all other fireworks are to maintain the global search ability. Additionally, we have analyzed the possibility to remove the rather time-consuming Gaussian mutation operator of EFWA.

From the result of our experimental evaluation we concluded the following observations:

1. The proposed dynFWA algorithm significantly improves the results of EFWA and reduces the runtime by more than 20 %.
2. Compared with SPSO2011, dynFWA achieves a better mean rank among 28 benchmark functions with similar computational cost.
3. The Gaussian mutation operator of EFWA is a necessary component of EFWA. However, removing this operator from dynFWA can reduce the runtime of dynFWA significantly without loss of optimization accuracy.

References

1. Y. Tan, Y. Zhu, Fireworks algorithm for optimization, in *Advances in Swarm Intelligence* (Springer, Berlin, 2010), pp. 355–364
2. S. Zheng, A. Janecek, Y. Tan, Enhanced fireworks algorithm, in *2013 IEEE Congress on Evolutionary Computation (CEC)* (IEEE, 2013), pp. 2069–2077
3. M. Clerc, J. Kennedy, The particle swarm—explosion, stability, and convergence in a multidimensional complex space. Trans. Evol. Comput. **6**(1), 58–73 (2002)
4. J.J. Liang, B.Y. Qu, P.N. Suganthan, A.G. Hernández-Díaz, Problem definitions and evaluation criteria for the CEC 2013 special session on real-parameter optimization (2013)
5. M. Zambrano-Bigiarini, M. Clerc, R. Rojas, Standard particle swarm optimisation 2011 at CEC-2013: a baseline for future PSO improvements, in *2013 IEEE Congress on Evolutionary Computation (CEC)* (2013), pp. 2337–2344, doi:10.1109/CEC.2013.6557848
6. M. Clerc, Standard particle swarm optimization, from 2006 to 2011. *Particle Swarm Central* (2011)

References

Chapter 8
Adaptive Fireworks Algorithm

The explosion amplitude in FWA is a key factor influencing the performance of Fireworks Algorithm, which needs to be controlled precisely. In this chapter, a new FWA algorithm called Adaptive fireworks algorithm is proposed by replacing the explosion amplitude operator in FWA with an adaptive method.

8.1 Motivation

Both FWA [1] and EFWA remain to be improved in many aspects. For example, neither way of calculating the explosion amplitude in the two algorithms are reasonable. Considering the explosion search in FWA and EFWA, the amplitude is a very important factor influencing its performance.

In order to improve the mechanism of calculating the amplitude of explosion in FWA and EFWA, an adaptive explosion amplitude is proposed by using the distance between the best firework and a certain selected individual as the explosion amplitude. We analyze the property of our adaptive amplitude and proved that it can adjust itself adaptively according to the search results. By applying adaptive explosion amplitude to the EFWA, a new algorithm called the adaptive fireworks algorithm (AFWA) is proposed.

8.2 Analysis of the Amplitude in FWA and EFWA

Considering the explosion search manner in the FWA and the EFWA, the amplitude of each firework is a fatal variable influencing the performance of the algorithm. As shown in Eq. (2.3), the amplitudes of other fireworks (except for the best one) are calculated according to the difference between their fitness and the best one's. However, the amplitude of the best firework is always 0 according to Eq. (2.3). In FWA, there is no other operator to deal with this problem, which means that the best

© Springer-Verlag Berlin Heidelberg 2015
Y. Tan, *Fireworks Algorithm*, DOI 10.1007/978-3-662-46353-6_8

firework will not contribute to the algorithm at all, despite its most numerous sparks. Note that according to Eq. (2.1), the better a firework's fitness is, the more sparks it generates. In FWA, The best fireworks takes the most but provides the least.

Thus, in EFWA [2], in order to make sure the best firework works, a minimal amplitude check is adopted, preventing the amplitude of the best firework from being 0. As is shown in Eqs. (6.1) and (6.3), the threshold of the amplitude is a nonlinear decreasing function of the generation number. In EFWA, the threshold is actually the amplitude of the best firework.

However, any preset amplitude, linear decrease or nonlinear decrease, whatever the parameters are set, cannot fit the evaluation function well: in some functions, it decreases too fast, and in others too slow. Decreasing too fast causes the search range converges early before the minima is reached. Decreasing too slow causes the search range is still too large to search precisely even when the minima is already within the search range. In either case, the algorithm performs badly.

The amplitudes of other fireworks in FWA and EFWA are calculated according to Eq. (2.3) adaptively, while the amplitude of the best firework still remains a big problem. Although it only influences local search of the algorithm, it is a key to the performance. The amplitude of the best firework needs to be adjusted automatically in order to fit all evaluation functions.

8.3 Adaptive Explosion Amplitude

In this section, an adaptive method is proposed, using already generated sparks to calculate the explosion amplitude of the best firework. We use the information obtained in this generation to calculate the amplitude of the best firework in the next generation. Considering the selection in EFWA, the best firework in next generation is the best individual found (could be a spark generated or a firework) in this generation. As we already know, all the amplitudes of other fireworks (in the next generation) are calculated according to the difference between their fitness and the best firework. The main problem is to get a reasonable amplitude of the best firework. For convenience, we only consider one firework within this section.

8.3.1 Principles

In order to calculate an adaptive amplitude for the best firework, we choose an individual and use its distance to the best individual, which is the firework in next generation, as the amplitude of the next explosion.

The individual we choose subjects to the following two conditions:

(1) Its fitness is worse than the best firework fitness of this generation.
(2) Its distance to the best individual (the firework of next generation) is minimal among all individuals subjecting to (1).
Namely,

$$\hat{s} = \arg\min_{s_i}(d(s_i, s^*)) \tag{8.1}$$

s.t.

$$f(s_i) > f(X) \tag{8.2}$$

where s_i stands for all sparks generated by the firework, s^* stands for the best individual among all sparks and the firework, X stands for the firework, d is a certain measurement of distance. Note that the algorithm always choose the best individual as the firework in next generation.

Condition (1) requires that the difference (in fitness) between this individual and the best individual is bigger than that between the firework and the best individual, i.e.,

$$f(\hat{s}) - f(s^*) > f(X) - f(s^*) \tag{8.3}$$

An intuitive understanding of this inequality is it assures that the scale on the domain of the evaluation function within the range $d(\hat{s}, s^*)$ is at least bigger than the improvement made in this generation. Our aim to find a better location $\tilde{s}$ in the next explosion such that

$$f(s^*) - f(\tilde{s}) > f(X) - f(s^*) \tag{8.4}$$

by estimating the potential range $d(\tilde{s}, s^*)$ with $d(\hat{s}, s^*)$. The algorithm makes a correspondence between the range and the domain. Some deeper consideration and its properties will be discussed after the complete algorithm is presented.

On the other hand, condition (2) helps to make sure the amplitude converges. If a farther individual subjecting to condition (1) is chosen, the amplitude could be, in the worst case, double the amplitude of this explosion. Under that circumstances, there is no guarantee the amplitude would not be locked on the maximum value (the range of the search space, for example). On the contrary, choosing the nearest individual is quite safe because if the function is regular locally and the sparks are numerous enough, the minimum distance would be at most slightly longer than the amplitude in this generation. Although we cannot promise the amplitude decreases every time, but with big iteration times and numerous sparks, it converges in general, as is shown in Fig. 8.1. It can be seen that at the early phase of the optimization, the explosion amplitude is usually very bigger, while at the later phase, the explosion amplitude tends to reduce, and to be quite small, thus to increase its local search ability.

To make a clear explanation of AFWA, in the following, we use figures to present the basic idea. In Fig. 8.2, the red dot with fitness 1.0 is the firework of this generation, and yellow dots are the sparks generated by it. Obviously, we will choose the yellow

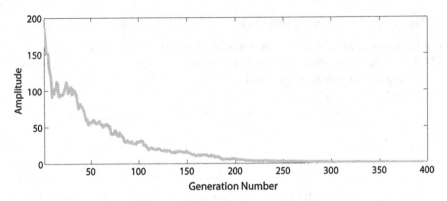

Fig. 8.1 Adaptive amplitude on sphere function

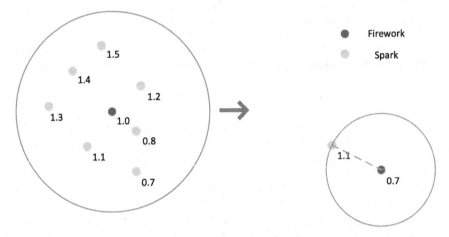

Fig. 8.2 An example of how adaptive amplitude is calculated

dot with fitness 0.7 as the firework in the next explosion. According to Eqs. (8.1) and (8.2), we choose the individual whose fitness is 1.1 as $\hat{s}$ and use its distance to the 0.7 individual (s^*) as the amplitude of next explosion. Without condition (1) we will choose the 0.8 individual, which makes the algorithm converges too fast. Without condition (1) we will choose the 1.5 individual, which makes the algorithm does not converge at all. If the evaluation function changes regularly, with numerous sparks, there is a good chance for us to find the individual with fitness 0.3 in the next explosion, if 0.7 is not close to local minima.

It is clear that this algorithm does not care either the scale of the range or the domain. Rather, it detects the relationship between the scale of the range and the scale of the domain adaptively without any more evaluation. Changing the scale of either will not influence the performance. In a sense, it provides the gradient information of the evaluation function.

Considering the way fireworks explode in EFWA, where they explode in each dimension independently, we use infinity norm as the distance measure, namely the maximum difference among all dimensions. Besides, in order to further slow down the convergence rate and improve the global search, the adaptive amplitude calculated above is multiplied by a certain coefficient (usually bigger than 1). Finally, considering the sparks of each explosion is limited, in order to minimize the influence of very bad luck (for example, every spark is worse than the firework, in which case the amplitude shrinks very fast, or on the contrary the amplitude doubled last time), we also adopt a simple smoothing mechanism, which uses the average of the amplitude calculated above and the amplitude of this generation as the amplitude.

8.3.2 Algorithm

The complete algorithm of calculating the amplitude is shown in Algorithm 8.1.

Algorithm 8.1 Calculate the adaptive amplitude for the firework of generation $g + 1$

1: $A(g + 1) \leftarrow UB - LB$
2: **for** $i = 1$ to n **do**
3:    **if** $||s_i - s^*||_\infty > A(g + 1)$ and $f(s_i) > f(X)$ **then**
4:       $A(g + 1) \leftarrow ||s_i - s^*||_\infty$
5:    **end if**
6: **end for**
7: $A(g + 1) \leftarrow \lambda \cdot A(g + 1)$
8: $A(g + 1) \leftarrow 0.5 \cdot (A(g) + A(g + 1))$
9: **return** $A(g + 1)$

In Algorithm 8.1, UB and LB stand for the upper bound and lower bound of the search space respectively, $s_1 \ldots s_n$ stands for all sparks generated by the firework in generation g, X stands for the firework in generation g, s^* stands for the best individual in generation g, namely the firework in generation $g + 1$.

The parameter λ has a great impact on the performance of the algorithm. In a sense, it controls the balance between global search and local search. If λ is too small, the adaptive amplitude converges too fast to a local minima without searching the neighborhood. While if it is too big, the adaptive amplitude does not converge. Generally speaking, as long as it converges stably, the bigger the better. In experiment, we usually use $\lambda = 1.3$ empirically.

The computational complexity of Algorithm 8.1 is $O(n)$, which means it does not add much cost. Actually, compared to other operators such as generating sparks, $O(n)$ is not dominant.

By using the information provided by these already calculated individuals efficiently, this algorithm returns a relatively accurate amplitude, within which the explosion process has a great chance to make a great improvement.

Specifically, when the amplitude is too long, the chance to find a better location become low. On the contrary, if the amplitude is too short, explosion algorithm can only obtain a comparatively small improvement. However, the proposed algorithm promises that there is likely to exists a location within the amplitude which brings a greater improvement than the last generation, so long as the function is relatively smooth in the small neighborhood.

8.3.3 Explanation

For most optimization problems, the search process by only one firework can be briefly divided into three stages (note that they are not strictly distinguished):

(1) Global search. At the beginning, the algorithm does not have any information about the evaluation function, so it has to explore globally to decide which region is comparatively promising for further exploitation. In our algorithm, the amplitude of the firework is set to the range of the search space at the beginning, and the sparks it generates will be distributed in the whole search space. In this way, the algorithm detects the rough information about which region is good and which is not. Then, the algorithm will choose the best individual among all the sparks and the firework. As far as the amplitude is concerned, it is really hard to predict whether it will become longer or shorter. There are two possible cases, roughly speaking, as shown in Fig. 8.3.

In the first case, $\hat{s}$ and s^* are on the different sides of the firework. To the limited information we have, we can assume in reason that the firework is in the same local region as $\hat{s}$ and s^*, and that the $d(\hat{s}, s^*)$ gives the longest estimate of the region's scale. In this case, the amplitude will most probably become longer and the s^* will walk a longest step toward the potential local best.

In the second case, $\hat{s}$ and s^* are on the same side of the firework. Recall that the fitness of $\hat{s}$ we use is worse than the firework, while s^* is better. To the limited

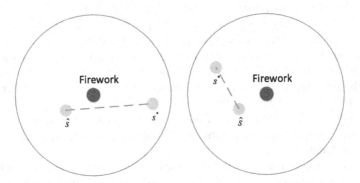

Fig. 8.3 *Left* $\hat{s}$ and s^* are on the different sides of the firework; *Right* $\hat{s}$ and s^* are on the same side of the firework

information we have, we can assume in reason that the firework is not in the same local region as $\hat{s}$ and s^*, and that the region containing s^* is more promising than that containing the firework, and that the $d(\hat{s}, s^*)$ gives the longest estimate of the region's scale. In this case, the amplitude will most probably become shorter, in order to give up the region of the firework and search more efficiently.

In summary, $d(\hat{s}, s^*)$ always gives the longest possible estimate of the scale of the local region containing s^* and keeps the search most efficient.

(2) Local search. When the firework enters a certain local region, which we can assume as a bowl region, the main task is to walk toward the bottom as fast as possible. When the amplitude is still much shorter than the distance between the firework and the bottom, we can assume that the neighborhood of the firework is monotonous, which means $\hat{s}$ and s^* will undoubtedly on the different sides of the firework. From the monotonic we can also assume that the s^* is very close to the border of the firework's amplitude, because there is no farther spark in its direction. Then, it is most likely that,

$$d(\hat{s}, s^*) > d(firework, s^*) \tag{8.5}$$

where, $\hat{s}$ is the best individual in current generation g and $\hat{s}$ is the individual calculated under Eq. (8.1). Figure 8.4 shows this situation.

As shown in Fig. 8.5, the experimental result also supports this conclusion.

In summary, the amplitude will become longer and longer in stage (2) to fasten the search as long as the firework is still far from the local minima.

(3) Refine search. At the end of the search, when the local minima is already within the amplitude of the firework, the algorithm need to search more precisely than in the above stages. In this case, the s^* is not the farthest spark, rather it is the spark closest to the local minima. So, in contrary to stage (2), the amplitude will most probably become shorter and shorter (shown in Fig. 8.6) and enable the algorithm to search more and more precisely, unless the local minima is not within the amplitude any longer, which brings the algorithm back to stage (2). Finally, as we have shown in Fig. 8.1, the amplitude converges after all, and the local minima is certainly reached.

The properties of our algorithm in all the three stages proves it a promising global-to-local search algorithm.

There are some extreme cases for the algorithm:

(a) It is possible that all the sparks' fitness is worse than the firework's. It is most likely to happen in stage (3). Usually, it implies that the firework is quite close to the local minima, while the amplitude is too large. In AEA algorithm, if it happens, $\hat{s}$ will just be the closest spark to the firework, and the amplitude in next generation will become shorter than this generation. When the amplitude is reduced to a reasonable size (there is certainly a better location in a very small neighborhood unless the local minima is already reached), the algorithm will go on to work normally.

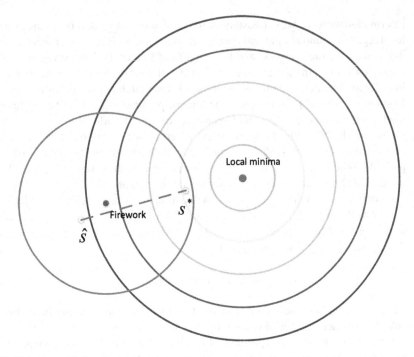

Fig. 8.4 Stage (2): Local search

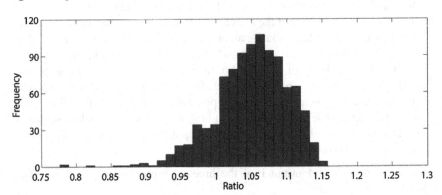

Fig. 8.5 A histogram of the ratio by which the amplitude increases at stage (2)

(b) It is possible that all the sparks' fitness is better than the firework's. It is most
    likely to happen in stage (1). According to Algorithm 8.1, the amplitude will be
    set to the range of the search space. It is still reasonable because the firework can
    be considered a local maxima in this case, which means search around it will be
    meaningless.

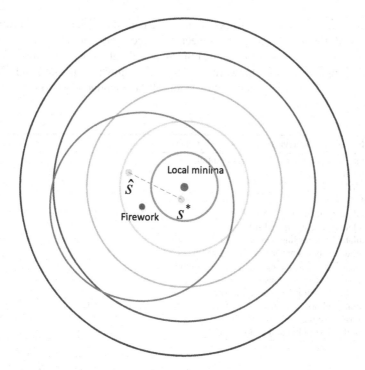

Fig. 8.6 Stage (3): Refine search

Therefore, even in the extreme cases, the algorithm still has a strong capability of error correction.

The explosion amplitude could be considered as the step size in the fireworks algorithm. To the best of our knowledge, the adaptive explosion amplitude we proposed here is a brand new method to control the step size in an evolutionary algorithm, which would find more potential application in other EC algorithms.

8.4 Adaptive Fireworks Algorithm

In this section, the adaptive explosion amplitude is applied to the best firework of enhanced fireworks algorithm.

At the beginning after initialization, we set the amplitude of the best firework to the range of the search space. Then, for each generation, we use AEA algorithm to calculate the amplitude of the best firework in the next generation.

When there are more than two fireworks adopted in the algorithm, the $\hat{s}$ we look for in Sect. 8.3 could be a spark as well as a firework.

The amplitudes of other fireworks are still unchanged as calculated according to Eq. (2.3), and the minimal amplitude check is unnecessary and are dropped in our algorithm since it is designed to control the amplitude of the best firework. Except for the amplitude of the best firework, we basically follow the operators in EFWA.

Algorithm 8.2 shows the complete version of the adaptive fireworks Algorithm [3].

Algorithm 8.2 Adaptive Fireworks Algorithm

1: randomly select m fireworks in the potential space
2: evaluate their fitness
3: $A^* \leftarrow UB - LB$
4: **repeat**
5:    calculate N_i according to Eq. (2.1)
6:    calculate A_i according to Eq. (2.3)
7:    calculate A^* according to Algorithm 8.1
8:    generate N_i sparks according to Algorithm 6.1
9:    generate Gaussian sparks according to Algorithm 2.2
10:    evaluate all sparks' fitness
11:    keep the best individual as a firework
12:    randomly choose other $m - 1$ fireworks among the rest individuals
13: **until** termination criteria is met
14: **return** the best individual and its fitness

In Algorithm 8.2, m is the number of fireworks, UB and LB stand for the upper bound and lower bound of the search space respectively, A_i stands for the amplitude of each firework, N_i stands for the number of sparks of each firework, A^* stands for the amplitude of the best firework.

Note that being different from the EFWA, m can be 1 in the AFWA, because the amplitude of the best firework is calculated independently.

We can see from Algorithm 8.2 that the number of parameters adopted in the AFWA is less than that in EFWA, since EFWA adopts 2 parameters in minimal amplitude check, which is now replaced by adaptive explosion amplitude.

8.5 Experiments

In order to illustrate and compare the performance of AFWA and EFWA, experiments on 28 CEC13's benchmark functions [4] were conducted. The introduction of the 28 functions is shown in Appendix A. In AFWA and EFWA, $m = 5$, $N_{min} = 2$, $N_{max} = 100$, $\hat{N} = 200$, $\hat{A} = 100$ and $NG = 5$. In AFWA, $\lambda = 1.3$. Besides, the results of SPSO2007 and SPSO2011, which is the latest version of the Standard PSO, were adopted as a baseline. The parameters of SPSO2007 are the same as [5], and the results of SPSO2011 are obtained directly from [6] and http://t.cn/8Fqg1rN. Evaluation times: 300,000, Run times: 51, Dimension: 30. Experiment environment: MATLAB2011b; Win 7; Intel Core i7-2600 CPU; 3.7GHZ; 8GB RAM.

The mean error of the four algorithms is presented in Fig. 8.7.

The mean ranking of the four algorithms' mean error, which represents the mean average error rank on 28 benchmark functions was calculated, shown in Table 8.1. For each function, each algorithm achieves the fitness error rank, then the average value is calculated for all functions to represent the performance for each algorithm.

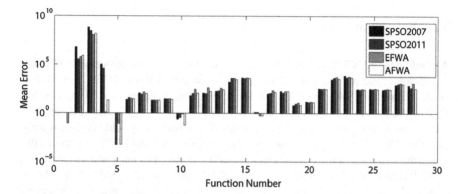

Fig. 8.7 Mean error on 28 functions

Table 8.1 Mean ranking of mean error on 28 functions

Algorithm	SPS02007 [5]	SPSO2011 [6]	EFWA [2]	AFWA
Mean ranking	2.4286	2.1786	3.3929	1.8929

Table 8.2 T-test results on AFWA versus EFWA

Function	1	2	3	4	5	6	7	8	9	10	11	12	13	14
H	1	0	0	1	1	1	1	1	1	1	1	1	1	1
Function	15	16	17	18	19	20	21	22	23	24	25	26	27	28
H	1	1	1	1	1	1	1	1	1	1	1	1	1	1

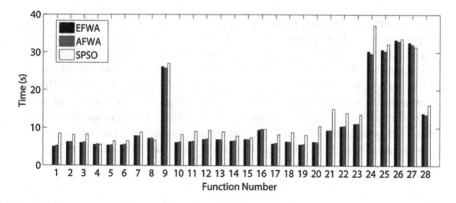

Fig. 8.8 Time consumed by each algorithm on 28 functions

A set of T-tests were also conducted to illustrate whether the improvement of AFWA over EFWA is significant. The null hypothesis is the results of the AFWA and those of the EFWA come from distributions with equal means. $H = 1$ indicates that the null hypothesis can be rejected at the 5 % level. Table 8.2 shows the H values.

Figure 8.8 shows the time consumed by each algorithm.

8.6 Discussion

According to Fig. 8.7, Tables 8.1 and 8.2, we can see that AFWA outperformed EFWA significantly, except for function 2 and function 3 where AFWA and EFWA performed almost the same, EFWA only beat AFWA on function 4 and function 18. AFWA and EFWA share almost all the parameters in common, so the significant difference in their performance proved that the new adaptive explosion amplitude is significantly effective. More precisely, since the global search in AFWA and EFWA are almost the same, the adaptive explosion amplitude actually improved its local search capability.

AFWA also performed much better than SPSO2007 and SPSO2011: AFWA beats SPSO2011 on 18 out of 27 functions and beats SPSO2007 on 16 out of 27 functions (except for function 1 where they are even). According to Table 8.1, judging from the overall performance, AFWA is the best algorithm among all the four algorithms.

In terms of computation cost, as is shown in Fig. 8.3, the time consumed by AFWA and EFWA were very close, which is less than SPSO. The computational cost of calculating the adaptive explosion amplitude is very low($O(n)$), compared to other operators such as generating sparks. In addition, according to [7], the computation cost of the EFWA is smaller than the PSO.

Lastly, it turns out from the performance and the computation cost that the adaptive fireworks algorithm is a very promising, efficient, and simple algorithm.

8.7 Summary

In this chapter, we analyzed the amplitudes of explosion in FWA and EFWA, and then proposed an adaptive explosion amplitude. The distance of the best firework and a certain individual subjecting to some conditions is employed as the amplitude of the explosion. We analyzed the property of the adaptive explosion amplitude and come to the conclusion that the adaptive explosion amplitude for firework explosion is a theoretically promising operator. Replacing with adaptive explosion amplitude, adaptive fireworks algorithm was proposed. According to the experimental results on CEC13's 28 benchmark functions, the performance is greatly improved: the AFWA not only outperforms EFWA but also beats SPSO2007 and SPSO2011 totally.

References

1. Y. Tan, Y. Zhu, Fireworks algorithm for optimization, in *Advances in Swarm Intelligence* (Springer, Berlin, 2010), pp. 355–364
2. S. Zheng, A. Janecek, Y. Tan, Enhanced fireworks algorithm, in *2013 IEEE Congress on Evolutionary Computation (CEC)* (IEEE, 2013), pp. 2069–2077
3. J. Li, S. Zheng, Y. Tan, Adaptive fireworks algorithm, in *IEEE Congress on Evolutionary Computation (CEC)* (2014), pp. 3214–3221
4. J.J. Liang, B.Y. Qu, P.N. Suganthan, A.G. Hernández-Díaz, Problem Definitions And Evaluation Criteria for the CEC 2013 Special Session On Real-parameter Optimization (2013)
5. D. Bratton, J. Kennedy, Defining a standard for particle swarm optimization, in *Swarm Intelligence Symposium, SIS 2007* (IEEE, 2007), pp. 120–127
6. M. Zambrano-Bigiarini, M. Clerc, R. Rojas, Standard particle swarm optimisation 2011 at CEC-2013: a baseline for future PSO improvements, in *2013 IEEE Congress on Evolutionary Computation (CEC)* (2013), pp. 2337–2344
7. Y. Tan, C. Yu, S. Zheng, K. Ding, Introduction to fireworks algorithm. Int. J. Swarm Intell. Res. (IJSIR) **4**(4), 39–70 (2013)

Chapter 9
Cooperative Fireworks Algorithm

In the previous studies on FWAs, researchers ignored the cooperation and interaction between the individual fireworks in the swarm, which are the most important core for any swarm intelligence algorithm. By incorporating a probabilistically oriented explosion mechanism (POEM) into the conventional FWA, a novel Cooperative Fireworks Algorithm (CoFWA, for short) is proposed to enhance the interactions among the individual fireworks in the swarm. In the CoFWA, the POEM mechanisms of sparks generation and fireworks selection are well designed to strengthen the cooperative capability of the individual fireworks in the CoFWA. It turns out by many experiments that the CoFWA significantly outperforms two most recent variants of FWA (i.e., EFWA and dynFWA) and SPSO2011 and shows a competitive performance against the state-of-the-art swarm intelligence algorithms.

9.1 Introduction

The Fireworks Algorithm (FWA) [1] is a newly developed successful swarm intelligence (SI) algorithm inspired by explosion of fireworks in real life. Like other swarm intelligence algorithm, it also aims at finding the position with the best (usually minimum) fitness in the search space. Inspired by real fireworks, the fireworks (i.e., individuals) in FWA are set off at the potential search space. In the initialization step, a few of fireworks are randomly chosen in the search space and their corresponding fitness is evaluated accordingly. Subsequently, the fireworks explode and then generate different types of sparks within their local areas. Finally, some fireworks are selected from the set of candidates which include either the newly generated sparks or the original fireworks for the next generation. In such a way, the algorithm continues this searching until a termination criterion (time, maximum number of iterations or fitness evaluations, or convergence) is reached. Since FWA has proven its efficiency in dealing with optimization problems, a lot of improvement

© Springer-Verlag Berlin Heidelberg 2015
Y. Tan, *Fireworks Algorithm*, DOI 10.1007/978-3-662-46353-6_9

works on FWA have been done and published, such as Enhanced Fireworks Algorithm (EFWA) [2], Adaptive Fireworks Algorithm (AFWA) [3], Dynamic Search in Fireworks Algorithm (dynFWA) [4] and just to name a few.

FWA consists of four parts, i.e., initialization, explosion, mutation, and selection.

In the part of explosion, the explosion amplitude and the population of the newly generated explosion sparks differ among the fireworks in order to ensure diversity and balance the global and local searches. Fireworks located at good positions can generate a large population of explosion sparks within a smaller range, i.e., with a small explosion amplitude. On contrary, fireworks located at positions with lower fitness values can only generate a smaller population within a larger range, i.e., with higher explosion amplitude.

In the selection process, the best firework is always kept, but the rest $m - 1$ fireworks are selected randomly.

In the framework of FWA, one can discover that the cooperation and interaction between the individuals in the swarm do not play an important role in the FWA. The interaction between fireworks is only reflected on the population of explosion sparks and the explosion amplitude, but the distribution of how the sparks are generated is simply regarded as a uniform distribution, which does not use any information from other fireworks. In particular, in the selection part, only the best firework is kept and the other $m - 1$ fireworks are randomly selected among sparks. We consider that this kind of selection mechanism makes the fireworks lose too much information from previous generations even though it guarantees the mutation of fireworks.

9.2 Probabilistically Oriented Explosion Mechanism (POEM)

As indicated by the name, the Probabilistically Oriented Explosion Mechanism (POEM) is a way with which a firework explodes in different directions with different probabilities. Unlike commonly used the uniform explosion with which firework explodes in all direction with same probability, POEM means a firework is able to explode in anisotropy with probability, in different directions with different probabilities, as illustrated by Figs. 9.1, 9.2 and 9.3.

In the POEM, a kind of interaction between each firework and the best firework in the current swarm is designed to determine the orientation of explosion of each firework. First of all, the best firework is called as the global guide (S_g) in the current swam. Second, when a firework ($S_l, (l = 1, \ldots, N, l \neq g)$) except the best one explodes, the sparks will not generate randomly in the amplitude and directions, instead, more sparks will be generated in the area in the directions of the global guide but fewer sparks will be generated in the directions which are away further from the global guide, as illustrated in Figs. 9.1, 9.2 and 9.3.

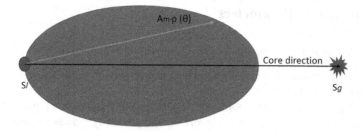

Fig. 9.1 Gaussian explosive way in POEM

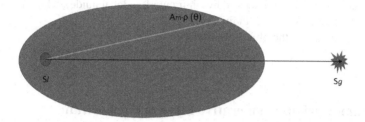

Fig. 9.2 Ellipsoid explosive way in POEM

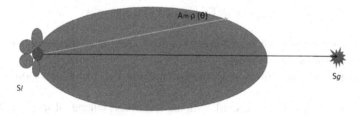

Fig. 9.3 Sinc explosive way in POEM

There are a number of firework explosive ways in POEM so long as they obey the above explanation for POEM. For example, one can have POEM with Gaussian explosive way, POEM with Ellipsoid explosive way and POEM with Sinc explosive way, as shown in Figs. 9.1, 9.2 and 9.3, respectively. By use of those kinds of anisotropic explosions for each firework, the cooperation and interaction among fireworks would be established and then strengthened further against the uniform explosion in FWA as before. As a result, a promising performance is expected. Sure, more evidences will be noted in the successive sections.

1. POEM with Gaussian explosive way
2. POEM with Ellipsoid explosive way
3. POEM with Sinc explosive way

9.3 Remarks on Parameters

There are following three parameters which should be explained in detail.

1. A_m: it is the amplitude of explosion which is inversely proportional to the fitness of the firework S_l, $(l = 1, \ldots, N)$.
2. $\rho(\theta)$: It is a point in polar coordinate system. It will take the maximum value in the (core) direction to the elite firework S_g and gradually fade as increase of the angle deviated from the core direction. It should be a normalized parameter like a probability.
3. M_l: number of sparks generated by lth firework (a local guide), which is proportional to the fitness of the lth firework.
4. S_g: the estimate of the global optimum.
5. S_l, $(l = 1, \ldots, N, l \neq g)$: the estimates of local optima by fireworks.

9.4 Framework of Cooperative Firework Algorithm (CoFWA)

There are following four steps in the Framework of Cooperative Firework Algorithm (CoFWA).

Step 1: Initialization—A few of fireworks in a swam, i.e., 5 fireworks, are generated randomly in the search space.

Step 2: POEM—A shower of sparks generated in POEM by explosion of each firework.

Step 3: Memory—The elite spark with highest fitness among all sparks and fireworks is selected as a global guide in the current swarm, denoted by S_g. The spark with highest fitness in those sparks generated in POEM by one firework survives, with a probability p_l (which is proportioned to the fitness), for next explosion, as a local guide, S_l, $(l = 1, \ldots, N, l \neq g)$, where N is the number of fireworks in the swarm, i.e., $N = 5$.

Step 4: Mapping Rule—there are two typical mapping rules to be exploited, i.e., Random Rule (If a spark is outside boundaries, it will be thrown back randomly) and Module Rule (taking module operation once a spark is outside boundaries).

Repeat Step 1–Step 4 until a termination criterion is reached.

It turns out that the CoFWA is a kind of swarm intelligence optimization algorithm which optimizes with a champion. In other words, S_g is the estimate of the global optimum, i.e., the champion, while S_l, $(l = 1, \ldots, N, l \neq g)$ are the estimates of local optima by fireworks.

9.5 A Kind of Realization of POEM

In order to strengthen the cooperation and interaction among fireworks of FWA, a mechanism called Probabilistically Oriented Explosion Mechanism (POEM) is adopted here. In the POEM, we design a kind of interaction between each firework and the best firework in the current generation to determine the orientation of explosion of each firework. We regard the best firework as the global guide (S_g) in the current swam. When a firework (S_l) except the best one explodes, the sparks will not generate randomly in the amplitude, instead, more sparks will be generated in the area which is closer to the global guide but fewer sparks will be generated in the area which is further from the global guide, as shown in Fig. 9.4. In order to achieve this goal, a special Gaussian mutation and a shift from the firework to the global guide is designed in POEM.

Specifically, there is no explosion orientation in FWA, as shown in Fig. 9.5. In order to enhance the cooperation between fireworks, at first, the sparks will be generated by a normal distribution with mean 0 and standard deviation b in the first dimension and standard deviation a in the next $d - 1$ dimension as illustrated in Fig. 9.6. Next, a matrix is calculated to rotate the orientation of the first dimension toward the global best firework, as shown in Fig. 9.7. Finally, a shift is calculated to move the sparks closer to the global best, as shown in Fig. 9.8.

A realization of the POEM can be specifically written as follows:

POEM Probabilistically Oriented Explosion Mechanism

---

  for S_l

    $shift = S_g - S_l$

    sample r_1 from $U(0, b)$

    sample r_j from $U(0, a)$, $j = 2, 3, \ldots, d$

    $\mathbf{r} = (r_1, r_2, \ldots, r_d)$

    $s_{li} \leftarrow S_l + \mathbf{M} * \mathbf{r} + \lambda * shift$, $i = 1, 2, \ldots, N_l$

  end for

Where $U(0, d)$ is the normal distribution with mean 0 and standard deviation d, N_l is the number of sparks of firework l, λ is a parameter which controls the shift from the firework to the global guide on a certain level, and $\mathbf{M}$ is the matrix to get the orientation from S_l to S_g. b is usually set to $\max\{|shift_i|, i = 1, 2, \ldots, d\}$, and a is equal to $\beta * b$ where $0 < \beta < 0.5$.

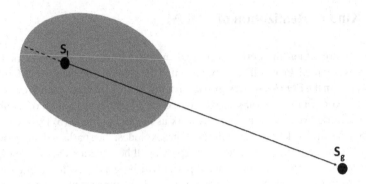

Fig. 9.4 Schematic diagram of POEM

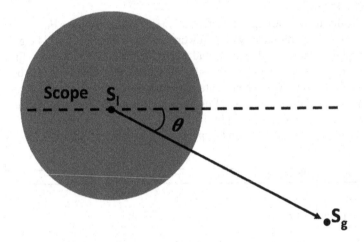

Fig. 9.5 There is no explosion orientation in FWA

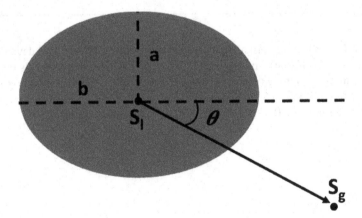

Fig. 9.6 Sparks generated by the normal distribution with zero mean and standard deviation b in the first dimension and a in the next $d - 1$ dimension

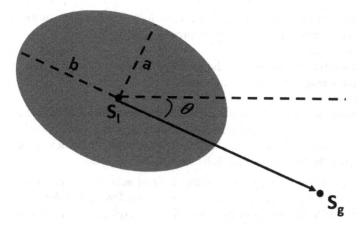

Fig. 9.7 Rotating the orientation of the first dimension toward the global best firework

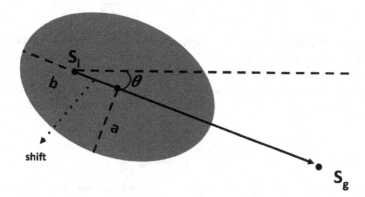

Fig. 9.8 A shift is calculated to make the sparks closer to the global best

9.6 The Proposed CoFWA

By incorporating the (POEM) into the original FWA, a novel Cooperative Fireworks Algorithm (CoFWA) is proposed here for enhancing the cooperations and interactions of the individual fireworks. Briefly, the CoFWA contains the following four main parts:

1. Initialization
   m fireworks are initialized randomly as the first generation.
2. Explosion
   Each firework's fitness is evaluated to determine the number of sparks for each firework. A firework with *higher* fitness can generate a *greater* number of sparks and a firework with *lower* fitness can only generate a *smaller* number of sparks. And then, each firework except the global guide firework generates a certain number of sparks by POEM. If the location of a spark is out of bound, it will be

randomly mapped into the search space. As for the global guide, it will generate the sparks the same way as the best firework in dynFWA.

3. Mapping rule
   Random Rule is used here. If the location of a spark is out of bound, it will be randomly mapped into the search space.

4. Selection
   The best spark of each firework in the current generation has remained as the firework for the next generation. In this selection algorithm, the fireworks selected in the current generation will help the global guide in the next generation to search the best fitness more efficiently in the search space.

The parts (2)–(4) are repeated until the maximal number of iterations.

Then we will briefly introduce some operators used in the CoFWA.

Like in FWA, the numbers of sparks that are exploded by each firework are calculated as follows:

$$N_l = \hat{N} \cdot \frac{y_{max} - f(S_l) + \xi}{\sum_{l=1}^{m}(y_{max} - f(S_l)) + \xi} \qquad (9.1)$$

where $\hat{N}$ is a parameter of controlling the total number of sparks generated by the m fireworks, $y_{max} = max(f(S_l))(l = 1, 2, 3, \ldots, m)$ is the maximum (worst) value of the objective function among the m fireworks, and ξ, which denotes the smallest constant in the computer, is utilized to avoid zero division error. In order to avoid the overwhelming effects of splendid fireworks, the number of sparks is bounded by

$$N_l = \begin{cases} N_{min} & \text{if } N_l < N_{min} \\ N_{max} & \text{if } N_l > N_{max} \\ N_l & \text{otherwise} \end{cases} \qquad (9.2)$$

where N_{min} and N_{max} are the lower bound and upper bound for the spark numbers. Algorithm 9.1 shows how sparks are generated in the CoFWA.

Algorithm 9.1 Sparks generated by a firework in CoFWA
---
for $l = 1$ to m do
  $shift = S_g - S_l$
  M is the matrix to get the orientation between S_l
and S_g
    if $S_g \neq S_l$ **then**
      for $i = 1$ to N_l **do**
        sample s_{li} from **POEM**
      end for
    else then
      for $i = 1$ to N_l **do**
        sample r from $U(0, 1)$
        $s_{li} \leftarrow S_l + r * scope,$
      end for
    end if
  end for
  return all the s_{li}

Where *scope* is the explosion amplitude of the global guide.
Algorithm 9.2 shows the selection.

Algorithm 9.2 Selecting fireworks for the next generation

---
for $l = 1$ **to** m
  $S_l \leftarrow \arg\min_{s_{li}} (f(s_{li}))$
end for
return S_l

Finally, the complete version of the CoFWA is shown in Algorithm 9.3.

Algorithm 9.3 Cooperative Fireworks Algorithm

Initialize m fireworks x_i and constant parameters
repeat
  Calculate sparks number and amplitude
  Generate explosion sparks by Algorithm 9.1
  Evaluate fitness of newly created sparks
  Select fireworks for next iteration by Algorithm 9.2
until termination

9.7 Convergence Theorem of CoFWA

In this section, we want to give a convergence proof for CoFWA.

At first, we will describe a proposition of global convergence criteria.

Given a function f from $\mathbb{R}^n$ to $\mathbb{R}$ and S a subset of $\mathbb{R}^n$. We seek a point z in S which minimizes f on S or at least which yields an acceptable approximation of the infimum of f and S. This proposition provides a definition of what a global optimizer must produce as output, given the function f and the search space S. The random search algorithm can be described as follows: given a random initial starting point in S, called $\mathbf{z}_0$. In the kth iteration, this algorithm requires a probability space $(\mathbb{R}^n, \mathcal{B}, \mu_k)$, where μ_k is a probability measure (corresponding to a distribution function on $\mathbb{R}^n$) on $\mathcal{B}$, and $\mathcal{B}$ is the σ-algebra of subsets of $\mathbb{R}^n$. Then a vector ξ_k will be generated from the sample space $(\mathbb{R}^n, \mathcal{B}, \mu_k)$. In the $k+1$th iteration, the point z_{k+1} equals to $D(\mathbf{z}_k, \xi_k)$ and μ_k updates to μ_{k+1}, where D is a function that constructs a solution to the problem.

Next, we will first give a local convergence proof for the CoFWA. Before that, we want to introduce a theorem:

H 1 $f(D(\mathbf{z}, \xi)) \leq f(\mathbf{z})$ and if $\xi \in S$, then $f(D(\mathbf{z}, \xi)) \leq f(\xi)$

H 2 *To any* $\mathbf{z}_0 \in S$, *there corresponds a* $\gamma > 0$ *and an* $0 < \eta \leq 1$ *such that:*

$$\mu_k[(\text{dist}(D(\mathbf{z}, \xi), R_\epsilon) \leq \text{dist}(\mathbf{z}, R_\epsilon) - \gamma) \text{ or}$$
$$(D(\mathbf{z}, \xi) \in R_\epsilon)] \geq \eta$$

for all k and all $\mathbf{z}$ *in the compact set* $\mathbf{L}_0 = \{\mathbf{z} \in S | f(\mathbf{z} \leq f(\mathbf{z}_0)\}.$ *Where* $R_\epsilon = \{\mathbf{z} \in S | f(\mathbf{z}) < \psi + \epsilon\},$ *and* $\psi = \inf(t : v[\mathbf{z} \in S | f(\mathbf{z} < t] > 0),$ $v[A]$ *is the Lebesgue measure on the set A.*

Theorem 9.1 *Suppose that f is a measurable function, S is a measurable subset of* $\mathbb{R}^n$ *and (H1) and (H2) are satisfied. Let* $\{\mathbf{z}_k\}_{k=0}^\infty$ *be a sequence generated by the algorithm. Then,*

$$\lim_{k \to \infty} P[\mathbf{z}_k \in R_\epsilon] = 1$$

where $P[\mathbf{z}_k \in R_\epsilon]$ *is the probability that at step k, the point* $\mathbf{z}_k$ *generated by the algorithm is in the optimality region,* $R_\epsilon.$

Proof Let $\mathbf{z}_0$ be the initial starting point generated randomly. Since L_0 is compact, there always exists an integer p such that (by assumption H2)

$$\gamma p > \text{dist}(\mathbf{a}, \mathbf{b}) \quad \forall \mathbf{a}, \mathbf{b} \in L_0.$$

By (H3) it follows that
$$P[\mathbf{z}_1 \in R_\epsilon] \geq \eta$$

and
$$P[\mathbf{z}_2 \in R_\epsilon] \geq \eta \times P[\mathbf{z}_1 \in R_\epsilon] \geq \eta^2.$$

There probabilities are disjoint, so repeated application (p times) of (H2) yields

$$P[\mathbf{z}_p \in R_\epsilon] \geq \eta^p$$

Applying (H2) another p times results in

$$P[\mathbf{z}_{2 \times p} \in R_\epsilon] \geq \eta^{2 \times p}$$

Hence, for $k = 1, 2, \ldots$

$$P[\mathbf{z}_{kp} \in R_\epsilon] = 1 - P[\mathbf{z}_{kp} \notin R_\epsilon] \geq 1 - (1 - \eta^p)^k.$$

Now (H1) implies that $\mathbf{z}_1, \ldots, \mathbf{z}_{p-1}$ all belong to L_0 and by the above it then follows that
$$P[\mathbf{z}_{kp+l} \in R_\epsilon] \geq 1 - (1 - \eta^p)^k \quad \text{for} \quad l = 0, 1, \ldots, p - 1.$$

This shows that all steps between kp and $(k+1)p$ satisfy (H2).

This completes the proof, since $(1 - \eta^p)^k$ tends to 0 as k goes to $+\infty$. $\square$

According to the above theorem, it remains to show that the CoFWA satisfies both (H1) and (H2) to prove local convergence. The proof will be first presented for unimodal optimization problems, after which the multimodal case will be discussed again.

The proof for the CoFWA starts by choosing the initial value.

$$\mathbf{x}_0 = \arg\max_{\mathbf{x}_i}\{f(\mathbf{x}_i)\}, \quad i \in 1 \ldots m,$$

where $\mathbf{x}_i$ represents the position of firework i. That is, $\mathbf{x}_0$ represents the worst firework, yielding the largest f value. Now define $L_0 = \{\mathbf{x} \in S \mid f(\mathbf{x}) \leq f(\mathbf{x}_0)\}$, the set of all points with f values smaller than that of the worst firework $\mathbf{x}_0$. It is assumed that all the fireworks lie in the same 'basin' of the function.

In the next section this will be extended to the general case with multiple basins.

From Algorithm 9.2, function D introduced in assumption H1 is defined for the CoFWA as

$$D(S_{g,k}, S_{l,k}) = \begin{cases} S_{g,k} & \text{if } f(\mathbf{x}_{l,k,i}) \geq f(S_{g,k}) \\ \mathbf{x}_{l,k,j} & \text{if } j = \arg\min_i(f(\mathbf{x}_{l,k,i})) \\ & \text{and} \quad f(\mathbf{x}_{l,k,j}) < f(S_{g,k}) \end{cases}$$

where $\mathbf{x}_{l,k,j}$, $j = 1, 2, \ldots$, are the sparks that the firework S_l generates in the kth iteration. The definition of D above clearly complies with (H1), since the S_g is monotonic by definition.

Then we go back to the generation of the sparks of the best firework.

$$\mathbf{x}_{g,k+1} = S_{g,k} + r * \rho_k, \quad \text{sample } r \text{ from } U(0, 1) \tag{9.3}$$

where ρ is the scope of the explosion of the best firework. In the dynFWA, stagnation is prevented by ensuring that $\rho > 0$ for all time steps. Note that S_g is always in L_0. It is possible, however, that $\mathbf{x}_{g,k+1} \notin L_0$, due to the cumulative effect of a growing $r * \rho_k$ vector, so that $S_{g,k} + r * \rho_k \notin L_0$. But $S_g \in M_k$ and $\in L_0$, so $S_g \in M_k \cap L_0$. This implies that $v[M_k \cap L_0] > 0$, so a new sampled spark arbitrarily close to S_g, and thus in L_0, can be generated. Using this fact, it is now possible to consider the local convergence property of the CoFWA.

If we assume that S is compact and has a nonempty interior, then L_0 will also be compact with a nonempty interior.

Further, L_0 will include the essential infimum, contained in the optimality region R_ϵ, by definition. Now R_ϵ is compact with a nonempty interior, thus we can define a ball B' centered at $\mathbf{c}'$ contained in R_ϵ, as shown in Fig. 9.6. Now pick the point $\mathbf{x}' \in \arg\max_{\mathbf{x}}\{\text{dist}(\mathbf{c}', \mathbf{x}) \mid \mathbf{x} \in L_0\}$, as illustrated in Fig. 9.9.

Let B be the hypercube centered at $\mathbf{x}'$, with sides of length $2(\text{dist}(\mathbf{c}', \mathbf{x}') - 0.5\rho)$.

Let C be the convex hull of $\mathbf{x}'$ and B'. Consider a line tangent to B', passing through $\mathbf{x}'$ (i.e one of the edges of C). This line is the longest such line, for $\mathbf{x}'$ is the point furthest from B'. This implies that the angle subtended by $\mathbf{x}'$ is the smallest such angle of any point in L_0. In turn, this implies that the volume $C \cap B$ is smaller than that of $C' \cap B$ for any other convex hull C' defined by any arbitrary point $\mathbf{x} \in L_0$.

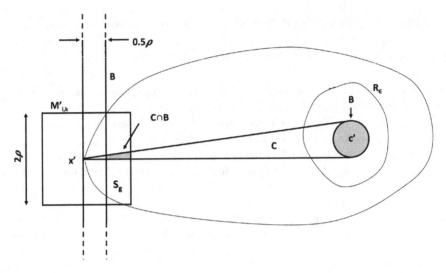

Fig. 9.9 Schematic diagram

Then for all **x** in L_0

$$\mu_k[(\text{dist}(D(S_{g,k}, S_{l,k}), R_\epsilon) \leq \text{dist}(S_{g,k}, R_\epsilon) - 0.5\rho] \geq \eta$$
$$= \mu[C \cap B] > 0 \qquad (9.4)$$

where μ_k is the uniform distribution on the hypercube centered at **x**, with side length 2ρ. It was shown above that the CFWA can provide such a hypercube.

Since $\mu[C \cap B] > 0$, the probability of selecting a new point **x** so that it is closer to the optimality region R_ϵ is always nonzero.

This is sufficient to show that the CoFWA complies with (H2).

So this completes the proof that the sequence values $\{S_{g,k}\}_{k=0}^{\infty}$ generated by the CoFWA will converge to the optimality region.

In sequel, we talk about function with multiple minima.

It was assumed above that L_0 was convex-compact. A non-unimodal function, with S including multiple minima, will result in a non-convex set L_0. Even if all the sparks are contained in the same convex subset, the CoFWA is not guaranteed to yield a point in the same convex subset as it started from, especially not during the earlier iterations. This is because the scope of explosion can yield a value larger than the diameter of the basin in which the firework currently resides. If the point found in a different convex subset yields a function value smaller than the current global best firework, then the algorithm will move its global firework to this new convex subset of L_0. This process could continue until the algorithm converges onto the essential

Table 9.1 List of mean fitness on the 28 benchmark function and mean fitness rank for SPSO2011, EFWA, dynFWA, and CoFWA

Fun.	SPSO2011	Rank	EFWA	Rank	dynFWA	Rank	CoFWA	Rank
1	$-1.4000E+03$	1	$-1.3999E+03$	4	$-1.4000E+03$	1	$-1.4000E+03$	1
2	$3.3719E+05$	1	$6.8926E+05$	3	$8.6937E+05$	4	$6.6407E+05$	2
3	$2.8841E+08$	4	$7.7586E+07$	2	$1.2317E+08$	3	$7.0544E+07$	1
4	$3.7543E+04$	4	$-1.0989E+03$	2	$-1.0896E+03$	3	$-1.0997E+03$	1
5	$-1.0000E+03$	1	$-9.9992E+02$	4	$-1.0000E+03$	1	$-1.0000E+03$	1
6	$-8.6210E+02$	3	$-8.5073E+02$	4	$-8.6995E+02$	2	$-8.7226E+02$	1
7	$-7.1208E+02$	2	$-6.2634E+02$	4	$-7.0010E+02$	3	$-7.4492E+02$	1
8	$-6.7908E+02$	3	$-6.7907E+02$	4	$-6.7910E+02$	1	$-6.7910E+02$	1
9	$-5.7123E+02$	3	$-5.6846E+02$	4	$-5.7587E+02$	2	$-5.8191E+02$	1
10	$-4.9966E+02$	3	$-4.9916E+02$	4	$-4.9995E+02$	1	$-4.9992E+02$	2
11	$-2.9504E+02$	3	$5.8198E+00$	4	$-2.9589E+02$	2	$-3.2202E+02$	1
12	$-1.9604E+02$	2	$3.9944E+02$	4	$-1.4222E+02$	3	$-2.4554E+02$	1
13	$-6.1406E+00$	2	$2.9857E+02$	4	$5.3830E+01$	3	$-7.5014E+01$	1
14	$3.8910E+03$	4	$2.7240E+03$	2	$2.9180E+03$	3	$2.4201+E03$	1
15	$3.9093E+03$	2	$4.4595E+03$	4	$4.0227E+03$	3	$3.312+E03$	1
16	$2.0131E+02$	4	$2.0063E+02$	3	$2.0058E+02$	2	$2.0019E+02$	1
17	$4.1626E+02$	2	$6.2461E+02$	4	$4.4261E+02$	3	$3.8921E+02$	1
18	$5.2063E+02$	2	$5.7361E+02$	3	$5.8782E+02$	4	$4.8429E+02$	1
19	$5.0951E+02$	3	$5.1022E+02$	4	$5.0726E+02$	2	$5.05986E+02$	1
20	$6.1346E+02$	2	$6.1466E+02$	4	$6.1328E+02$	1	$6.1402E+02$	3
21	$1.0088E+03$	1	$1.1178E+03$	4	$1.0102E+03$	3	$1.0098E+03$	2
22	$5.0988E+03$	3	$6.3181E+03$	4	$4.1262E+03$	1	$4.1705E+03$	2
23	$5.7313E+03$	3	$7.5809E+03$	4	$5.6526E+03$	2	$4.2643E+03$	1
24	$1.2667E+03$	2	$1.3452E+03$	4	$1.2729E+03$	3	$1.2599E+03$	1
25	$1.3993E+03$	3	$1.4426E+03$	4	$1.3970E+03$	2	$1.3822E+03$	1
26	$1.4861E+03$	3	$1.5461E+03$	4	$1.4607E+03$	2	$1.4011E+03$	1
27	$2.3046E+03$	3	$2.6210E+03$	4	$2.2804E+03$	2	$2.0597E+03$	1
28	$1.8013E+03$	3	$4.7651E+03$	4	$1.6961E+03$	1	$1.7010E+03$	2
Mean rank		2.57		3.67		2.25		1.25

infimum contained in its convex subset, at which point it will no longer be able to get out of the convex subset if the diameter of the subset is greater than 2ρ. By the end, the global firework converges to the local minimum of the current convex subset. □

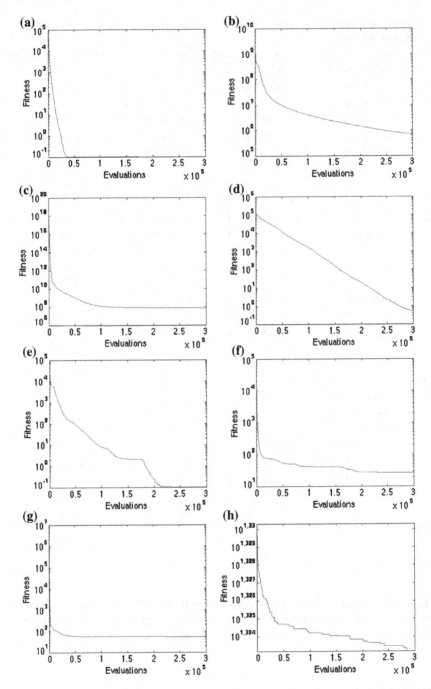

Fig. 9.10 Convergence curves of CoFWA on 28 Benchmark functions (f1–f8) of IEEE CEC 2013 averaged over 30 simulation runs. **a** $f1$, **b** $f2$, **c** $f3$, **d** $f4$, **e** $f5$, **f** $f6$, **g** $f7$, **h** $f8$

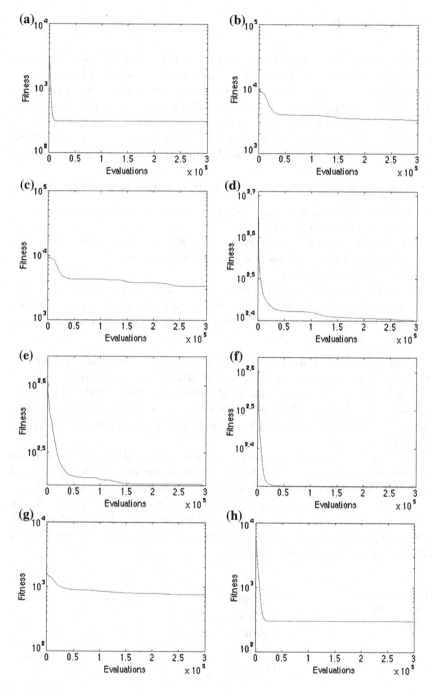

Fig. 9.11 Convergence curves of CoFWA on 28 Benchmark functions (f21–f28) of IEEE CEC 2013 averaged over 30 simulation runs. **a** $f21$, **b** $f22$, **c** $f23$, **d** $f24$, **e** $f25$, **f** $f26$, **g** $f27$, **h** $f28$

9.8 Experiments

The experimental platform is MATLAB 2013a, running under Windows 8.1 on an Intel Core i5 CPU with 2.4 GHz and 4 GB RAM. The parameters in CoFWA are set as follows. The number of fireworks is set to 5 and the maximum number of explosion sparks is 100. The parameter λ is 0.34. All other parameters for dynFWA and all EFWA parameters are identical to [2, 4], SPSO2011 parameters are listed in [5, 6]. To validate the performance of the proposed CoFWA algorithm, we used the IEEE CEC'2013 benchmark suite that includes 28 different benchmark functions as listed in [7].

For the algorithms in experiments, the detailed descriptions are given below.

- EFWA—the enhanced fireworks algorithm which is an efficient improvement of the conventional fireworks algorithm.
- dynFWA—the dynamic search fireworks algorithm which used a dynamic explosion amplitude for the firework at the current best location.
- CoFWA—as described above.
- SPSO2011—the most recent SPSO variant. Compared to earlier versions of SPSO, it features an improved velocity update by exploiting the idea of rotational invariance for the velocity update instead of sequential dimension-by-dimension update in the previous SPSO versions.

In this chapter, every algorithm runs 51 times on each benchmark optimization function and the number of fitness function evaluations is set to $3 * 10^5$.

Table 9.1 presents the experimental results of four algorithms (SPSO2011, EFWA, dynFWA, CoFWA) on the 28 benchmark functions. According to Table 9.1, it turns out that the CoFWA outperformed EFWA and dynFWA as well as SPSO2011 significantly.

In order to show the convergence performance of the CoFWA, Fig. 9.10 plotted the convergence curves of the CoFWA on benchmark functions f1–f8, and Fig. 9.11 plotted the convergence curves of the CoFWA on benchmark functions f21–f28.

According to the results in Table 9.1, we can rank the four algorithms in terms of average mean ranks. As a result, Table 9.2 shows the ranking among CoFWA with EFWA and dynFWA as well as SPSO2011. It can be seen from Table 9.2 that CoFWA ranks the first among the four algorithms.

Table 9.2 Ranking of SPSO2011, EFWA, dynFWA, and CoFWA

Algorithms	Ranking
CoFWA	1
dynFWA	2
SPSO2011	3
EFWA	4

9.9 Conclusion

In this chapter, the shortcomings of FWA which do not have the efficient mechanism of interaction among fireworks were pointed out. To deal with this critical issue, a probabilistically oriented explosion mechanism (POEM) is designed to enhance the cooperation in a swarm of fireworks. Based on the detailed analysis, a new algorithm called Cooperative Fireworks Algorithm (CoFWA) was proposed and its convergence was also proven. The convergence proof guaranteed that CoFWA is theoretically efficient and the experiments gave a strong evidence that the CoFWA performs much better than the EFWA and the dynFWA as well as SPSO2011.

In future, we will try to reduce the constant parameters and set them automatically according to the environment.

References

1. Y. Tan, Y. Zhu, Fireworks algorithm for optimization, in *Advances in Swarm Intelligence* (Springer, Berlin, 2010), pp. 355–364
2. S. Zheng, A. Janecek, Y. Tan, Enhanced fireworks algorithm, in *2013 Congress on Evolutionary Computation (CEC)* (IEEE, 2013), pp. 2069–2077
3. J. Li, S. Zheng, Y. Tan, Adaptive Fireworks Algorithm, in *IEEE Congress on Evolutionary Computation (CEC)* (2014), pp. 3214–3221
4. S. Zheng, A. Janecek, Y. Tan, Dynamic Search in Fireworks Algorithm, in *IEEE Congress on Evolutionary Computation (CEC)* (2014), pp. 1–7
5. M. Clerc, Standard Particle Swarm Optimization (2006–2011)
6. M. Zambrano-Bigiarini, M. Clerc, R. Rojas, Standard Particle Swarm Optimisation 2011 at CEC-2013: A Baseline for Future PSO Improvements, in *IEEE Congress on Evolutionary Computation (CEC)* (2013), pp. 2337–2344
7. J.J. Liang, B.Y. Qu, P.N. Suganthan, A.G. Hernández-Díaz, Problem definitions and evaluation criteria for the CEC 2013 special session on real-parameter optimization (2013)

Chapter 10
Hybrid Fireworks Algorithms

Fireworks algorithm has a broad research area and is suitable for combination with other algorithms to produce a new hybrid algorithm. This chapter focuses on hybrid fireworks algorithms, including Fireworks Algorithm with Differential Mutation (FWA-DM), Hybrid Fireworks Optimization Method with Differential Evolution Operators (FWA-DE), Culture Fireworks Algorithm (CFWA), and Hybrid Biogeography-Based Optimization and Fireworks Algorithm (BBO_FWA).

10.1 FWA-DM

FWA consists of explosion operator, mutation operator, mapping rule, and selection strategy [1]. FWA mutation operation is not limited to Gaussian mutation, but can also be the differential mutation. The operation of mutation enhances the diversity of FWA. Differential mutation (DM) operator utilizes the difference information between individuals and improves the interaction capability of populations. FWA-DM can improve the performance of FWA.

For fireworks algorithm, mutation operator is a very important operator. A firework is able to search both its vicinity and beyond area. The experimental results of FWA is better if the mutation operator is utilized. Each firework mutates to enhance the diversity of the population. In addition, this search mechanism can be replaced by other mutation mechanism to further enhance the search performance of FWA.

In this section, the differential mutation operator is first introduced. Then, the process of utilizing differential mutation operator in FWA is exhibited. The figure of the process is also shown, along with the details of the new algorithm.

© Springer-Verlag Berlin Heidelberg 2015
Y. Tan, *Fireworks Algorithm*, DOI 10.1007/978-3-662-46353-6_10

10.1.1 Differential Mutation Operator

Operator: DM/best/1/exp. In this operator, DM means the differential mutation operator and the word best indicates that the best one is kept for the mutation. Number 1 means the number of difference vectors used and the abbreviation 'exp' stands for an exponent recombination. The formula for this operator is as follows:

$$X_{i1}^k = X_B^k + F * (X_{i2}^k - X_{i3}^k), \tag{10.1}$$

where X_{i1}^k means the k dimension of the target individual and F is the scale factor generally between 0 and 2 [2]. X_B^k is the k dimension of the current best individual, while X_{i2}^k and X_{i3}^k are two distinguished random individuals on their k dimensions. The left part of Fig. 10.1 represents Gaussian mutation operator and the right part shows the details of DM operator.

10.1.2 Applying DM to EFWA

After introducing EFWA and DM, respectively, it is important and right time to apply DM to EFWA and thus compensate the shortcomings of EFWA.

Assume N denotes the number of individuals in a population for EFWA, which does not change during the optimization process. At first, N individuals are selected randomly and should lie in the feasible space. The N individuals form a population and the population is marked as POP1. Next, for each individual, a spark is produced around it within a certain amplitude.

$$A = A_{\min} * rand(0, 1), \tag{10.2}$$

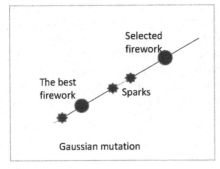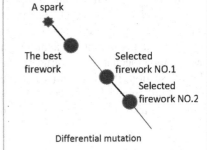

Gaussian mutation

Differential mutation

Fig. 10.1 The difference between GM and DM

$$A_{\min} = A_{init} - (A_{init} - A_{final})$$
$$*\sqrt{(2 * evals_{\max} - t) * t/evals_{\max}}, \qquad (10.3)$$

where A means the amplitude of each firework, while $A_{\min}$ decreases in the way of nonlinear. $rand(0, 1)$ generates random number from 0 to 1. A_{init} and A_{final} are constants, representing the initial and final amplitudes of the explosions. Parameter t stands for the number of iterations so far and parameter $evals_{\max}$ is the maximum function evaluation times.

Then, each new generated spark is compared with its corresponding individual. The one with better fitness value is kept and used to form a new population with N individuals marked as POP2. Finally, the DM operator is applied to POP2 and a new population is generated as POP3.

To select the individuals for next generation and continue the evolutionary process, the individuals in population POP3 are compared with individuals at the correspondence places in population POP2. The better ones are selected and passed down to the next generation, forming a new population POP1. The iteration of FWA-DM continues till the terminate condition is met.

The process of applying DM to EFWA is drawn in Fig. 10.2. The first row represents population POP1 with N individuals. The second row shows the generated explosion sparks after applying EFWA to POP1. After comparing the sparks in the first row with explosion sparks in the second row, better sparks are chosen and displayed in the third row. Then DM operator is used and population POP3 is produced. The better individuals between population POP2 and POP3 are selected for next iteration as a new population POP1. It is obvious that since DM is introduced, the communication of individuals is enhanced. As a result, the diversity of the population is guaranteed. In summary, the algorithm of FWA-DM [3] is given below.

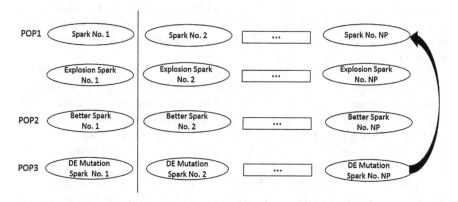

Fig. 10.2 The process of applying DM to enhanced FWA

Algorithm 10.1 The process of FWA-DM

1: randomly generate N individuals as POP1
2: **while** terminal condition is not met **do**
3:    generate N sparks from POP1 as explosion sparks
4:    choose better individuals as POP2
5:    apply DM operator and generate POP3
6:    choose better individuals between POP2 and POP3 as a new POP1
7: **end while**

It can be seen from Algorithm 10.1 that FWA-DM is simple and easy to implement.

FWA-DM provided a brand new way to solve function optimization problems by introducing DM operator to improve the performance of EFWA. Experimental results on 30 benchmark functions of CEC 2014 proved that FWA-DM could solve many function optimization problems effectively, as it outperformed EFWA on most functions.

10.2 FWA-DE

FWA is a relatively new swarm-based metaheuristic for global optimization. The algorithm is inspired by the phenomenon of fireworks display and has a promising performance on a number of benchmark functions. However, in the sense of swarm intelligence, the individuals including fireworks and sparks are not well-informed by the whole swarm. We develop an improved version of the FWA by combining with differential evolution (DE) operators: mutation, crossover, and selection. At each iteration of the algorithm, most of the newly generated solutions are updated under the guidance of two different vectors that are randomly selected from highly ranked solutions, which increases the information sharing among the individual solutions to a great extent. Experimental results show that the DE operators can improve diversity and avoid prematurity effectively, and the hybrid method outperforms both the FA and the DE on the selected benchmark functions.

For a D-dimensional optimization, the fitness value of a solution is determined by values of all components, and a solution that has discovered the region corresponding to the global optimum in some dimensions may have a low fitness value because of the poor quality in the other dimensions [4]. Thus, many population-based optimization methods, including DE, comprehensive learning PSO [4], fully informed PSO [5], enable the individuals to make use of the beneficial information in the swarm more effectively to generate better quality solutions.

In FWA, after obtaining the set R of all fireworks and sparks, the locations for new fireworks are selected based on distance to other locations in R so as to keep diversity of the swarm. Here we introduce the DE operators to the FWA to improve the diversification strategy.

In the hybrid algorithm, after obtaining the set R of locations, we first sort the locations in decreasing order of fitness, and create a set S of p candidate locations which are randomly selected from the top $2p$ locations in R. Afterward, we apply the standard DE process to the new solution set S: for each vector $\mathbf{x}_i \in S$, generate a mutant vector $\mathbf{v}_i$, mix the components of $\mathbf{x}_i$ and $\mathbf{v}_i$ to obtain a trial vector $\mathbf{u}_i$, and replace $\mathbf{x}_i$ with $\mathbf{u}_i$ in case that $\mathbf{u}_i$ is better.

After the DE process, we check that whether the new best result solution $\mathbf{x}_1^* \in S$ is worse than the first (best) one $\mathbf{x}^* \in R$. If it is so, we replace a randomly selected solution $\mathbf{x} \in S$ with $\mathbf{x}^*$, and thus make the best solution of the swarm will never degrade in the next generation. The DE procedure is shown as follows:

1. Let S be the empty set;
2. Sort R in decreasing order of vector fitness, and let x^* be the top one solution in R;
3. If $|R| > 2p$, truncate the length of R to $2p$, i.e., maintain the top $2p$ locations in R;
4. Randomly select p locations from R and add them to S, where the selection probability of each $x \in R$ is $f(x)/\sum_{z \in R} f(z)$;
5. Apply the DE mutation, crossover, and selection operators to each solution in S;
6. Let $\mathbf{x}_1^*$ be the best solution in R; if $f(\mathbf{x}_1^*) < f(\mathbf{x}^*)$, randomly select a solution $\mathbf{x} \in S$ and replace it with $\mathbf{x}^*$.

Generally speaking, the above DE procedure helps to improve the algorithm in the following aspects:

- For high quality (top $2p$) vectors in R, each of them has an opportunity to influence the new fireworks for the next generation, in terms of the roulette wheel selection and DE selection operations.
- For candidate fireworks in S, each of them has an opportunity to be informed by existing high-quality vectors at each dimension, in terms of the DE mutation and crossover operations.
- In particular, the DE mutation operator makes the difference of two random vectors acts as a search direction for the third one [6]; in comparison with large-amplitude explosion and distance-based selection used in the FWA, the mutation operation is more effective in improving the probability of obtaining the global optimum, whilst requiring less computational cost.

Since the DE operators are introduced into the algorithm for improving the diversity of solutions, the values of control parameters including m, $\hat{A}$, s_{max}, and s_{min} can be decreased, which will compensate the computational cost of DE operations to a certain extent. According to our analysis, in most conditions, the specific ("bad") spark generation procedure can also contribute to the result solution quality of the hybrid algorithm, but its contribution is much less than that in the standard FA. Thus, the specific procedure can either be remained or discarded in the hybrid algorithm; if it is remained, the parameter values mentioned above can be further decreased to save computational cost.

10.3 CFWA

The problem of the digital filter design is a multiparameter optimization problem. This section presents a joint objective function to design finite impulse response (FIR) digital filters and infinite impulse response (IIR) digital filters, and a cultural firework algorithm is proposed to implement filter designs. The design of the filter is transformed into the constrained optimization problem, and the cultural firework algorithm is used to search optimal value of filter design parameters in the parameter space with parallel search. The cultural firework algorithm (CFWA) is a multidimensional search algorithm for optimization of real numbers, using mechanisms of cultural evolution to update the locations of cultural sparks [7].

10.3.1 Design of Digital Filter

The goal of the fitness function is to evaluate the status of each cultural firework. In FIR and IIR digital filter design based on CFWA, the optimization goal of firework location is the minimization of the following objective function:

$$f(x) = \begin{cases} \alpha E_F + \beta E_I, x \in s.t \\ \delta [\alpha E_F + \beta E_I], x \notin s.t \end{cases} \tag{10.4}$$

where α and β are two parameters, satisfying $\alpha + \beta = 1$, $\alpha \in \{0, 1\}$, $\beta \in \{0, 1\}$. E_F and E_I are the objective of the total squared error in frequency domain of FIR and IIR digital filter, $s.t.$ denotes the constraint condition of vector x, δ is positive constant which is limited to $\delta > 1$.

What is following is the procedure to implement CFWA.

10.3.2 CFWA Implementation

The procedure for implementing CFWA is given in Algorithm 10.2.

Briefly, the CFWA is proposed for designing FIR and IIR filters and can converge around the optima quickly, which demonstrated FWA was successfully applied to the real-world applications like design of digital filters.

Algorithm 10.2 The process of CFWA

1: According to design requirement, select values of α, β and δ, where δ is equal to 0 for a non-constraint problem.
2: Randomly select an initial population of the q candidate solutions within the given domains, and initialize belief space.
3: Evaluate the performance scores of population space by a given objective function.
4: Select the p initial locations from the q locations.
5: Set off cultural fireworks at the p locations.
6: Calculate objective function of the new locations.
7: According to the acceptance function to select excellent locations, and update the belief space.
8: Select the top q different locations from the 2q cultural sparks which include the cultural sparks or fireworks of the current and previous generation for the next generation (iteration).
9: If it has not met the termination condition (the termination condition is set as maximum iteration times in general), then back to step 4; else the algorithm stops.

10.4 BBO_FWA

The key idea here is to introduce the migration operator of BBO to FWA, so as to enhance information sharing among the population, and thus improve solution diversity and avoid premature convergence. A migration probability is designed to integrate the migration of BBO and the Gaussian mutation operator of FWA, which cannot only reduce the computational burden, but also achieve a better balance between solution diversification and intensification. The Gaussian explosion of the enhanced FWA (EFWA) is reserved to keep the high exploration ability of the algorithm.

10.4.1 Biogeography-Based Optimization (BBO)

Borrowing ideas from biogeographic evolution over space and time, BBO [8] is another population-based heuristic for optimization problems. In BBO, each solution in the population is analogous to "habitats" or "islands," the solution components are analogous to a set of suitability index variables (SIVs), and the fitness of the solution is analogous to the species richness or habitat suitability index (HSI) of the island. The method mainly works on the principle of immigration and emigration of the species from one island to another, and therefore evolves the islands to find better solutions to the problem. BBO has proven itself a competitive method to other well-known heuristics on a wide set of problems [8–12].

A distinct feature of BBO is its migration operator, which indicates that high HSI islands have a high species emigration rate μ and low HSI islands have a high species immigration rate λ. The migration rates are functions of the HSI value or fitness of the islands. λ_i and μ_i of each island X_i are calculated as follows (but there are also other nonlinear migration models can be used in [8, 13]):

$$\lambda_i = I \left(\frac{f_i - f^{\min}}{f^{\max} - f^{\min}} \right), \tag{10.5}$$

$$\mu_i = E \left(\frac{f^{\max} - f_i}{f^{\max} - f^{\min}} \right), \tag{10.6}$$

where I and E are the maximum possible immigration rate and emigration rate, respectively, which are typically both set to 1.

At each time, the migration operator migrates a SIV from an emigrating island to an immigrating island, which are probabilistically selected according to the emigration and immigration rates of the islands. Algorithm 10.3 shows the basic procedure of a probably migration operation on an island X_i.

Algorithm 10.3 The migration operation in BBO

---

1 **for** $k = 1$ to D **do**
2   **if** $rand() < \lambda_i$ **then**
3   Select an emigrating island X_j with probability $\propto \mu_j$;
4          $X_{i,k} \leftarrow X_{j,k}$;

10.4.2 A Hybrid Biogeography-Based Optimization and Fireworks Algorithm (BBO_FWA)

For a high-dimensional optimization problem, the fitness value of a solution is codetermined by its component values of all dimensions. A solution that has discovered the region corresponding to the global optimum in some dimensions may have a low fitness value because of the poor quality in the other dimensions. Thus, some well-known population-based evolutionary algorithms, including differential evolution (DE) [2], comprehensive learning PSO [4], fully informed PSO [5], enable the individuals to make the utmost use of the beneficial information in the population and thus perform a very effective search.

In the original FWA, the individuals in the population never directly interacts with each other. EFWA makes a slight improvement on Gaussian mutation operator to let some individuals learn from the best individual found so far. On the other hand, FWA used a distance-based metric for selecting individuals in less crowded regions to the next generation so as to keep diversity. But such a selection operator is computational expensive, and thus EFWA turns to a random selection operator.

In the hybrid algorithm, we employ a diversification strategy that integrates the BBO's migration mechanism to FWA. In fact, the migration operator of BBO and the Gaussian mutation operator of FWA both have their advantages and disadvantages:

- The migration operator contributes greatly to the information sharing between different individuals by making low HSI islands probably learning from high HSI ones. It is also computational cheap (which requires only one function evaluation at each time, while the explosion requires s_i evaluations).
- The Gaussian mutation operator provides a good balance between exploration and exploitation. In particular, when a high-quality firework is nearby the global optimum, the explosion enables an intensive local search around the optimum.

To combine their advantages while reducing their disadvantages as much as possible, we introduce a migration probability, denoted by ρ, to the hybrid algorithm. Each firework X_i has a probability of ρ to apply the migration operator, and a probability of $(1 - \rho)$ to explode.

Since the migration operator helps to enhance the information sharing and increase the solution diversity, and the Gaussian explosion also utilizes the information of the global best, here we do not use the elitism method that always put the best-known individual to the new population. Algorithm 10.4 presents an overview of the hybrid BBO_FWA.

Algorithm 10.4 The hybrid BBO_FWA

---

1 Randomly initialize a population P of n fireworks;
2 **while** (stop criterion is not met) **do**
3   **for each** firework $X_i \in P$ **do**
4   **if** $rand(0, 1) < \rho$ **then**
5   use Algorithm 10.3 to perform migration on X_i;
6   **else**
7   use Lines 5-12 of Algorithm 1 to produce sparks;
8   **for** $j = 1$ **to** M_g **do**
9   select a random firework X_i;
10   use Lines 15-18 of Algorithm 1 to produce a spark;
11   add the new individuals to P;
12   randomly select n individuals for P;
13   update A_k^{min} and the migration rates;
14 **return** the best individual found so far.

In general, for problems with complex objective functions, we prefer to set a larger value of the probability ρ which allows a smaller number of function evaluations to reduce the computational burden. A larger ρ can also enhance the solution diversity and thus improve the exploration ability for multimodal functions. In contrast, a smaller ρ is more suitable for those functions whose optima are often located in very narrow or sharp ridges, since the Gaussian mutation operator can diverse the search along different directions and thus decrease the chance of skipping the optima. Empirically, the value of ρ can range from 0.5 to 0.8 to achieve an obvious performance improvement over both BBO and FWA/EFWA.

10.4.3 Discussions

FWA is a metaheuristic method inspired by the phenomenon of fireworks explosion, and has received much interest in recent years. FWA has drawbacks of high computational cost and lacking of information sharing among the population. A hybrid algorithm BBO_FWA is proposed here, which integrates the migration operator of BBO with the explosion operator of FWA based on a migration probability, and thus effectively increases the solution diversity without harming the exploitation ability of FWA.

BBO_FWA uses a fixed migration probability p which is easy to implement. However, as indicated by the numerical experiments, the parameter value needs to be fine-tuned to obtain the best results on different problems. Experimental results on selected benchmark functions show that the hybrid BBO_FWA has a significant performance improvement in comparison with both BBO and EFWA.

10.5 Summary

As a novel swarm intelligence algorithm, FWA can hybridize with other SI algorithms to produce high effective hybrid algorithms. In this chapter, FWA-DM is first introduced and the experimental results are given in detail. Second, FWA-DE is stated as FWA and DE hybrid and generated an high effective hybrid algorithm, where DE algorithm is used to improve the selection strategy in FWA. Third, CFWA is introduced as the locations and fitness values of fireworks are kept as knowledge in a library. The individuals can learn from the library and improve themselves. At last, BBO_FWA is presented. The immigration and emigration of biogeography-based optimization is introduced to improve the performance of FWA. Hence, all in one, FWA is very much suitable for hybridization with other algorithms and produces a lot of effective hybrid algorithms.

References

1. Y. Tan, Y. Zhu, Fireworks algorithm for optimization, in *Advances in Swarm Intelligence* (Springer, Berlin, 2010), pp. 355–364
2. R. Storn, K. Price, Differential evolution-a simple and efficient heuristic for global optimization over continuous spaces. J. Glob. Optim. **11**(4), 341–359 (1997)
3. C. Yu, J. Li, Y. Tan, Improve enhanced fireworks algorithm with differential mutation. In *2014 Conference on IEEE System, Man, and Cybernetics (SMC)*. IEEE (2014)
4. J.J. Liang, A.K. Qin, P.N. Suganthan, S. Baskar, Comprehensive learning particle swarm optimizer for global optimization of multimodal functions. IEEE Trans. Evol. Comput. **10**(3), 281–295 (2006)
5. R. Mendes, J. Kennedy, J. Neves, The fully informed particle swarm: simpler, maybe better. IEEE Trans. Evol. Comput. **8**(3), 204–210 (2004)

6. Y.-C. Lin, K.-S. Hwang, F.-S. Wang, Co-evolutionary hybrid differential evolution for mixed-integer optimization problems. Eng. Optim. **33**(6), 663–682 (2001)
7. H. Gao, M. Diao, Cultural firework algorithm and its application for digital filters design. Intern. J. Model., Identif. Control **14**(4), 324–331 (2011)
8. D. Simon, Biogeography-based optimization. IEEE Trans. Evol. Comput. **12**(6), 702–713 (2008)
9. A. Bhattacharya, P.K. Chattopadhyay, Biogeography-based optimization for different economic load dispatch problems. IEEE Trans. Power Syst. **25**(2), 1064–1077 (2010)
10. U. Singh, H. Kumar, T.S. Kamal, Design of Yagi-Uda antenna using bio-geography based optimization. IEEE Trans. Antennas Propag. **58**(10), 3375–3379 (2010)
11. I. Boussad, A. Chatterjee, P. Siarry, M. Ahmed-Nacer, Biogeography-based optimization for constrained optimization problems. Comput. Oper. Res. **39**(12), 3293–3304 (2012)
12. Y.-J. Zheng, H.-F. Ling, H.-H. Shi, H.-S. Chen, S.-Y. Chen, Emergency railway wagon scheduling by hybrid biogeography-based optimization. Comput. Oper. Res. **43**, 1–8 (2014)
13. H. Ma, D. Simon, Blended biogeography-based optimization for constrained optimization. Eng. Appl. Artif. Intell. **24**(3), 517–525 (2011)

Part III
Advanced Topics

FWA was first proposed for single-objective optimization problems and has been widely studied based on the CPU platform. After years of study, FWA has extended far beyond this conventional domain. In this part, we present some advanced topics on FWA.

Multi-objective optimization problems are universal and much more complicated than their single-objective counterparts. Chapter 11 will introduce the applications of FWA on multi-objective problems. In Chapter 12, with the help of the well-known S-metric, a kind of hyper-volume indicator, an S-metric multi-objective fireworks algorithm (MOFWA) is proposed for efficiently solving multi-objective optimization problems. Combinatorial optimization problems are of great importance in the real world and many combinatorial problems are proved to be NP hard and thus difficult to be solved exactly. After some modifications, FWA can be applied to combinatorial optimization problems. Chapter 13 comes up with a discrete FWA for tackling Traveling Salesman Problem (TSP).

GPU is a game-changing force in the domain of High Performance Computing (HPC). Thanks to GPU's parallelism and great computational power, swarm intelligence algorithms are able to fully exploit their inherent parallelism. Building swarm intelligence algorithms on GPU platform is an increasingly important and popular research topic. Chapter 14 describes a GPU-based FWA. As will see, GPU-based FWA can achieve great speedup compared to CPU-based implementation. The enormous acceleration implies that FWA is capable of applying to problems of greater scale in more different domains.

Chapter 11
FWA for Multiobjective Optimization

This chapter is to present some research works of FWA for multiobjective optimization, of which this is a successful instance like the multiobjective fireworks algorithm (MOFWA) proposed by Zheng et al. in [1] for oil crop fertilization, which takes into consideration not only crop yield and quality but also energy consumption and environmental effects. The variable-rate fertilization (VRF) is a key aspect of prescription generation in precision agriculture, which typically involves multiple criteria and objectives. To solve the problem efficiently, a hybrid multiobjective fireworks optimization algorithm (MOFWA) is proposed to evolve a set of solutions to the Pareto optimal front by mimicking the explosion of fireworks. Especially, MOFWA uses the concept of Pareto dominance for individual evaluation and selection, and combines differential evolution (DE) operators to increase information sharing among the individuals. The proposed MOFWA outperforms some state-of-the-art methods on a set of real-world VRF problems.

11.1 Introduction

Variable-rate fertilization (VRF) decision problem is concerned with specifying the dosage of each type of fertilizer for the crop(s) in each field. Traditional VRF problems are typically described as a single-objective model maximizing crop yield or the rate of yield to cost. Nevertheless, different VRF solutions may cause very different impacts not only on crop yield, but also on crop quality, soil quality as well as other aspects of the environment. In modern precision agriculture systems, VRF decision should take both economical and ecological effects into consideration and thus should be modeled as an optimization problem with multiple objectives and constraints.

In a multiobjective optimization problem (MOP), there are often conflicts among various objectives and it is impossible to find a single solution that is optimal in terms of all objectives. Therefore, it is preferable to search for a set of Pareto optimal or non-dominated solutions, i.e., there is no solution that is better than another solution

© Springer-Verlag Berlin Heidelberg 2015
Y. Tan, *Fireworks Algorithm*, DOI 10.1007/978-3-662-46353-6_11

in all objectives. Multiobjective models are first introduced to irrigation decision in agricultural practices. Raju and Kumar [2] studied the irrigation problem with three conflicting objectives including net benefits, agricultural production, and labor employment. They employed a constraint method that successively optimizes each individual objective, while all the others are constrained. Chen et al. [3] presented a fuzzy irrigation problem model which divides the period of growth into multiple stages and considers two objectives including the crop yield and the risk of shrivel. They used a weighted function to combine the two objectives, and applied a multi-dimensional dynamic programming approach to solve the problem stage by stage. However, methods like the constraint and the weighted sum transform the problem to a single-objective one and do not simultaneously evolve all the objectives. Kilic and Anac [4] considered a problem of increasing the benefit from production, increasing the total area irrigated and reducing the water loss, and used a mathematical programming method to solve it, but the three objectives have a close relationship and generally do not conflict each other.

In recent years, multiobjective models and algorithms have been introduced for fertilization decision problems. Wang and Zhang [5] presented a model of multi-objective rice fertilization,and employed a linear approximation iteration method to solve the problem. However, such programming methods are only efficient for small-sized problems, which have motivated many researchers to use more effective heuristic algorithms. When solving a generalized fertilization problem, Yuan et al. [6] applied the genetic algorithm (GA) and introduced a reduction factor into the fitness function to improve convergence efficiency, but they did not employ the multiobjective evolution strategy in the algorithm. Some researchers (e.g., [7]) also used neural network approaches for fertilization decisions, in which fertilizer variables are used as network inputs fertilizer and objective values are used as network outputs; nevertheless, the topology design and optimal weights search remain a main difficulty in algorithm implementation.

Multiobjective evolutionary algorithms (MOEA) have been considered as one of the most effective approaches to MOP, mainly because they deal simultaneously with a set of possible solutions (the so-called population) which allows to find several members of the Pareto optimal set in a single run of the algorithm, and their applicabilities are less susceptible to the shape or continuity of the Pareto front [8]. A variety of Pareto-based MOEA have been proposed in the last decade, including the non-dominated sorting genetic algorithm (NSGA) [9] and NSGA-II [10], the strength Pareto evolutionary algorithm (SPEA) [11] and SPEA2 [8], the Pareto archived evolution strategy (PAES) [12], the Pareto differential evolution algorithm (PDE) [13], the non-dominated sorting particle swarm optimizer (NSPSO) [14], the constrained nonlinear multiobjective optimization immune algorithm (CNMOIA) [15], the coevolutionary particle swarm optimization algorithm (CCPSO) [16], the dominating tree-based multiobjective evolutionary algorithm (DTEA) [17], the adaptive multiobjective particle swarm optimization algorithm (MO-TRIBES) [18], the hybrid MOEA with two crossover operators [19], the multiobjective endocrine particle swarm optimization algorithm (MOEPSO) [20], etc. These methods have attracted much attention among researchers and shown promising results for many practical problems [21, 22].

For example, Niknam et al. [23] applied an MOEA for volt/var control in distribution networks, which simultaneously and effectively minimize electrical energy losses, voltage deviations, total electrical energy costs, and total emissions of renewable energy sources and grid. In [24] Ahmadi et al. used an MOEA for optimal design of a poly generation energy system, which considers minimizing total cost rate of the system while maximizing the system energy efficiency. Studies on the application of MOEA in the field of sustainable and renewable energy can also be found in [24–28]. However, the studies of MOEA in fertilization decision problems are very few. The only report we found was that by Reddy and Kumar [29], where an MODE approach was proposed for the simultaneous evolution of optimal cropping pattern and operation policies for a multi-crop irrigation reservoir system, and the result provides a wide spectra of Pareto optimal solutions and gives sufficient flexibility to select the best irrigation planning and reservoir operation strategy.

Fireworks optimization algorithm (FWA) [30] is a relatively new heuristic method inspired by the phenomenon of fireworks explosion. The algorithm selects a certain number of locations in the search space, each for exploding a firework to generate a set of sparks (new solutions); the fireworks and sparks with good performance (fitness) are chosen as the locations for the next generations fireworks, and the evolutionary process continues until a desired result is obtained. Numerical experiments on a number of benchmark functions show that FWA can converge to a global optimum with a much smaller number of function evaluations than that of particle swarm optimizers [31, 32]. Recently, Zheng et al. [33] develop an improved version of FWA by introducing differential evolution (DE) operators [34] to improve the population diversity. However, to our best knowledge, there is still no research on the application of fireworks optimization in multiobjective problems.

The motivation for MOFWA derives from precision agriculture practices on oil crop production in East China. The government and the agronomists try to guide the farm managers to achieve balance not only between expected profits and potential risks but also between short-term profits and long-term profits. We establish multiobjective optimization models for oil crop fertilization problems, and try to solve the problems based on different heuristic methods, among which we find that the multiobjective fireworks optimization algorithm (MOFWA) has more advantages over other methods on the efficiency and effectiveness. The proposed MOFWA employs a problem-specific strategy for generating the initial population, uses the concept of Pareto dominance for individual evaluation and selection, combines the DE operators including mutation, crossover, and selection to increase the information sharing among the population, and thus significantly decreases the computational cost and improves the quality of result solution set.

Note that majority parts of the chapter are excerpted from Ref. [1] on behalf of Prof. Yujun Zheng.

Table 11.1 The parameters representation

m	Number of fields
n	Number of types of fertilizer
a_i	Average gradient of field i
d_i	Average plant density of field i
p_j	Unit price of fertilizer j
x_{ij}	Dosage of fertilizer j in field i
x_{ij}^0	Inherent quantity of fertilizer j in field i
y_{ij}	Residual of fertilizer j in field i
Y	Function for estimating the crop yield
Q	Function for evaluating the crop quality
C	Function for estimating the cost of fertilization
E	Function for estimating the energy consumption of fertilization
R	Function for estimating the residual fertilizer

11.2 Problem Model

Different VRF decisions significantly affect the production of oil crops and the ecological conditions of environment. Besides the yield of crops and the cost of fertilizers, here we are also concerned with overall crop quality, energy consumption, and residual fertility.

The problem is to determine a dosage matrix $x = (x_{ij})_{m \times n}$, i.e., the dosage of each fertilizer in each field. For a given field i, the expected crop yield can be estimated by a fertilizing effect function $Y_i(x)$. There are a number of models for describing the relationship between yield and fertilizer, among which the most widely used form of Y_i is a quadratic function as follows [35]:

$$Y_i(X) = \sum_{j=1}^{n}\sum_{k=1}^{n} a_{ijk}\hat{x}_{ij}\hat{x}_{ik} + \sum_{j=1}^{n} b_{ij}\hat{x}_{ij} + c_i, \qquad (11.1)$$

where a_{ijk} is the quadratic regression coefficient, b_{ij} is the simple regression coefficient, c_i is the constant coefficient, and $\hat{x}_{ij} = x_{ij} + x_{ij}^0$, i.e., the sum of the dosage of fertilizer x_{ij} and the inherent fertility x_{ij}^0 in the field (Table 11.1).[1]

A crop usually has a number of quality criteria, e.g., protein content, oil content, crop density, etc. In many cases, a majority of such criteria are not contradictory to each other, and thus we can establish a fertilizing effect function to evaluate the

---

[1] For the fields with the same or similar soil conditions, their equations can have the same coefficients. This is also the case for some coefficients in the following computational equations.

comprehensive quality index. Such a function can also be modeled by a quadratic regression equation as follows:

$$Q_i(x) = \sum_{j=1}^{n}\sum_{k=1}^{n} a'_{ijk}\hat{x}_{ij}\hat{x}_{ik} + \sum_{j=1}^{n} b'_{ij}\hat{x}_{ij} + c'_i, \tag{11.2}$$

where a'_{ijk}, b'_{ij} and c'_i are regression coefficients.

In case there are indeed conflicts between some quality criteria, then we can establish a set of quality subindices, e.g., $Q'Q''$, etc., and construct a computational model for each of them. Given unit price p_j for each fertilizer j, the total fertilizer cost can be computed as:

$$C(x) = \sum_{i=1}^{m}\sum_{j=1}^{n} p_j x_{ij}. \tag{11.3}$$

And we use the following empirical formula to estimate the energy consumption of crop fertilization based on the gradient α_i and plant density d_i of each field i (where λ is a constant coefficient):

$$E_i(x) = \lambda \sum_{j=1}^{n} \alpha_i d_i^{1/3} x_{ij}. \tag{11.4}$$

For agricultural soil, the residual fertility is an important criterion to evaluate the soil quality. A simple formula to estimate the residual of fertilizer j in field i is as follows [36]:

$$y_{ij} = \mu x_{ij} + v_j x_{ij}^0 \tag{11.5}$$

where μ_j and v_j are coefficients relevant to the characteristics of the fertilizer.[2] Therefore, the homogeneity of residual fertilizer j in the planting area can be computed as:

$$R_j(x) = \sum_{i=1}^{m}(y_{ij} - \overline{y})^2 \tag{11.6}$$

where $\overline{y} = (\sum_{i=1}^{m} y_{ij})/m$. Based on above analysis, we model the multiobjective optimization VRF problem for oil crop production as follows:

$$max\, Q(x) = \sum_{i=1}^{m} Y_i(x)Q_i(x) \tag{11.7}$$

---

[2]If the soil conditions vary greatly from the different fields, we may need to define the coefficients μ_{ij} and v_{ij} for not only different kinds of fertilizer but also different fields. However, the goal is to improve the homogeneity of fertilizer content among the fields with the same or similar soil conditions.

$$minC(x) = \sum_{i=1}^{m}\sum_{j=1}^{n} p_j x_{ij} \qquad (11.8)$$

$$minE(x) = \sum_{i=1}^{m} E_i(x) \qquad (11.9)$$

$$s.t. \sum_{i=1}^{m} Y_i(x) \geq Y_L \qquad (11.10)$$

$$\sum_{i=1}^{m}\sum_{j=1}^{n} p_j x_{ij} \leq C^{\cup} \qquad (11.11)$$

$$\sum_{j=1}^{n} R_j(x) \leq R^{\cup} \qquad (11.12)$$

$$x_{ij} \geq 0, \forall i \text{ and } j \qquad (11.13)$$

where Y^L is the lower limit of the total crop yield, $C^{\cup}$ is the upper limit of total fertilizer cost, and $R^{\cup}$ is the upper limit of the fertilizer residual homogeneity index. In general, (11.7) to (11.13) constitute a nonlinear multiobjective programming problem.

11.3 MOFWA for VFR Problem

11.3.1 Fitness Assignment Strategy

There are a variety of fitness assignment strategies for MOEA, and in the MOFWA we employ a strategy based on the Pareto strength used in SPEA2 [8]. That is, for an individual solution x_i in the population P or in the non-dominated solution archive NP, a strength value s(xi) is calculated according to the number of other individuals it dominates:

$$s(x_i) = |x_j \in P \cup NP|x_i \succ x_j| \qquad (11.14)$$

where $\succ$ denotes the Pareto dominance relation. And the raw fitness value of x_i is determined by the strengths of its dominators:

$$r(\mathbf{x}_i) = \sum_{(\mathbf{x}_j \in P \cup NP)(\mathbf{x}_j \succ \mathbf{x}_i)} s(\mathbf{x}_j) \qquad (11.15)$$

It should be noted that the fitness is to be minimized here. A density value of x_i is then calculated by (11.16) and incorporated into the fitness by (11.16):

$$d(x_i) = \frac{1}{\delta_k(x_i)} \tag{11.16}$$

$$f(x_i) = r(x_i) + d(x_i) \tag{11.17}$$

where $_k(x_i)$ is the distance from x_i to its kth nearest individual, and k is typically set to equal to the square root of sample size $P \cup N$ [37].

In addition, if a solution violates the problem constraints (11.10) to (11.12), we compute the degrees of constraint violations, respectively.

$$u_Y(x_i) = \begin{cases} Y^L - \sum\limits_{i=1}^{m} Y_i(\mathbf{x}_i) & \text{if } \sum\limits_{i=1}^{m} Y_i(\mathbf{x}_i) \le Y^L, \\ 0 & \text{else.} \end{cases} \tag{11.18}$$

$$u_c(x_i) = \begin{cases} C(\mathbf{x}_i) - C^U & \text{if } C(\mathbf{x}_i) \ge C^U, \\ 0 & \text{else.} \end{cases} \tag{11.19}$$

$$u_c(x_i) = \begin{cases} \sum\limits_{j=1}^{n} R_j(\mathbf{x}_i) - R^U & \text{if } \sum\limits_{j=1}^{n} R_j(\mathbf{x}_i) \ge R^U, \\ 0 & \text{else} \end{cases} \tag{11.20}$$

and multiply the fitness value by a penalty coefficient $p(x_i)$ as follows.

$$p(x_i) = w_1 u_Y(x_i) + w_2 u_C(x_i) + w_3 u_R(x_i) \tag{11.21}$$

where w_1, w_2 and w_3 are three predefined weights.

11.3.2 Evolutionary Strategies

As most MOEA, MOFWA maintain two collections of solutions: a population P and a non-dominated solution archive NP, and at each algorithm iteration selects non-dominated solutions from P to update NP. However, in MOFWA such update is done twice at each iteration. In detail, for each generation of population P, we explode its individuals (fireworks) to generate a number of sparks according to the method described in Sect. 3.1, compute their fitness values, and then select those non-dominated sparks to update NP.

Afterward, we randomly select p solutions from all the fireworks and sparks, where the selection probability of a solution is proportional to its fitness value. For each selected solution x_i, the DE operators [34] are applied as follows:

1. (Mutation) generate a mutant solution v_i by adding the weighted difference between two randomly selected solutions to a third one:

$$v_i = x_{r1} + \gamma(x_{r2} - x_{r3}) \qquad (11.22)$$

   where random indexes $r_1, r_1, r_1, \in \{1, 2, \ldots, p\}$ and coefficient $\gamma > 0$.
2. (Crossover) generate a trial solution u_i by mixing the components of the mutant solution and the original one, where each jth component of u_i is determined as follows:

$$u_i^j = \begin{cases} v_i^j & \text{if } rand(0, 1) < C_r \text{ or } j = r(i), \\ x_i^j & \text{else} \end{cases} \qquad (11.23)$$

   where Cr is the crossover probability ranged in $(0, 1)$ and $r(i)$ is a random integer within $(0, N]$ for each i.
3. (Selection) choose the better one for the next generation by comparing the trial solution with the original one:

$$x_i = \begin{cases} u_i & \text{if } f(u_i) \le f(x_i), \\ x_i & \text{else} \end{cases} \qquad (11.24)$$

At the last step, if the trial solution u_i is selected, we test whether it is a non-dominated solution in current generation and, if so, use it to update NP.

11.3.3 The MOFWA Framework

Typically, the algorithm termination condition can be a maximum number of iterations (generations) or a maximum CPU time. Besides, we can also set a maximum number of non-improvement iterations, i.e., the algorithm stops if it cannot find a new non-dominated solution after certain iterations.

The basic algorithm flow is shown in Fig. 11.1. Next we will discuss some implementation details on initial population generation and non-dominated archive maintenance.

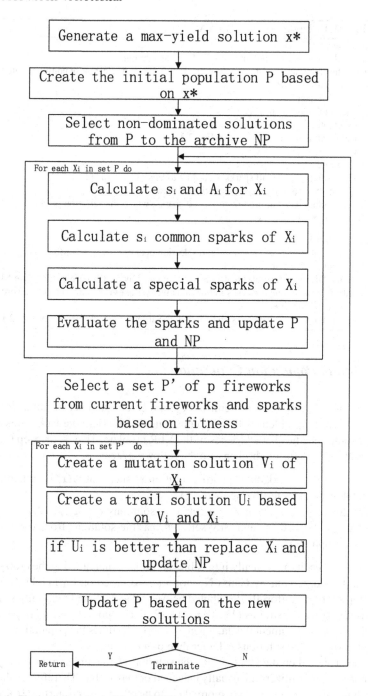

Fig. 11.1 The basic flowchart of the MOFWA

Algorithm 11.1 The proposed MOFWA.

1: Initialization
2:     Randomly generate a population P of p feasible solutions.
3:     Create the empty non-dominated solution archive NP, and select those non-dominated solutions from P to update NP.
4: Iterative improvement
5:     For each individual x_i in P do:
6:         Calculate s_i for x_i according to Eq. (2.1).
7:         Calculate A_i for x_i according to Eq. (2.3).
8:         Generate s_i sparks of x_i.
9:         Generate a specific spark of x_i.
10:        Compute fitness for all sparks according to Eqs. (11.14) to (11.21).
11:        Update NP based on the new solutions (sparks).
12:        Select p solutions from the fireworks and sparks, where the selection probability of each solution x_i is $f(\mathbf{x}_i) / \sum_{j \in P} f(\mathbf{x}_j)$
13:        For $i = 1$ to p do:
14:            Apply the mutation, crossover, and selection operators to x_i according to Eqs. (11.22), (11.24) and get a trial solution u_i.
15:            If the DE result indicates that x_i is to be replaced by u_i, then use u_i to update NP.
16:        Update P by including the best solution and other $p1$ ones randomly selected according to distance-based probability.
17:        If the termination condition is satisfied, then the algorithm stops; else go to step (2.1).

11.3.4 Initial Population Generation

For many MOEA, it may be difficult to obtain a certain number of initial feasible solutions when the problem is heavily constrained [38]. Here we tackle this issue based on some specific characteristics of the VFR problem. That is, the step (1.1) of the MOFWA can be further broken into the following steps:

1.1.1 For $i = 1$ to m, work out the partial solution x_i^* such that $Y_i(x_i^*)$ is maximized, and thus obtain a fertilizing matrix $(x_{ij}^*)_{m \times n}$.
1.1.2 Let $\triangle_{ij} = (x_{ij}^*)/q$, generate a set of q solutions $(\triangle_{ij})_{m \times n}$, $(2\triangle_{ij})_{m \times n}, \ldots,$ $(q\triangle_{ij})_{m \times n}$, and then randomly select p feasible solutions from them as the initial population P.

Step (1.1.1) results a (typically infeasible) solution x^* maximizing the crop yield, which can be worked out by Gauss Newton method based on Eq. (11.1). Next we generate q solutions uniformly distributed between 0 and x^*, and randomly select p solutions satisfying constraints (11.10) to (11.12) to compose the initial population. According to our experimental data, when q is about 45 times the population size p, the number of feasible solutions is larger than p in most cases.

Our idea is based on the fact that, for the solution x* maximizing the crop yield, the proportions of fertilizers are usually balanced or reasonable. Therefore, solutions with such fertilizer proportions are more likely to be effective, and starting from such a population canal so improve (but not critically affect) the algorithm performance.

11.3.5 Non-dominated Archive Maintenance

According to our experiments, the size of archive NP may increase rapidly during the search process, and thus it is reasonable to limit the size of NP, especially for large problem instances. In general, when NP reaches the size limit, a new non-dominated solution can be inserted into NP and an archive one will be removed only if the diversity of NP can be potentially improved by doing that, because the preservation and improvement of diversity of the population is crucial not only to avoid losing potentially efficient solutions but also avoid premature convergence [39]. In MOFWA we use the minimum pairwise distance metrics [40] which is of low computational cost. During the search process, whenever NP reaches the size limit, we mark two solutions x_a, $x_b \in NP$, the Euclidean distance between which is the minimum among all pairs in NP, i.e.,

$$dis(x_a, x_b) = min_{x, x \in NP, \wedge x \neq x'} dis(x, x'). \tag{11.25}$$

And the following procedure is applied for possible inclusion of a new solution x if $|NP| = |NP|^U$:

1. If x is dominated by any $x_i \in NP$, then x is discarded.
2. Else if x dominates some $x_i \in NP$, then remove those x' and insert x.
3. Else if $dis(x_a, x_b) < min_{x, x \in NP, \wedge x \neq x_a} dis(x, x')$, then remove x_a and insert x.
4. Else if $dis(x_a, x_b) < min_{x, x \in NP, \wedge x \neq x_b} dis(x, x')$, then remove x_b and insert x.
5. Else choose a closest $x_1 \in NP$ to x; if $min_{x, x \in NP, \wedge x' \neq x_1} dis(x, x') < dis(x, x_1)$, then remove x1 and insert x.
6. Else discard x.

11.4 Computational Experiments

In this section, we assess the performance of the proposed MOFWA on a set of VRF problems for three kinds of oil crops including Brassic anapus (BN or rapeseed), Canarium album (CA or olive), and Camellia oleifera (CO). The BN VRF problem considers two kinds of fertilizers, i.e., N and P, and the CA and CO VRF problem considers three kinds of fertilizers, i.e., N, P, and K. We conduct extensive experiments by comparing with competitive algorithms on the rest two kinds of crops, and then apply the MOFWA to a real-world CO VRF problem in East China.

11.4.1 Comparative Experiments on BN and CA VRF Problems

We test the MOFWA on the BN and CA VRF problems in comparison with four other MOEA: NSGA-II [10], PDE [13], DE-MOC [41], and NSPSO [14]. The mathematical models for crop yield and/or quality estimation are based on the results from [42, 43]. The experiments are conducted on a computer of $4\times$ Intel Xeon X3430 $2.4\,$GHz processors and $4 \times 2\,$GB memory. For all the algorithms, the upper limit of the archive size is set to 20, and the maximum number of generations is set to $100\,n\sqrt{m}$. Moreover, if the algorithm cannot find a new non-dominated solution after 300 continuous generations, it is also stopped. The population size is set to 30 for the MOFWA, 200 for the NSGA-II, and 100 for PDE, DE-MOC, and NSPSO. The penalty coefficient computed by Eq. (11.27) applies to all the competitive algorithms. For the MOFWA, the DE-related parameters are set as those in [34], and other parameter settings are given in Table 11.2, which are suggested by our preliminary empirical data. For each crop, we generate 10 problem instances with swarm size m ranges from 20 to 800. On each instance, each algorithm is run 30 times with different random seeds. The performance metrics include:

1. The CPU time $t(s)$
2. The hypervolume HV [44] of the result solution set.
3. The coverage c [44], which is a comparative metric on the result sets obtained by different algorithms. Let S1 and S2 be two solution sets, then the coverage $c(S_1, S_2)$ is defined as the fraction of solutions in S_2 that are (strictly) dominated by a solution in S_1:

$$c(S_1, S_2) = \frac{||x_2 \in S_2|\exists x_1 \in S_1 : x_1 \succ x_2||}{|S_2|}. \tag{11.26}$$

In the comparative experiments, we calculate the coverage of the result set of MOFWA over that of the four competitive algorithms, denoted by c_1, c_2, c_3, and c_4 respectively, and the coverage of the result sets of the four competitive algorithms over that of MOFWA, denoted by c'.

The experimental results of the BN and CA problems are summarized, respectively in Tables 11.3 and 11.4, and the variations of CPU time with the problem size are illustrated, respectively in Figs. 11.2 and 11.3. As we can see, for relatively small-sized problems ($mn \leq 200$), all the algorithms achieve nearly the same Pareto front, and MOFWA consumes a little more CPU time than NSGA-II, PDE, and

Table 11.2 Parameter values of our algorithm for the test problems (x_{UB}, x_{LB} denote the upper and lower search bounds)

Parameter	s_m	s_{min}	s_{max}	$\hat{A}$	q	w_1	w_2	w_3
Value	25	2	20	$min_{1 \leq k \leq D} \frac{x_{UB,k} - x_{LB,k}}{7}$	$5p$	4.8	1.5	2.0

Table 11.3 Computational results on the BN VFR problem instances, where the values are averaged over 30 independent runs

m	NSGA-II			PDE			DE-MOC			NSPSO			MOFWA					
	t	HV	C'	t	HV	C'	t	HV	C'	t	HV	C'	t	HV	C1	C2	C3	C4
20	0.47	6.40E+0 (0.00E+0)	0	0.33	6.40E+0 (0.00E+0)	0	0.54	6.40E+0 (0.00E+0)	0	0.7	6.40E+0 (0.00E+0)	0	0.62	6.40E+0 (0.00E+0)	0	0	0	0
50	1.22	3.60E+1 (0.00E+0)	0	1.1	3.60E+1 (0.00E+0)	0	1.29	3.60E+1 (0.00E+0)	0	1.77	3.60E+1 (0.00E+0)	0	1.45	3.60E+1 (0.00E+0)	0	0	0	0
100	3.89	1.50E+3 (0.00E+0)	0	3.14	1.30E+3 (0.00E+0)	0	3.44	1.50E+3 (0.00E+0)	0	5.67	1.50E+3 (0.00E+0)	0	4.13	1.50E+3 (0.00E+0)	0	0	0	0
150	9.49	1.70E+4 (−3.20E+3)	0	7.33	1.20E+4 (−2.40E+3)	0	7.89	1.70E+4 (−3.90E+3)	0	13.3	1.90E+4 (0.00E+0)	0	10.3	1.90E+4 (0.00E+0)	0	0.13	0	0
200	17.7	6.60E+5 (−1.90E+5)	0	15.6	3.90E+5 (−8.80E+4)	0	17	5.50E+5 (−1.70E+5)	0	28.9	7.10E+5 (−2.00E+5)	0	20.4	7.70E+5 (−1.70E+5)	0.1	0.17	0.1	0.05
300	65.3	2.70E+7 (−6.60E+6)	0	52.8	2.00E+7 (4.5+6)	0	61.1	2.40E+7 (−6.20E+6)	0	98.3	3.50E+7 (−6.10E+6)	0	63.3	4.00E+7 (−5.90E+6)	0.15	0.21	0.17	0.1
400	158	8.20E+8 (−1.90E+8)	0	141	4.40E+8 (−9.20E+7)	0	150	6.60E+8 (−1.60E+8)	0	262	1.10E+9 (−2.20E+8)	0	149	2.20E+9 (−3.10E+8)	0.18	0.2	0.18	0.15
500	417	2.40E+9 (−5.10E+8)	0	394	9.80E+8 (−2.50E+8)	0	434	2.40E+9 (−4.20E+8)	0	631	2.80E+9 (−4.90E+8)	0.03	370	5.90E+9 (−4.70E+8)	0.2	0.33	0.24	0.13
600	1064	7.20E+10 (−1.70E+10)	0	855	4.80E+10 (−8.50E+9)	0	989	7.00E+10 (−1.50E+10)	0	1375	9.30E+10 (−1.90E+10)	0	869	1.70E+11 (−2.10E+10)	0.33	0.48	0.37	0.15
800	3107	4.50E+12 (−9.90E+11)	0	2737	3.20E+12 (−5.60E+11)	0	3225	4.90E+12 (−1.00E+12)	0	4619	7.00E+12 (−1.30E+12)	0.07	3021	9.60E+12 (−9.80E+11)	0.55	0.75	0.5	0.21

For HV the numbers in the brackets are standard deviations

Table 11.4 Computational results on the CA VFR problem instances, where the values are averaged over 30 independent runs

m	NSGA-II			PDE			DE-MOC			NSPSO			MOFWA						
	t	HV	C'	t	HV	C'	t	HV	C'	t	HV	C'	t	HV	C'	$C1$	$C2$	$C3$	$C4$
20	0.78	2.00E+1 / 0.00E+0	0	0.52	2.00E+1 / 0.00E+0	0	0.75	2.00E+1 / 0.00E+0	0	1.01	2.00E+1 / 0.00E+0	0	0.97	2.00E+1 / 0.00E+0	0	0	0	0	0
50	1.9	4.90E+2 / 0.00E+0	0	1.65	4.90E+2 / 0.00E+0	0	1.81	4.90E+2 / 0.00E+0	0	2.44	4.90E+2 / 0.00E+0	0	2.17	4.90E+2 / 0.00E+0	0	0	0	0	0
100	5.47	1.90E+4 / -3.60E+5	0	4.94	1.60E+4 / -3.30E+5	0	5.3	2.00E+4 / -4.00E+5	0	6.32	2.30E+4 / 0.00E+0	0	5.7	2.30E+4 / 0.00E+0	0	0.1	0.17	0.1	0
150	14.2	6.80E+5 / -1.30E+5	0	12.7	4.30E+5 / -1.10E+5	0	14.7	7.00E+5 / -1.50E+5	0	16.9	9.70E+5 / 0.00E+0	0	14.2	9.70E+5 / 0.00E+0	0	0.1	0.2	0.1	0
200	30.4	3.00E+7 / -4.90E+6	0	24.9	2.20E+7 / -3.80E+6	0	33	3.20E+7 / -5.40E+6	0	39.9	4.80E+7 / -5.90E+6	0	29.1	5.70E+7 / 0.00E+0	0	0.2	0.25	0.17	0.15
300	111	3.40E+9 / -5.20E+8	0	94.6	2.10E+9 / -3.90E+8	0	121	3.80E+9 / -4.90E+8	0	177	4.10E+9 / -3.90E+8	0	104	5.10E+9 / -3.10E+8	0	0.18	0.29	0.18	0.17
400	316	2.60E+11 / -4.90E+10	0	269	1.80E+11 / -3.50E+10	0	330	2.90E+11 / -4.80E+10	0	420	3.10E+11 / -3.70E+10	0	293	4.30E+11 / -4.00E+10	0.05	0.26	0.38	0.2	0.17
500	904	8.80E+12 / -2.40E+12	0	732	6.70E+12 / -1.20E+12	0	959	9.60E+12 / -2.60E+12	0	1330	1.40E+13 / -1.60E+12	0	846	1.90E+13 / -1.70E+12	0.12	0.5	0.57	0.33	0.17
600	2881	2.30E+15 / -3.80E+14	0	2254	1.80E+15 / -3.10E+14	0	3022	2.80E+15 / -3.90E+14	0	3789	3.90E+15 / -4.30E+14	0	2652	5.50E+15 / -5.20E+14	0	0.45	0.83	0.4	0.25
800	5963	2.00E+18 / -3.90E+17	0	4807	9.60E+17 / -3.10E+17	0	6300	2.80E+18 / -4.20E+17	0	6480	3.70E+18 / -4.40E+17	0	5902	4.80E+18 / -4.80E+17	0.09	0.48	0.92	0.4	0.22

For HV the numbers in the brackets are standard deviations

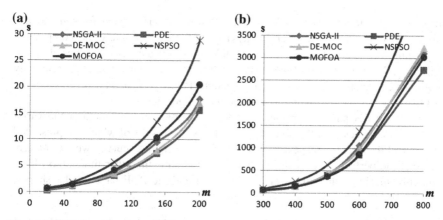

Fig. 11.2 CPU time (s) variation with problem size m on the BN VFR problem: **a** $300 \leq m \leq 800$, **b** $m \leq 200$

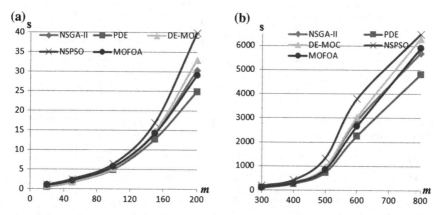

Fig. 11.3 CPU time (s) variation with problem size m on the CA VFR problem: **a** $300 \leq m \leq 800$, **b** $m \leq 200$

DE-MOC. However, with the increase of the problem size, MOFWA shows more and more performance advantages over the other three algorithms. For the problem instances with $mn > 400$, MOFWA always obtains higher quality solution set (in terms of hypervolume and coverage) with less computational time, in comparison with NSGA-II, DE-MOC, and NSPSO. PDE is faster than MOFWA, but the quality of its result sets is far less than that of the MOFWA.

In particular, the coverage metrics show that the result solutions obtained by the MOFWA are never dominated by those obtained by the NSGA-II, PDE, and DE-MOC, given that the values of c' are always zero in the corresponding table columns of the three algorithms. Only on large-sized problems, NSPSO occasionally obtains a few solutions that dominate some result solutions of the MOFWA, but the coverage value is always smaller than C4, and it should be noted that NSPSO consumes much more computational cost than MOFWA. On the contrary, except for some small-sized problem instances, there are always a part of result solutions of the competitive algorithms dominated by those of the MOFWA, as indicated by the values of c_1, c_2, c_3, and c_4. For very large problem instances ($mn > 1000$), those values are relatively high, which means that the distance between the Pareto front and the result set of the MOFWA is much closer than those of the other algorithms.

For those BN and CA problems where $m \geq 200$, we also monitor the variations of HV values with the algorithm iterations, the results of which are presented in Figs. 11.4 and 11.5 respectively. From the convergence curves we can clearly see that the performance of the MOFWA overwhelms the other three algorithms. Roughly speaking, the curves of the NSGA-II and the PDE have similar shapes,

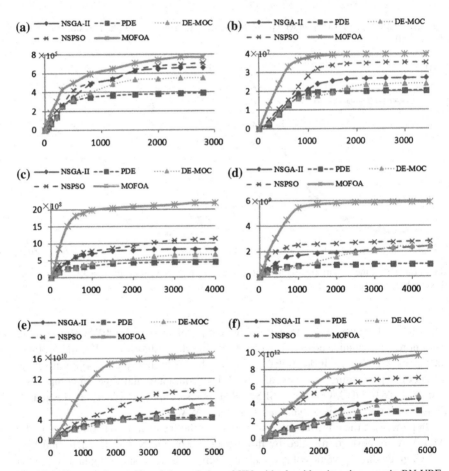

Fig. 11.4 Comparative results of the variation of HV with algorithm iterations on six BN VRF problem instances. The HV values are averaged over 30 simulation runs. **a** $m = 200$, **b** $m = 300$, **c** $m = 400$, **d** $m = 500$, **e** $m = 600$, **f** $m = 800$

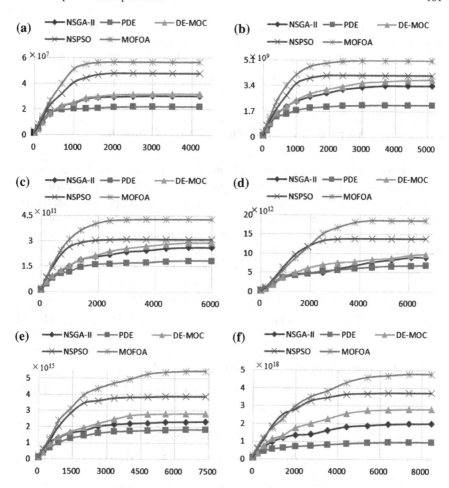

Fig. 11.5 Comparative results of the variation of HV with algorithm iterations on six CA VRF problem instances. The HV values are averaged over 30 simulation runs. **a** $m = 200$, **b** $m = 300$, **c** $m = 400$, **d** $m = 500$, **e** $m = 600$, **f** $m = 800$

both are smooth and grow slowly toward the final results. The PDE has the lowest performance among the algorithms, which indicates that its search capability is very limited on the test problems. The performance of the NSPSO is the best in the three competitive algorithms, and its convergence curves grow very fast; in particular, during the early algorithm iterations, the NSPSO occasionally achieve better results than the MOFWA on some large problem instances (for example, see Figs. 11.4f and 11.5d, f). Therefore,we can conclude that the NSPSO has a good search capability, but it is easy to be trapped in the local optima and thus converge prematurely.

By comparison, we found that the MOFWA also converges fast, but it can effectively avoid the local optima and guide the search toward the Pareto fronts. In summary, on the test VFR problems, the proposed MOFWA has significant performance advantages over the other competitive algorithms.

11.4.2 Case Study of a Real-World CO VRF Problem

We apply the MOFWA to a real-world Camellia oleifera VRF problem in Yichun District, Jiangxi Province, East China. The plant area is 585 ha and can be partitioned into two parts: 375 ha in sunny slope and 210 ha in semi-sunny slope, which are further divided into 40 fields and 24 fields, respectively, according to their gradients. The crop quality is evaluated based on the percentage of unsaturated fatty acid in fruits. According to our field experiments, the yield and quality estimation model of the sunny and semi-sunny areas are, respectively, as follows:

$$
\begin{aligned}
Y_a = {} & 850.2 + 9.3x_1 + 6.1x_2 + 14.88x_3 + 6.68x_1x_2 \\
& + 2.04x_2x_3 - 3.06x_1x_3 - 0.61x_1^2 - 4.12x_2^2 - 0.35x_3^2
\end{aligned}
\tag{11.27}
$$

$$
\begin{aligned}
Y_b = {} & 661.8 + 19.02x_1 + 26.46x_2 + 1.49x_3 - 4.78x_1x_2 \\
& + 1.6x_2x_3 + 3.08x_1x_3 - 3.98x_1^2 - 1.65x_2^2 - 0.67x_3^2,
\end{aligned}
\tag{11.28}
$$

$$
\begin{aligned}
Q_a = {} & 59.9 + 0.37x_1 + 4.94x_2 + 0.62x_3 + 0.52x_1x_2 \\
& - 0.04x_2x_3 + 0.04x_1x_3 - 0.31x_1^2 - 0.47x_2^2 - 0.02x_3^2,
\end{aligned}
\tag{11.29}
$$

$$
\begin{aligned}
Q_b = {} & 49.1 + 0.9x_1 + 2.25x_2 + 1.18x_3 + 0.39x_1x_2 \\
& - 0.03x_2x_3 + 0.1x_1x_3 - 0.27x_1^2 - 0.35x_2^2 - 0.04x_3^2,
\end{aligned}
\tag{11.30}
$$

where Y_a and Y_b stand for the yields of the sunny area and the semi-sunny area respectively, and Q_a and Q_b stand for the corresponding quality objectives. And the mathematical formulation of the real-world problem is as follows (where a_i denoted the area of the field i):

$$min \ Q(X) = \sum_{i=1}^{40} a_i Y_a(X) Q_a(x) + \sum_{i=41}^{64} a_i Y_b(x) Q_b(x) \tag{11.31}$$

$$min \ C(x) = \sum_{i=1}^{64} \sum_{j=1}^{3} p_j x_{ij} \tag{11.32}$$

$$min \ E(x) = \sum_{i=1}^{6} 4 E_i(x) \tag{11.33}$$

s.t.

$$\sum_{i=1}^{64} Y_i(x) \geq Y^L \tag{11.34}$$

$$\sum_{i=1}^{64} \sum_{j=1}^{3} p_j x_{ij} \leq C^U \tag{11.35}$$

$$\sum_{j=1}^{3} R_j(x) \leq R^U \tag{11.36}$$

$$x_{ij} \geq 0, \forall i = 1, 2, \ldots, 64. \ \forall j = 1, 2, 3. \tag{11.37}$$

The unit price of the fertilizer N, P, and K are respectively 6.3, 3.5, and 4.6 (RMB/kg). The gradient and plant density d of each field are presented in Table 11.5.

The organization has a precision agriculture management software which employs a multiobjective random search (MORS) algorithm based on [45] for VFR optimization and decision. The software has been used for three years and demonstrated its contribution to crop production. We respectively execute the MORS and the MOFWA algorithms for five times on this problem. The MORS consumes 48 s in average and its best result set consists of 10 solutions. The MOFWA uses only 5.1 s in average and its best result set consists of 6 solutions. Table 11.6 presents the objective function values of the result solutions.

It is not difficult to observe that none of the solutions obtained by the MOFWA are dominated by those obtained by the random search. On the contrary, except solution ♯7, all the other solutions obtained by the MORS are dominated by the result of the MOFWA. For example, the MOFWAs solution ♯1 dominates the random search solution ♯1, the MOFWAs ♯3 dominates the MORSs ♯8 and ♯9, and the MOFWAs ♯5 dominates the random search ♯2 ♯6. That is, the coverage of the result set of MOFWA to that of MORS is 88.9 %, and the coverage of MORS to MOFWA is 0. After empirical analysis, the decision-maker believes that solutions ♯2, ♯3, and ♯5 of the MOFWA (but none of the random search) are preferred VRF solutions that achieve excellent balance between the objectives, and takes solution ♯3 as the final

Table 11.5 The input data of the real-world CO VRF problem: field gradients and plant densities

Fields	1	2	3	4	5	6	7	8	9	10	11	12
$\alpha_i(o)$	8.24	9.05	8.6	8.29	9.32	8.1	7.63	9.81	10.69	8.8	8.43	8.91
d_i (per ha)	1227.4	1306.2	1256	1275.4	1323.5	1280.3	1249.7	1290.1	1310.7	1218.6	1283.4	1311.9
Fields	13	14	15	16	17	18	19	20	21	22	23	24
$\alpha_i(o)$	9.53	9.14	8.22	9.17	11.26	10.8	8.95	9.79	8.68	11.05	9.36	9.6
d_i (per ha)	1287.1	1277.3	1265.2	1300.1	1331.5	1298.6	1257	1310.3	1190.4	1359.2	1281	1286.7
Fields	25	26	27	28	29	30	31	32	33	34	35	36
$\alpha_i(o)$	9.86	9.45	10.7	11.47	10.44	9.72	10.66	11.02	11.19	9.85	10.26	10.03
d_i (per ha)	1309.4	1270.8	1367.3	1385	1333.9	1259.4	1351	1403.8	1388.9	1327.6	1340	1315.3
Fields	37	38	39	40	41	42	43	44	45	46	47	48
$\alpha_i(o)$	11.22	10.37	10.64	11.08	16.9	16.45	15.41	17.52	16.1	17.05	15.79	16.18
d_i (per ha)	1364.6	1311.1	1315	1335.6	1540	1524.9	1500.7	1623.9	1474.8	1588.8	1510	1525.2
Fields	49	50	51	52	53	54	55	56	57	58	59	60
$\alpha_i(o)$	15.32	14.66	14.78	16.36	15.71	16.08	14.8	15.68	16.33	15.03	15.69	15.85
d_i (per ha)	1507.8	1502.5	1495.8	1540.3	1527.2	1550.2	1477	1542.4	1562.1	1446	1526.5	1528
Fields	61	62	63	64								
$\alpha_i(o)$	16.15	16.93	15.7	15.56								
d_i (per ha)	1564.4	1611.5	1540.6	1506.2								

Table 11.6 The non-dominated solution sets obtained by the old random search algorithm and by the MOFWA for the real-world CO VRF problem

Algorithm	Optimized solutions
MORS	# 1(56.2, 86.5, 32.9)
	#2(54.8, 84.4, 33.2)
	#3(54.4, 85.7, 31.0)
	# 4(54.1, 85.2, 32.1)
	# 5(53.8, 80.3, 34.0)
	# 6(53.5, 84.0, 33.8)
	# 7(52.8, 84.9, 30.7)
	# 8(52.6, 83.5, 31.9)
	# 9(52.5, 82.7, 32.1)
MOFWA	# 1(57.3, 84.9, 31.4)
	# 2(56.7, 82.9, 31.2)
	# 3(56.4, 82.6, 31.5)
	# 4(56.1, 81.3, 33.4)
	# 5(55.6, 83.6, 30.8)
	# 6(54.8, 85.1, 30.5)

The solutions are represented by their objective values (Y, C, E)

decision which suggests to put more emphasis on the crop quality for sunward fields and put more emphasis on the crop yield for non-sunward fields. Furthermore, having seen the effectiveness of the new algorithm, the organization decides to integrate the MOFWA into the precision agriculture management software.

11.5 Discussion

From the computational experiment results, we can see that the proposed MOFWA can obtain high-quality solutions for the VRF problem:

- For small-sized problem instances where $mn \leq 200$, the MOFWA has a much higher probability of reaching the Pareto optimal front of the problem than other state-of-the-art multiobjective optimization algorithms.
- For relatively middle-sized and large-sized instances, the result solution sets obtained by the MOFWA are always better than the result sets obtained by other competitive algorithms. The larger the instance size, the more obvious the advantage of the MOFWA is.

Among the stochastic optimization algorithms used in the comparative experiments, MOFWA exhibits the best performance in terms of convergence speed and robustness, and its performance is not very sensitive to the sizes of problem instances. This is mainly because the common explosion of high-quality fireworks provides an excellent capability of local search, and specific explosion of low quality helps to diversify the search. Moreover, the introduction of DE operators enhances the algorithms capability of exploration. The effective combination of the exploration and exploitation makes the MOFWA intrinsically a robust stochastic search method [46]. On the other hand, the enhanced diversification mechanism also provides good capability of avoiding early convergence to local optima. In general, we can conclude that the proposed MOFWA is mostly appropriate for the test VRF problem instances studied in the chapter.

However, the fireworks explosion method employed in the MOFWA has its drawbacks. At each algorithm iteration, each individual (firework) in the population generates a certain number of new solutions (sparks). Such a number is set in the range of 2–20 in our algorithm, and thus the required number of function evaluations (NFE) of MOFWA is much larger than other algorithms with the same population size. Due to the complexity of the nonlinear objective functions and constraint functions of the VRF problem, this make the MOFWA consume high computational resources. To mitigate the effect, the population size of the MOFWA is set much smaller than that of other algorithms, and we fine-tune the control parameters of the MOFWA to make it generate appropriate number of sparks at different stages of evolution.

Another drawback is that the number of control parameters to be manually tuned in the MOFWA is more than other typical evolutionary algorithms such as GA and DE. Moreover, the integration with DE operator increases the number of parameters. This will bring more burdens when apply the MOFWA to a new problem.

Nevertheless, as a new heuristic method for multiobjective optimization, the MOFWA offers a great performance advantage in solving complicated VRF problem instances. Based on our experiments, the recommended parameter values of the algorithm are provided for the proposed VRF problem. And we believe that the algorithm can be improved and applied to many other kinds of optimization problems further.

11.6 Summary

Fireworks algorithm is a swarm intelligence algorithm that has a promising performance on many global optimization problems. However, the study of its application in multiobjective optimization is very few. An efficient MOFWA algorithm was presented for oil crop VRF problems, which uses a problem-specific strategy for generating the initial population, uses the concept of Pareto dominance for individual evaluation and selection, and combines the DE operators to increase the information sharing and thus diversify the search. The algorithm has been successfully applied to a number of VRF problems and demonstrated its efficiency and effectiveness.

This research provides a novel insight into the practicability of the fireworks algorithm heuristic in multiobjective optimization. Except the strategy of initial population generation, most features of the MOFWA could be adapted or extended to solve other multiobjective optimization problems.

References

1. Y.-J. Zheng, Q. Song, S.-Y. Chen, Multiobjective fireworks optimization for variable-rate fertilization in oil crop production. Appl. Soft Comput. **13**(11), 4253–4263 (2013)
2. K.S. Raju, D.N. Kumar, Multicriterion decision making in irrigation planning. Agric. Syst. **62**(2), 117–129 (1999)
3. C. Shouyu, Fuzzy optimization of multi-dimensional multi-objective dynamic programming and its application to farm irrigation. J. Hydraul. Eng. **4**, 33–38 (2002)
4. M. Kilic, S. Anac, Multi-objective planning model for large scale irrigation systems: method and application. Water Resour. Manag. **24**(12), 3173–3194 (2010)
5. Y. Wang, D. Zhang, The optimization model of multi-objective fertilization of rice seedbed. J. Biomath. **18**(4), 467–472 (2002)
6. Y. Yuan, L. Mao, L. Lujiu, Z. Guobing, C. Xi, W. Li, Algorithm of fertilization model based on intelligent computing. Trans. Chin. Soc. Agric. Eng. **12**, 2008 (2008)
7. Y. Helong, D. Liu, G. Chen, B. Wan, S. Wang, B. Yang, A neural network ensemble method for precision fertilization modeling. Math. Comput. Model. **51**(11), 1375–1382 (2010)
8. C.A.C. Coello, D.A. Van Veldhuizen, G.B. Lamont, *Evolutionary Algorithms for Solving Multiobjective Problems*, vol. 242 (Springer, Berlin, 2002)
9. N. Srinivas, K. Deb, Muiltiobjective optimization using nondominated sorting in genetic algorithms. Evol. Comput. **2**(3), 221–248 (1994)
10. K. Deb, A. Pratap, S. Agarwal, T. Meyarivan, A fast and elitist multiobjective genetic algorithm: NSGA-II. IEEE Trans. Evolut. Comput. **6**(2), 182–197 (2002)
11. E. Zitzler, L. Thiele, Multiobjective evolutionary algorithms: a comparative case study and the strength Pareto approach. IEEE Trans. Evol. Comput. **3**(4), 257–271 (1999)
12. J.D. Knowles, D.W. Corne, M-PAES: a memetic algorithm for multiobjective optimization, in *Proceedings of the 2000 Congress on Evolutionary Computation*, vol. 1 (IEEE, 2000), pp. 325–332
13. H.A. Abbass, R. Sarker, C. Newton, PDE: a pareto-frontier differential evolution approach for multi-objective optimization problems, in *Proceedings of the 2001 Congress on Evolutionary Computation*, vol. 2 (IEEE, 2001), pp. 971–978
14. X. Li, A non-dominated sorting particle swarm optimizer for multiobjective optimization, in *Genetic and Evolutionary Computation GECCO 2003* (Springer, Berlin, 2003), pp. 37–48
15. Z. Zhang, Immune optimization algorithm for constrained nonlinear multiobjective optimization problems. Appl. Soft Comput. **7**(3), 840–857 (2007)
16. C.K. Goh, K.C. Tan, D.S. Liu, S.C. Chiam, A competitive and cooperative co-evolutionary approach to multi-objective particle swarm optimization algorithm design. Eur. J. Oper. Res. **1**, 42–54 (2010)
17. C. Shi, Z. Yan, Z. Shi, L. Zhang, A fast multi-objective evolutionary algorithm based on a tree structure. Appl. Soft Comput. **10**(2), 468–480 (2010)
18. P.K. Tripathi, S. Bandyopadhyay, S.K. Pal, An adaptive multi-objective particle swarm optimization algorithm with constraint handling, in *Handbook of Swarm Intelligence*, ed. by B.K. Panigrahi, Y. Shi, M.-H. Lim. Adaptation, Learning, and Optimization, vol. 8 (Springer, Berlin, 2011), pp. 221–239
19. W.K. Mashwani, A. Salhi, A decomposition-based hybrid multiobjective evolutionary algorithm with dynamic resource allocation. Appl. Soft Comput. **12**(9), 2765–2780 (2012)
20. D. Chen, F. Zou, J. Wang, A multi-objective endocrine PSO algorithm and application. Appl. Soft Comput. **11**(8), 4508–4520 (2011)
21. A. Zhou, Q. Bo-Yang, H. Li, S.-Z. Zhao, P.N. Suganthan, Q. Zhang, Multiobjective evolutionary algorithms: a survey of the state of the art. Swarm Evol. Comput. **1**(1), 32–49 (2011)
22. S. Chen, Y. Zheng, C. Cattani, W. Wang, Modeling of biological intelligence for SCM system optimization. Comput. Math. Methods Med. **2012**, 30 (2011)
23. T. Niknam, M. Zare, J. Aghaei, Scenario-based multiobjective Volt/Var control in distribution networks including renewable energy sources. IEEE Trans. Power Deliv. **27**(4), 2004–2019 (2012)

24. P. Ahmadi, M.A. Rosen, I. Dincer, Multi-objective exergy-based optimization of a polygeneration energy system using an evolutionary algorithm. Energy **46**(1), 21–31 (2012)
25. T. Niknam, H. Zeinoddini Meymand, H. Doagou Mojarrad, An efficient algorithm for multiobjective optimal operation management of distribution network considering fuel cell power plants. Energy **36**(1), 119–132 (2011)
26. T. Niknam, A. Kavousifard, S. Tabatabaei, J. Aghaei, Optimal operation management of fuel cell/wind/photovoltaic power sources connected to distribution networks. J. Power Sources **196**(20), 8881–8896 (2011)
27. P. Ahmadi, I. Dincer, Thermodynamic and exergoenvironmental analyses, and multi-objective optimization of a gas turbine power plant. Appl. Therm. Eng. **31**(14–15), 2529–2540 (2011)
28. Y.-J. Zheng, S.-Y. Chen, Y. Lin, W.-L. Wang, Bio-inspired optimization of sustainable energy systems: a review. Math. Probl. Eng. **2013**, 28 (2013)
29. M. Janga Reddy, D. Nagesh Kumar, Evolving strategies for crop planning and operation of irrigation reservoir system using multi-objective differential evolution. Irrig. Sci. **26**(2), 177–190 (2008)
30. Y. Tan, Y. Zhu, Fireworks algorithm for optimization, in *Advances in Swarm Intelligence* (Springer, Berlin, 2010), pp. 355–364
31. J. Kennedy, R. Eberhart et al., Particle swarm optimization, in *Proceedings of IEEE international conference on neural networks*, vol. 4 (Perth, Australia, 1995), pp. 1942–1948
32. Y. Tan, Z.M. Xiao, Clonal particle swarm optimization and its applications, in *IEEE Congress on Evolutionary Computation. CEC 2007* (2007), pp. 2303–2309
33. Y.J. Zheng, X.L. Xu, H.F. Ling, A hybrid fireworks optimization method with differential evolution. Neurocomputing (2012)
34. R. Storn, K. Price, Differential evolution-a simple and efficient heuristic for global optimization over continuous spaces. J. Glob. Optim. **11**(4), 341–359 (1997)
35. J.L. Yu, *Agricultural Experiments with Polydesign* (Beijing University of Agriculture Press, Beijing, 1993)
36. R.K. Ru, *Soil Plant Nutrition Principles and Fertilization* (Chemical Industry Press, Beijing, 1998)
37. B.W. Silverman, *Density Estimation for Statistics and Data Analysis* (Chapmanand Hall, London, 1986)
38. Z. Cai, Y. Wang, A multiobjective optimization-based evolutionary algorithm for constrained optimization. IEEE Trans. Evol. Comput. **10**(6), 658–675 (2006)
39. M.S. Alam, M.M. Islam, X. Yao, K. Murase, Diversity guided evolutionary programming: a novel approach for continuous optimization. Appl. Soft Comput. **12**(6), 1693–1707 (2012)
40. J. Hájek, A. Szöllös, J. Šístek, A new mechanism for maintaining diversity of pareto archive in multi-objective optimization. Adv. Eng. Softw. **41**(7), 1031–1057 (2010)
41. W. Gong, Z. Cai, A multiobjective differential evolution algorithm for constrained optimization, in *IEEE Congress on Evolutionary Computation. CEC 2008* (IEEE World Congress on Computational Intelligence, 2008), pp. 181–188
42. Q.-Y. Guo, Z.-Y. LI, X.-W. Tu, Plant nutritional aspects and effects of fertilizer application in rapeseed in red-yellow soil of South China. Fertilizer application of double-low rapeseed cultivar, zhongshuang no. 7 in red paddy soil. Chin. J. Oil Crop Sci. **1**, 011 (2001)
43. L. Yankun, W. Huishan, H.J. Zhuang, Z. Lin, J.L. Yongye, Preliminary report on fertilization trial of canarium album. Guangdong For. Sci. Technol. **5**, 004 (2007)
44. E. Zitzler, K. Deb, L. Thiele, Comparison of multiobjective evolutionary algorithms: empirical results. Evol. Comput. **8**(2), 173–195 (2000)
45. J. Wei-yi Qian, Y. Yang, H.W. Yang, W. Jin-xia, A new random group search algorithm for solving the multi-objective programming problems. J. Liaoning Norm. Univ. Nat. Sci. **30**(2), 141 (2007)
46. D. Karaboga, B. Basturk, On the performance of artificial bee colony (ABC) algorithm. Appl. Soft Comput. **8**(1), 687–697 (2008)

Chapter 12
S-Metric-Based Multi-objective Fireworks Algorithm

This chapter is to present how to apply FWA to solving multi-objective optimization problems with the help of a hypervolume indicator such as S-metric, then proposes a S-metric multi-objective fireworks algorithm (S-MOFWA). The S-metric is a frequently used quality measure for solution sets comparison in evolutionary multi-objective optimization algorithms (EMOAs), which is also used to evaluate the contribution of a single solution among the solution sets. Traditional multi-objective optimization algorithms usually perform a $(\mu + 1)$ strategy and update the external archive one by one, while the proposed S-MOFWA performs a $(\mu + \mu)$ strategy, thus converging faster to a set of Pareto solutions by three steps: (1) Exploring the solution space by mimicking the explosion of fireworks; (2) Performing a simple selection strategy for choosing the next generation of fireworks according to their S-metric; (3) Utilizing an external archive to maintain the best solution set ever found, with a new archive definition and a novel updating strategy, which can update the archive with μ solutions in a single process. The detailed comparison results with NSGA-II, SPEA2, and PESA2 demonstrate the efficiency of the proposed S-MOFWA.

12.1 Introduction

A decision vector a is said to dominate a vector $b(a \prec b)$, iff $f_i(a) \leq f_i(b)$ for all i and $f_j(a) \leq f_j(b)$ for at least one j with $i, j \in 1, \ldots, n, f : \mathbb{R}^m \rightarrow \mathbb{R}^n$ and $a, b \in \mathbb{R}^m$. The set of non-dominated decision vectors in $\mathbb{R}^n$ is called a Pareto (optimal) set. The corresponding image under f in the solution space is called the Pareto front. The multi-objective optimization problem requires to get a best Pareto front.

Multi-objective optimization problem has two or more conflicting objectives to optimize simultaneously. Lack of prior knowledge about the objectives, we usually investigate into a vast number of solutions and reserve the non-dominated ones, i.e., a Pareto solution set, as the approximation of the true Pareto optimal set. This idea contributes directly to the rapid development of evolutionary multi-objective optimization algorithm (EMOA) [1].

© Springer-Verlag Berlin Heidelberg 2015
Y. Tan, *Fireworks Algorithm*, DOI 10.1007/978-3-662-46353-6_12

In Multiple Single Objective Pareto Sampling (MSOPS) [2, 3], the relationship between conflicting objectives was quantified as weights added to each objective. Despite unknown, when enough trials of ways to combine and adjust the weights were explored, enough non-dominated solutions would be found to make a good spread of the Pareto front.

During the evolution of EMOA, there are two crucial points to consider, i.e., convergence and diversity. Most EMOAs consider them apart. For example, in NSGA-II [4], the non-dominated sorting is performed first to classify all the solutions into hierarchical dominated levels as their measure of convergence, then a crowding distance will be calculated as their measure of diversity. In SPEA2 [5], the fitness of each solution is composed by two parts: dominated strength which represents the convergence and density metric which represents the diversity.

Different from the above methods, indicator-based methods choose a single metric to represent both convergence and diversity. IBEA [6] defines a binary quality indicator and each solution needs to be compared with every other solution in order to get a single fitness. The comparison times required is as much as that for non-dominated sorting. *S*-metric [7] is another indicator that represents both convergence and diversity, by calculating the hypervolume of dominated space. A great amount of attention was paid to this metric since its proposal. Nicola Beume and Carlos did a detailed analysis on the complexity of computing the hypervolume indicator [8]. Knowles and Corne [9] gave an example for fast calculating the *S*-metric.

In this chapter, a multi-objective fireworks algorithm based on *S*-metric is presented [10]. The explosion amplitudes of fireworks and the numbers of sparks are calculated according to each firework's *S*-metric, the next generation of fireworks is selected by their *S*-metric and an external archive with fixed size is used to maintain the best solution set found ever by a novel updating strategy.

12.2 *S*-Metric

The *S*-metric or hypervolume indicator was proposed by Zitzler and Thiele [7] for the first time. It can be viewed as the *size of the space covered* or *size of dominated space*. The *S*-metric is a frequently used quality measure for solution sets comparison in the field of evolutionary multi-objective optimization algorithms (EMOAs). Besides, *S*-metric can also be used to evaluate the contribution of a single solution among the solution set. Let Λ denote the Lebesgue measure, then the *S*-metric for a solution set $M = \{m_1, m_2, \ldots, m_i, \ldots m_n\}$ is defined as [11]:

$$S(M) := \Lambda \left(\bigcup_{m \in M} \{x | m \prec x \prec x_{ref}\} \right) \tag{12.1}$$

where, $\prec$ denotes the dominance relationship, i.e., $a \prec b$ means b is dominated by a. The x_{ref} is a reference point dominated by all valid solutions in the solution set.

Each solution in set M contributes a part to this total S-metric, so the S-metric for each solution is defined as

$$S(m_i) = \Delta S(M, m_i) := S(M) - S(M \setminus \{m_i\}).$$
(12.2)

Fig. 12.1 S-metric.
a S-metric for the set.
b S-metric for a solution

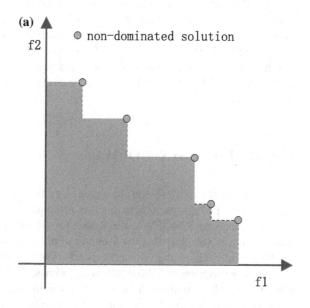

(a)

non-dominated solution

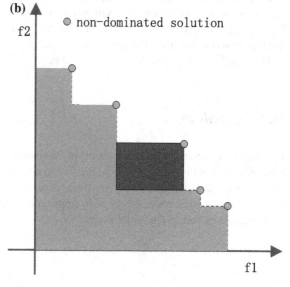

(b)

non-dominated solution

The S-metric for a solution m_i can be seen as the size of space dominated by m_i, but not dominated by any other solutions in the set. A large value means m_i is important in set M, while zero indicates that m_i must be dominated by some other solutions in set M.

Figure 12.1 shows the meaning of S-metric and the difference between S-metric for set and S-metric for a single solution. It can be seen that S-metric for a solution has no relation with the choice of a reference point. A solution with larger S-metric value lies further away from its nearest neighbors, which is coincident with the density metric of SPEA2 [5] in the meaning of diversity. Besides, a nonzero value guarantees that the solution is non-dominated, coincident with convergence meaning, because according to the definition in Eq. (12.2), the S-metric for a solution dominated by others is zero.

12.3 The Proposed S-MOFWA

As for multi-objective problems, there are more than one objective function, f_1, f_2, etc. But for an EMOA, the fitness function is unique.

In the proposed S-MOFWA, the fitness for an individual is its S-metric in the current solution set. Note that the S-metric of one solution is merely relevant to its neighbors and could change when the members of the current solution set change.

A brief description of S-MOFWA is as follows: At first, a number of fireworks will be randomly initialized in the search space. In each iteration, we calculate the numbers of explosion sparks and amplitudes of explosion for each firework. Then the explosion and Gaussian mutation will be performed to generate a number of sparks. Finally, fireworks for next iteration are selected from the candidates. An external archive is used to maintain the best solution set.

The details of each step can be described below.

12.3.1 Initialization

The algorithm randomly selects N points in the search space as the first generation of fireworks. As it is the first generation, there is no need to calculate or set the objective function values. Here, we simply set their S-metric the same values, so they have the same sparks numbers and explosion amplitudes, because we do not hurry to tag them as good or bad in the initial stage.

12.3.2 Calculation of the Explosion Sparks Number and Explosion Amplitude

As mentioned above, to make a contrast among the fireworks, the firework with better fitness will have larger number of explosion sparks and smaller explosion amplitude, while the firework with worse fitness will have smaller number of sparks and bigger explosion amplitude. To ensure this principle, the number of sparks z_i and explosion amplitude A_i for each firework x_i are calculated by Eqs. (12.3) and (12.4), which are determined by the S-metric solely.

$$z_i = M_e * \frac{S(x_i) + \varepsilon}{\sum_{i=1}^{N} (S(x_i)) + \varepsilon}, \tag{12.3}$$

$$A_i = \hat{A} * \frac{S_{\max} - S(x_i) + \varepsilon}{\sum_{i=1}^{N} (S_{\max} - S(x_i)) + \varepsilon}, \tag{12.4}$$

where $S_{\max} = \max(S(x_i))$, $i = 1, 2, \ldots, N$. M_e and $\hat{A}$ are two constant parameters to control the sparks number and amplitude. To bound the values z_i to a proper range, two other constants $a, b \in [0, 1]$ was also needed.

$$z_i = \begin{cases} round(aM_e) & \text{if } z_i < aM_e, \\ round(bM_e) & \text{if } z_i > bM_e, \\ round(z_i) & \text{otherwise.} \end{cases} \tag{12.5}$$

Note that the best solution, namely the solution with a max S-metric, obtains a very small explosion amplitude which is close to zero according to Eq. (12.4). This may be acceptable in the final stage of the evolution, but that is unreasonable for the early stage. The nonlinearly decreasing amplitude threshold $A_{\min}$ introduced in [12] is also employed here.

12.3.3 Explosion

For each firework x_i, z_i sparks will be generated according to Algorithm 12.1. This explosion procedure can be seen as the exploration among the search space.

Algorithm 12.1 Explosion

1: **for** $j = 1 \to z_i$ **do**
2:    initialize the location of the explosion spark: $\hat{x}_j = x_i$
3:    **for** each dimension k of x_i **do**
4:      **if** $rand(0, 1) < 0.3$ **then**
5:        $\hat{x}_j^k = x_i^k + A_i * rand(-1, 1)$
6:      **if** $\hat{x}_j^k$ out of bounds **then**
7:        $\hat{x}_j^k = U(x_{min}^k, x_{max}^k)$
8:      **end if**
9:      **end if**
10:    **end for**
11: **end for**

12.3.4 Gaussian Mutation

To increase the diversity of generated explosion sparks swarm, the algorithm will also generate a number of special Gaussian sparks through a process called Gaussian mutation. Algorithm 12.2 describes the calculation process of Gaussian sparks.

Algorithm 12.2 Gaussion mutation

1: initialize the location of the Gaussian spark: $\tilde{x}_i = x_i$
2: **for** each dimension k of x_i **do**
3:    **if** $rand(0, 1) < 0.5$ **then**
4:      $\tilde{x}_i^k = x_i^k * normrnd(1, 1)$
5:      **if** $\tilde{x}_i^k$ out of bounds **then**
6:        $\tilde{x}_i^k = U(x_{min}^k, x_{max}^k)$
7:      **end if**
8:    **end if**
9: **end for**

12.3.5 Fireworks Selection and Archive Updating

12.3.5.1 Calculation of S-Metric

In conventional FWA, the fireworks for next generation are selected according to their fitness and diversity measure [13]. As for *S*-MOFWA, the *S*-metric is calculated according to their values of objective functions and is also used as selection criteria. In case of two objectives, we sort the sparks in descending order according to the values of the first objective function f_1. After wiping away the points that are dominated, we will get a sequence sorted in ascending order according to the second objective function f_2 (Those points which lose in both two objective functions are dominated). As shown in Fig. 12.2a, the *S*-metric for solution m_i is calculated as in Eq. (12.6)

Fig. 12.2 *S*-metric.
a Calculation of *S*-metric.
b The increase of neighbors'
S-metric after wiping away a
solution

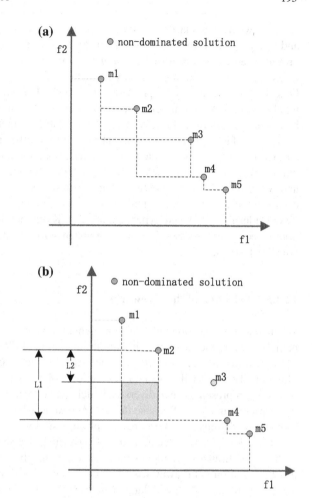

$$S(m_i) = (f_1(m_i) - f_1(m_{i-1})) * (f_2(m_i) - f_2(m_{i+1})). \qquad (12.6)$$

Noted that the calculation of *S*-metric does not depend on the order of the objectives to be considered. All the objectives are of coordinate importance. In the process of sorting the sparks by one objective, we can easily find out and erase those who are dominated by others. Then the *S*-metric of a non-dominated spark is determined by its nearest neighbors in each objective. The nearest neighbors restrict the space which is only dominated by this spark, but not dominated by any other sparks. The volume of this space is the so-called hypervolume, namely the *S*-metric.

For two objectives problem, in each iteration, after the sorting process we can calculate all the sparks' *S*-metric one by one. So the complexity is $O(n\log n)$. For three or more objectives problem, the complexity is $O(n^{d/2}\log n)$ due to Overmars and Yap [14]. Here, n is the number of sparks and d is the number of objectives.

Here, we need to clarify that the calculation of S-metric can be of high efficiency and is also suitable for more objective problems. Boris Naujoks and Nicola Beume presented an algorithm for calculating S-metric in three-objective space [15]. The LebMeasure algorithm described by Fleischer [16] and the Hypervolume by Slicing Objectives (HSO) algorithm proposed by Knowles [17] and Zitzler [18] compute the whole S-metric for all dimensional objectives. Furthermore, many researches have been done to improve the efficiency of calculating S-metric. While and Hingston presented a fast algorithm for calculating S-metric [19]. Then Bradstreet proposed an incremental version to calculate S-metric. On the basis of those improvements, many new S-metric-based EMOAs have been invented, such as SMS-EMOA [20] and MOPSOhv [21]. In addition, Ponweiser solved the multi-objective problems on a limited budget using S-metric [22], Beume accelerated S-metric calculation by considering dominated hypervolume [23], Kukkonen improved pruning of non-dominated solutions based on crowding distance for bi-objective optimization problems [24], so on.

12.3.5.2 Selection of the Fireworks

When selecting the fireworks of next generation, there are two keys to consider, namely convergence and diversity. In the conventional FWA, we choose the best one solution at first to guarantee convergence and lead the swarm developing in the right direction. Then we will select the rest $N-1$ solutions according to their crowding metric as a representation of diversity, which guaranteeing the algorithm will not drop into a premature stage. Because the best N individuals are usually generated by the same firework, thus located in a nearby place, resulting in a low diversity. So the other $N-1$ fireworks are selected according to crowding metric and located far away from the first one firework. While, as mentioned previously, a large S-metric guarantees that the solution is far away from its nearest neighbors and it is non-dominated. In another word, a solution with large S-metric points out a good region which has been explored too little. Our job is to strengthen the exploration in the suggested region. So there is no need to introduce another unnecessary diversity measure and we simply choose the best N solutions as the fireworks of next generation according to their S-metric, namely the fitness.

12.3.5.3 Updating of the External Archive

In conventional archive strategy, external archive is just a place for storing non-dominated solutions. When a new individual was generated, it would be examined whether it can be put into the archive or replace someone. This was done one by one, which means the external archive needs to be updated a thousand times if one thousand new individuals were generated.

The external archive always keeps a fixed number of solutions (assume the fixed number is K), which differs from the conventional grow-up strategy. These

K solutions are chosen from the candidates pool which includes the sparks generated by fireworks explosion and Gaussian mutation and the old archive. The selected K solution needs to ensure that they obtain a maximum S-metric in all the K-sets.

We need to point out that the solutions in this archive are not necessarily non-dominated. Noted that in the early stage of evolution, not enough non-dominated solutions should have been found. But the external archive always keeps K solutions. Inevitably, there are many dominated solutions existing in the archive, but this does not matter. Along with the iteration, the S-metric of the solution set in the archive will increase step by step. Fleischer proved that a finite solution set with the theoretic maximum of S-metric comes necessarily from the true Pareto front [16].

Figure 12.3 demonstrates the evolvement of the external archive on function KUR of S-MOFWA. The size of archive is fixed as $K = 100$. In the first 20 iterations, most of the solutions were dominated by those points close to the true Pareto front. After hundreds of iterations, nearly all of them became non-dominated, and the updates concentrate in the areas very close to the true Pareto front.

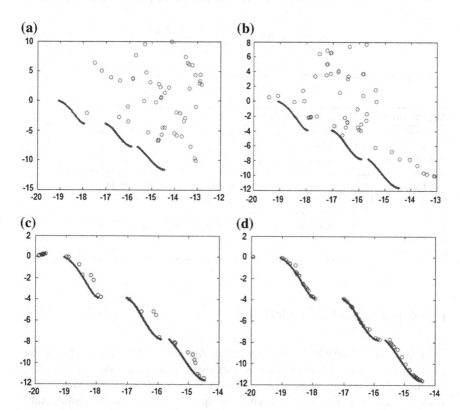

Fig. 12.3 MOFWA evolution process on KUR. **a** Iteration = 10. **b** Iteration = 20. **c** Iteration = 250. **d** Iteration = 300

In each iteration, we will choose K solutions to build up the new archive. This sounds like a combinational problem. However, one solution's S-metric is solely determined by its nearest neighbors, which indicates that wiping off a solution only influences the S-metric of a few solutions. Here, we keep a proper ratio (about 1:1) between the size of external archive and the total number of sparks, wipe off the worst one solution in the candidates set and fix its neighbors' S-metric. This procedure repeats iteratively, eventually remaining K solutions become the new archive.

As shown in Fig. 12.2b, after wiping off the solution m_3, its nearest neighbors m_2 and m_4 have a proportional increase in their S-metric, while the others remain unchanged. Take m_2 for instance, the ratio of new $S(m_2)$ to the old S-metric is L_2/L_1. So a simple update will be enough.

$$S(m_2) = S(m_2) * \frac{f_2(m_2) - f_2(m_4)}{f_2(m_2) - f_2(m_3)}, \tag{12.7}$$

Moreover, the corresponding operator also needs to be performed on m_4, another nearest neighbor:

$$S(m_4) = S(m_4) * \frac{f_1(m_4) - f_1(m_2)}{f_1(m_4) - f_1(m_3)}, \tag{12.8}$$

The procedure of updating the external archive is described in Algorithm 12.3.

Algorithm 12.3 Updating strategy for the external archive

1: note Q as the sparks generated by fireworks explosion and Gaussian mutation
2: note A as the external archive
3: note C as the candidates pool, set $C = Q \cup A$
4: calculate S-metric for each candidate in C /* *a minimum heap may be useful here.* */
5: **while** $|C| > K$ /* *K is the size of the external archive, a fixed number* */ **do**
6:    $r \leftarrow \arg\min_{m \in C}(S(m))$ /* *detect element of C with the lowest S-metric* */
7:    $C \leftarrow C \backslash \{r\}$ /* *eliminate detected element* */
8:    fix the S-metric values of nearest neighbors of r according to Eqs. (12.7) and (12.8)
9: **end while**
10: $A \leftarrow C$

12.3.6 Framework of S-MOFWA

The framework of S-MOFWA is described in Algorithm 12.4. First of all, we randomly initialize a set of fireworks in the search space, then calculate each firework's objective function value. Fitness value, namely the S-metric is calculated according to their objective function values. Then the sparks number and explosion amplitude are calculated by each firework's S-metric. After the explosion and Gaussian mutation, a group of sparks or candidate solutions are generated. When the external archive is updated, a set of solutions are selected as the next generation of fireworks and the algorithm goes into the next iteration.

Algorithm 12.4 Framework of S-MOFWA

1: randomly initialize N fireworks in the solution space
2: initialize external archive A as the set of the initial fireworks
3: iteration $p \leftarrow 0$
4: **while** terminal conditions are not met **do**
5:    $p \leftarrow p + 1$
6:    update the current evaluation times t
7:    update the nonlinearly decreasing amplitude threshold A_{min}
8:    **for** $j = 1 \rightarrow N$ **do**
9:      calculate objective function values of each firework x_i
10:    **end for**
11:    calculate each fireworks' S-metric according to Eq. (12.6)
12:    calculate each fireworks' sparks number and explosion amplitude according to Eqs. (12.3) and (12.4) respectively
13:    perform the firework explosion as Algorithm 12.1 for each firework
14:    perform the Gaussian mutation as Algorithm 12.2 for each firework
15:    calculate each candidate's S-metric /* *candidates refers to the sparks obtained by explosion and Gaussion mutation, and the solutions in the current archive* */
16:    select N best candidates as the next generation of fireworks, according to their S-metric
17:    update the external archive A
18: **end while**
19: output the external archive A

12.4 Experiments

To validate the performance of the proposed S-MOFWA, experiments on benchmark suite which contains six test functions were designed. Moreover, the performance comparison with other three well-known algorithms (NSGA-II, SPEA2 and PESA2) are also conducted.

12.4.1 Experimental Setup

The test problems used to compare different algorithms are chosen from a set of significant studies in the area of multi-objective optimization. Schaffer's problem (SCH) and Kursawe's problem (KUR) are presented by Veldhuizen [25]. ZDT1, ZDT2, ZDT3, ZDT6 are selected from the six test functions suggested by Zitzler [26]. All above problems have two objective functions and none of them have constraints. The properties of these functions are listed in Table 12.1.

For setting the parameters, the number of fireworks $N = 10$, size of external archive $K = 100$, total number of sparks $M_e = 100$, constants $a = 0.05$, $b = 0.4$, the mutation rate for Gaussian mutation $r_1 = 0.5$ and for explosion $r_2 = 0.3$, evaluation times $evals_{max} = 2,00,000$, the same for the comparison algorithms. Each function runs 20 repetition times. The rest of parameters for the fireworks are

Table 12.1 Benchmark functions (Dim. = dimension)

	Dim.	Range	Optimal locations	Property
SCH	1	$[-1000, 1000]$	$x \in [0, 2]$	Convex
KUR	3	$[-5, 5]$	Refer [25]	Non-convex
ZDT1	30	$[0, 1]$	$x_1 \in [0, 1], x_j = 0, j \neq 1$	Convex
ZDT2	30	$[0, 1]$	$x_1 \in [0, 1], x_j = 0, j \neq 1$	Non-convex
ZDT3	30	$[0, 1]$	$x_1 \in [0, 1], x_j = 0, j \neq 1$	Convex, disconnected
ZDT6	10	$[0, 1]$	$x_1 \in [0, 1], x_j = 0, j \neq 1$	Non-convex, nonuniformly spaced

identical to [12]. For setting the parameters of NSGA-II, SPEA2 and PESA2, they are set according to the source code of [**2011 Durillo Jmetal**].[1]

The experimental platform: Win8; Intel Core Duo $E4500$ CPU; 2.2 GHz; 2 GB RAM. The proposed *S*-MOFWA was run on MATLAB2012*b* and the comparison algorithms were run on Eclipse-Java.

12.4.2 Evaluation Criteria

We take the convergence measure and covered space measure to characterize the performance of one multi-objective optimization algorithm.

The **convergence measure** is the average distance to the closest point of the true Pareto front as used in [27]. The smaller the distance is, the closer the solution set lies to the true Pareto front. Usually, we uniformly select hundreds of points from the true Pareto front as a approximation. So this measure is calculated as the average distance to the closest point in the selected approximate set.

The **covered space measure** is a relative ratio of the covered hypervolume, for estimation of the diversity of the spread. The denominator could be the size of space covered by the true Pareto front, or the size of a fixed feasible function space. Here, we take the latter for simple calculation. The maximum values are limited by the different shapes of test functions, so the comparative values are only of meaning within the same test functions. Of course, a larger value indicates the spread is better distributed.

12.4.3 Experimental Results

The results for average convergence measure and covered space measure of *S*-MOFWA, NSGA-II, SPEA2, and PESA2 are listed on Tables 12.2 and 12.3, repec-

---

[1] jmetal: http://sourceforge.net/projects/jmetal/.

Table 12.2 Convergence measure

	ZDT1		ZDT2		ZDT3		SCH		KUR		ZDT6	
	Mean	Std	Mean	Std	Mean	Std	Mean	Std	Mean	Std	Mean	Std
S-MOFWA	**9.10E−04**	**1.00E−04**	**8.00E−04**	**0.00E+00**	**3.90E−03**	4.00E−04	**3.20E−03**	**2.00E−04**	**9.00E−03**	1.10E−03	**2.80E−03**	**1.00E−04**
NSGAII	1.40E−03	2.00E−04	1.10E−03	1.00E−04	4.60E−03	3.00E−04	9.90E−03	7.70E−03	1.31E−02	1.40E−03	2.80E−03	2.00E−04
SPEA2	1.30E−03	1.00E−04	8.00E−04	1.00E−04	4.70E−03	**2.00E−04**	1.19E−02	8.10E−03	1.01E−02	**7.00E−04**	2.80E−03	1.00E−04
PESA2	1.60E−03	3.00E−04	1.30E−03	4.00E−04	4.50E−03	4.00E−04	8.50E−03	7.40E−03	1.61E−02	2.40E−03	1.57E−02	2.05E−02

Table 12.3 Covered space measure

	ZDT1		ZDT2		ZDT3		SCH		KUR		ZDT6	
	Mean	Std	Mean	Std	Mean	Std	Mean	Std	Mean	Std	Mean	Std
S-MOFWA	0.6536	1.03E−02	0.3279	1.50E−03	**0.7795**	3.00E−04	**0.8303**	**1.70E−03**	**0.3558**	8.70E−03	**0.3227**	**0.00E+00**
NSGAII	0.6602	3.00E−04	0.3272	4.00E−04	0.7785	**1.00E−04**	0.811	5.87E−02	0.3435	**6.20E−03**	0.3205	2.00E−04
SPEA2	**0.6616**	**1.00E−04**	**0.3285**	**1.00E−04**	0.7789	1.00E−04	0.8001	5.05E−02	0.3446	6.98E−03	0.3224	2.00E−04
PESA2	0.6252	2.49E−02	0.3092	9.60E−03	0.7703	9.20E−03	0.8082	5.37E−02	0.3516	9.82E−03	0.3131	7.80E−03

Table 12.4 p-values for convergence measure (The values in bold indicate that MOFWA is significantly better compared with the other algorithms)

vs NSGA-II	8.857E−05	8.900E−05	1.030E−04	5.170E−04	1.030E−04	2.959E−01
vs SPEA2	8.900E−05	2.509E−02	8.900E−05	1.400E−04	1.162E−03	6.542E−01
vs PESA2	8.900E−05	1.200E−04	2.930E−04	3.185E−03	8.900E−05	2.190E−04

Table 12.5 p-values for covered space measure (The values in bold indicate that MOFWA is significantly better compared with the other algorithms)

vs NSGA-II	7.314E−02	4.550E−03	8.900E−05	7.932E−02	1.630E−04	8.900E−05
vs SPEA2	1.507E−03	5.016E−01	1.200E−04	7.189E−03	1.630E−04	8.900E−05
vs PESA2	3.380E−04	8.900E−05	1.325E−03	2.277E−02	2.190E−04	8.900E−05

tively. And the T-test results are listed in Tables 12.4 and 12.5. In addition, we also plot the final archive in the objective space on the benchmark functions of S-MOFWA (Fig. 12.4).

12.4.4 Discussion

For convergence measure, the proposed S-MOFWA performs best on all the test functions, specifically discussed by the type of functions as follows.

12.4.4.1 Convex Functions

On convex functions ZDT1 and SCH, the proposed S-MOFWA significantly outperforms the other three with smallest average convergence measure and standard deviation, both ability and stability are proven.

12.4.4.2 Non-convex Functions

On non-convex function ZDT2, the proposed S-MOFWA and SPEA2 both get the best convergence measure, and S-MOFWA shows a little superiority in the standard deviation. On function KUR, S-MOFWA performs best and SPEA2 ranks the second.

12.4.4.3 Disconnected Functions

On disconnected function ZDT3, S-MOFWA beats the other three in average convergence measure but the standard deviation ranks the last.

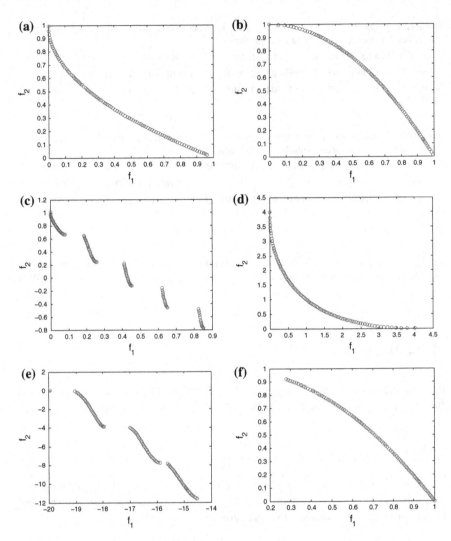

Fig. 12.4 Plots of the final archive of returned by S-MOFWA in the objective space. **a** ZDT1.
b ZDT2. **c** ZDT3. **d** SCH. **e** KUR. **f** ZDT6

12.4.4.4 Nonuniformly Spaced Functions

On nonuniformly spaced function ZDT6, S-MOFWA, NSGA-II, and SPEA2 get the
best average convergence measure, while PESA2 does not obtain a value of the same
order of magnitude.

For covered space measure, the S-MOFWA wins on four functions among all the
six functions. On ZDT1 and ZDT2, SPEA2 gets the best average covered space with
the smallest standard deviations, which means SPEA2 distributes the solutions better

along the non-dominated front. The density metric involved in the fitness assignment of SPEA2 may contribute a lot to this ability. On the other four functions, *S*-MOFWA finds a better spread than other algorithms.

12.5 Environmental/Economic Power Dispatch Problem

12.5.1 Problem Statement

The environmental/economic power dispatch (EED) problem is to minimize two competing objective functions, fuel cost and emission, while satisfying several equality and inequality constraints. Generally, the problem can be formulated as follows.

Minimization of Fuel Cost: The generators' cost curves are represented by quadratic functions with sine components. The superimposed sine components represent the rippling effects produced by the steam admission valve openings. The total fuel cost $F(P_G)$ can be expressed as

$$F(P_G) = \sum_{i=1}^{N} \left(a_i + b_i P_{G_i} + c_i P_{G_i}^2 + |d_i \sin \left[e_i \left(P_{G_i}^{\min} - P_{G_i} \right) \right]| \right), \quad (12.9)$$

where N is the number of generators, a_i, b_i, c_i, d_i, e_i are the cost coefficients of the ith generator, and P_{Gi} is the real power output of the ith generator. P_G is the vector of real power outputs of generators and defined as

$$P_G = [P_{G_1}, P_{G_2}, \ldots, P_{G_N}]^T. \quad (12.10)$$

Minimization of Emission: The atmospheric pollutants such as sulfur oxides SO_x and nitrogen oxides NO_x caused by fossil-fueled thermal units can be modeled separately. However, for comparison purposes, the total emission $E(P_G)$ of these pollutants can be expressed as

$$E(P_G) = \sum_{i=1}^{N} (10^{-2}(\alpha_i + \beta_i P_{G_i} + \gamma_i P_{G_i}^2) + \varsigma_i \exp(\lambda_i P_{G_i})), \quad (12.11)$$

where $\alpha_i, \beta_i, \gamma_i, \varsigma_i, \lambda_i$ are coefficients of the ith generator emission characteristics.

12.5.2 Evaluation Criterion

For a real-world problem, it is impossible to calculate the average distance to the true Pareto front, because we have no idea about the true Parato front. The only thing

Table 12.6   Evaluation results

Evaluation criteria			
Criteria (S-MOFWA) versus NSGA-II		Criteria (S-MOFWA) versus SPEA2	
Mean	Standard deviation	Mean	Standard deviation
0.6859	0.0221	0.7325	0.0447
Criteria1 (NSGA-II) versus S-MOFWA		Criteria1 (SPEA2) versus S-MOFWA	
Mean	Standard deviation	Mean	Standard deviation
0.3141	0.0112	0.2675	0.0104

we can do is to compare two algorithms' results (Pareto sets). Here we take each algorithm's portion in the Pareto set of the union of two algorithms' results as the evaluation criterion.

12.5.3 Results and Discussion

It can be seen that, in the union of two algorithms' results, most non-dominated solutions come from S-MOFWA while less part come from the contrastive algorithm (NSGA-II or SPEA2). This indicates the Pareto front found by S-MOFWA has a better spread than the other two algorithms (Table 12.6).

For the EED problem, S-MOFWA obtained a more economic and more environmental friendly solution than the solutions obtained by the other two algorithms.

12.6 Summary

This chapter presented the S-metric based multi-objective fireworks algorithm (S-MOFWA). As an indicator-based EMOA, the S-metric was introduced to guide the iteration process of FWA. The explosion numbers of sparks and explosion amplitudes are calculated according to each fireworks' S-metric. And the fireworks for next iteration are selected by the values of their S-metric. The selection for next generation and explosion to generate sparks were quite simpler compared to the conventional FWA. A novel archive strategy was employed to keep the best solution set, which reduced the labor in traditional archive strategy.

The experimental results on benchmark functions demonstrate that the proposed S-MOFWA outperforms the three well-known algorithms, i.e., NSGA-II, SPEA2 and PESA2, in terms of the measures of convergence and covered space. On most test functions, S-MOFWA can find a better spread and gets closer to the true Pareto front than the three competing algorithms.

References

1. K. Deb, *Multi-objective Optimization Using Evolutionary Algorithms*, vol. 16 (Wiley, Chichester, 2001)
2. E.J. Hughes, MSOPS-II: A general-purpose many-objective optimiser, in *IEEE Congress on Evolutionary Computation, CEC 2007* (IEEE, 2007), pp. 3944–3951
3. E.J. Hughes, Multiple single objective pareto sampling, in *The 2003 Congress on Evolutionary Computation, CEC'03*. vol. 4 (IEEE, 2003), pp. 2678–2684
4. K. Deb, A. Pratap, S. Agarwal, T.A.M.T. Meyarivan, A fast and elitist multiobjective genetic algorithm: NSGA-II. IEEE Trans. Evol. Comput. **6**(2), 182–197 (2002)
5. E. Zitzler, M. Laumanns, L. Thiele, SPEA2: Improving the strength pareto evolutionary algorithm (2001)
6. E. Zitzler, S. Knzli, Indicator-based selection in multiobjective search. *Parallel Problem Solving from Nature-PPSN VIII* (Springer, Berlin, 2004), pp. 832–842
7. E. Zitzler, L. Thiele, Multiobjective optimization using evolutionary algorithms: comparative case study, in *Parallel Problem Solving from Nature, PPSN V* (Springer, Berlin, 1998), pp. 292–301
8. N. Beume, C.M. Fonseca, M. López-Ibáñez, L. Paquete, J. Vahrenhold, On the complexity of computing the hypervolume indicator. IEEE Trans. Evol. Comput. **13**(5), 1075–1082 (2009)
9. J. Knowles, D. Corne, Properties of an adaptive archiving algorithm for storing nondominated vectors. IEEE Trans. Evol. Comput. **7**(2), 100–116 (2003)
10. L. Lang, Z. Shaoqiu, T. Ying, S-metric based multi-objective fireworks algorithm, in *The 2015 IEEE Congress on Evolutionary Computation*, vol. 2 (IEEE, 2015), pp. 1257–1264
11. M. Emmerich, N. Beume, B. Naujoks, An emo algorithm using the hypervolume measure as selection criterion. *Evolutionary Multi-criterion Optimization* (Springer, Berlin, 2005), pp. 62–76
12. S. Zheng, A. Janecek, Y. Tan, Enhanced fireworks algorithm, in *2013 IEEE Congress on Evolutionary Computation (CEC)* (IEEE, 2013), pp. 2069–2077
13. Y. Tan, Y. Zhu, Fireworks algorithm for optimization, in *Advances in Swarm Intelligence* (Springer, Berlin, 2010), pp. 355–364
14. H. Gazit, New upper bounds in Klee's measure problem. SIAM J. Comput. **20**(6), 1034–1045 (1991)
15. B. Naujoks, N. Beume, M. Emmerich, Multi-objective optimisation using s-metric selection: application to three-dimensional solution spaces, in *The 2005 IEEE Congress on Evolutionary Computation*, vol. 2 (2005) pp. 1282–1289
16. M. Fleischer, The measure of pareto optima applications to multi-objective metaheuristics, *Evolutionary Multi-criterion Optimization* (Springer, New York, 2003), pp. 519–533
17. J.D. Knowles, Local-search and hybrid evolutionary algorithms for Pareto optimization. Ph.D. thesis, University of Reading (2002)
18. E. Zitzler, Hypervolume metric calculation, *Computer Engineering and Networks Laboratory (TIK)* (Zürich, 2001)
19. L. While, P. Hingston, L. Barone, S. Huband, A faster algorithm for calculating hypervolume. IEEE Trans. Evol. Comput. **10**(1), 29–38 (2006)
20. N. Beume, B. Naujoks, M. Emmerich, SMS-EMOA: multiobjective selection based on dominated hypervolume. Eur. J. Oper. Res. **181**(3), 1653–1669 (2007)
21. I.C. Garcia, C.A. Coello Coello, A. Arias-Montano, MOPSHOhv: a new hypervolume-based multi-objective particle swarm optimizer, in *2014 IEEE Congress on Evolutionary Computation (CEC)* (IEEE, 2014), pp. 266–273
22. W. Ponweiser, T. Wagner, D. Biermann, M. Vincze, Multiobjective optimization on a limited budget of evaluations using model-assisted *S*-metric selection. *Parallel Problem Solving from Nature–PPSN X* (Springer, 2008), pp. 784–794
23. N. Beume, G. Rudolph, Faster S-metric calculation by considering dominated hypervolume as Klee S measure problem (2006)

24. S. Kukkonen, K. Deb, Improved pruning of non-dominated solutions based on crowding distance for bi-objective optimization problems, in *Proceedings of the World Congress on Computational Intelligence (WCCI-2006)* (IEEE Press, Vancouver, 2006), pp. 1179–1186

25. D.A. Van Veldhuizen, Multiobjective evolutionary algorithms: classifications, analyses, and new innovations. Technical report, DTIC Document (1999)

26. E. Zitzler, K. Deb, L. Thiele, Comparison of multiobjective evolutionary algorithms: empirical results. Evol. Comput. **8**(2), 173–195 (2000)

27. K. Deb, M. Mohan, S. Mishra, A fast multi-objective evolutionary algorithm for finding well-spread pareto-optimal solutions. KanGAL report 2003002 (2003)

Chapter 13
Discrete Firework Algorithm
for Combinatorial Optimization Problem

Considering the excellent performance of FWA for real parameter optimization problems, a novel FWA variant was proposed for tackling discrete optimization problems, especially for Travelling Salesman Problem (TSP). We first give a brief introduction to TSP, followed by the detailed description of the discrete fireworks algorithm (DFWA). The DFWA remains the basic framework of FWA and introduces some major changes in explosion operator, selection strategy and mutation operator, respectively. Specifically, exploration operator is redefined and mutation operator and selection strategy are modified according to the properties of discrete problems. In explosion operator, every firework is able to accept a worse solution and generate a spark with lower fitness, which refer to the mechanism of simulated annealing. However, the controlling parameter θ changes with the feedback of optimization process rather than time. In addition, this version of DFWA appropriately changed its behavior of the local search method to suit in the framework of FWA. A lot of experimental results demonstrated that DFWA is very efficient and effective for TSP, which sheds new light on more and more discrete combinatorial optimization problems.

13.1 Travelling Salesman Problem

13.1.1 Problem Statement

In essence, TSP is a discrete optimization problem whose aim is to find the minimal tour length. After starting from the original city, a salesman needs to go through all cities and then goes back to the original city. The tour length is defined as the total length between every adjacent cities in the visiting sequence.

TSP can be described formally as follows:

Definition 13.1 Given N cities$\{c_1, c_2, c_3, \ldots, c_N\}$, the distance between every two cities $c_i c_j$ denoted by $d(c_i, c_j)$, the Travelling Salesman Problem is to find a

© Springer-Verlag Berlin Heidelberg 2015
Y. Tan, *Fireworks Algorithm*, DOI 10.1007/978-3-662-46353-6_13

permutation $x = (x_1, x_2, x_3, \ldots, x_n)$, $x_i \in \{1, 2, 3, \ldots, N\}$ to minimize the tour length of x:

$$L(x) = \sum_{i=1}^{N-1} d(c_{x_i}, c_{x_{i+1}}) + d(c_{x_n}, c_{x_1}) \tag{13.1}$$

s.t.

$$\forall i \in \{1, 2, \ldots, N\}, \exists j \in \{1, 2, \ldots, N\}, x_j = i. \tag{13.2}$$

In this chapter, only the symmetric TSP is considered. For symmetric TSP, every two cities is connected and their distance satisfies

$$d(c_i, c_j) = d(c_j, c_i), \forall 1 \leq i, j \leq N. \tag{13.3}$$

TSP emerges in diverse domain from X-ray crystallography to VLSI chip fabrication [1]. An important real-world application of TSP is from the logistics industry, where minimizing the tour length can significantly reduce the delivery cost and gain more economic profit. Hence, the research on TSP optimization is worth wide attention.

Although the underlying principle of TSP is really simple, the solution space grows exponentially with the number of cities increasing. Taking 42 cities for example, if we enumerate all possible routes to calculate the minimal length, the number of feasible routes is $41! = 3.3^{49}$, which is intractable. It has been proved that TSP is a NP-complete problem [2]. There is no effective algorithm for solving TSP exactly. In general, there are two alternatives to address TSP: (1) turning to heuristics that merely find suboptimal tours, and (2) developing optimization algorithms that work well in most "real-world" cases, rather than worst-case instances. Driven by the insatiable real-world need, the scale of TSP to be solved increases rapidly, from the relatively humble 318 cities [3] all the way up to 2392 [4] and 7393 [5]. In fact, the last case took one of the most powerful cluster for more than 3 years.

In recent decades, a large number of swarm intelligent algorithms have been applied to TSP, such as genetic algorithm (GA), particle swarm optimization (PSO) as well as ant colony optimization (ACO) method, etc. Relying heavily on heuristic information, swarm intelligence can obtain good (though often suboptimal) solutions within acceptable time. Mhlenbein et al. [6] applied GA to TSP and obtained reasonable solutions on two instances with 442 cities and 531 cities, respectively. However, the solutions are far from the optimum. Braun [7] proposed a GA variant (dubbed improved GA) specifically for TSP. For TSP of small scale (less than 442 cities), the proposed method can easily obtain the best result very quickly. For TSP with 666 cities, the error rate computed by the improved genetic algorithm is only 0.04 %, which was a great breakthrough at that time.

In 2004, Clerc [8] proposed a discrete version of particle swarm optimization (DPSO). DPSO was applied to TSP in [9] where a local search mechanism was introduced. The performance of DPSO is comparable to the improved genetic algorithm and better in a few cases.

Ant algorithm is one of best algorithms for addressing TSP. The typical ant algorithm—ant colony system was first introduced by Dorigo and Gambardella [10] in 1997. The experimental results shown that ant colony system is originally appropriate for TSP. The original ACS obtains error rate of 0.68 % in 198-city instance, while the improved ACS with local search mechanism can obtain 0.11 % error in 318-cite instance. Notice that this result is still poorer than best versions of GA [11, 12]. A modified version of ACS—Max Min Ant System (MMAS) [13] significantly improves the performance of ACS. It can easily solve problems with the number of cities less than 100 and obtain 0.04 % error rate in 198-city instance and 1.93 % error rate in rat783, respectively. If local search mechanism is introduced, the error rate with more than 1000 cities (i.e., fl1157 instance) by Max Min Ant System is about 0.2 %, which is state-of-the-art result for TSP.

In summary, swarm intelligence algorithms are effective methods for solving TSP. With the number of cities less than 400, several improved algorithms exploiting local search mechanism can find the optimum with high probability.

13.2 Discrete Firework Algorithm

Conventional FWA is designed for real parameter optimization, where explosion operator and mutation operator are utilized for searching in large space and the selection strategy determines the fireworks in next generation. Because of the complexity in discrete function space, the completely random mutation strategy utilized in the FWA is not likely to have well performance in discrete space. Hence, we particularly introduced two novel operators which are quite different those of conventional FWA. In addition, we modified the definition of explosion amplitude as well to suit the discrete application. The framework of discrete fireworks algorithm (DFWA) remains the same with conventional FWA. In the following, we will discuss discrete fireworks algorithm for TSP in detail.

In TSP, the solution of DFWA is defined as city sequence $x = (x_1, x_2, x_3, \ldots, x_n)$ where x_i is the identifier of ith city. In travelling salesman problem, each city needs one and only one visit by the salesman. The target function is $L(x)$ Eq. (13.1).

Obviously, the goal of DFWA is:

$$\min_x L(x) \qquad (13.4)$$

s.t. Eq. (13.2).

13.2.1 Framework of DFWA

The framework of discrete fireworks algorithm is similar to that of the conventional fireworks algorithm. We keep the explosion operator, mutation operator, and

Fig. 13.1 Framework of discrete FWA

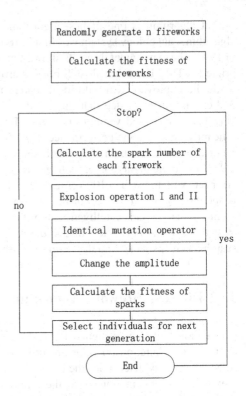

selection strategy but remove the mapping rule in the DFWA. For explosion operator, two explosion operations, i.e. explosion operation I and explosion operation II, are introduced here. For mutation operator, we put forward a new uniform mutation operator. Meanwhile, selection strategy of discrete firework algorithm is specially changed for travelling salesman problem. Because the offspring produced by DFWA would not exceed the boundary of definition domain, the mapping rule is unnecessary anymore. The framework of discrete fireworks algorithm is listed in Fig. 13.1. Note that the whole framework is consistent with conventional fireworks algorithm.

As shown in this figure, the discrete fireworks algorithm consists of general components in conventional FWA. At first, a feasible initial population is randomly generated, then sparks emerge by explosion operator and mutation operator. In the end of this iteration, some individuals among fireworks and sparks will be selected into next generation according to selection strategy Algorithm 13.1. Now we are going to introduce each component in next several sections.

Algorithm 13.1 Discrete Fireworks Algorithm

1: randomly generate N fireworks $x^{(i)}$
2: **while** stop criterion is not satisfied **do**
3:    generate sparks for each firework by explosion operator
4:    generate sparks for each firework by identical mutation operator
5:    change the amplitude
6:    select N fireworks into next generation
7: **end while**

13.2.2 Explosion Operator

The explosion operator is an important component in FWA, which is the major mechanism that drives local search and global search. Thanks to the smoothness in continuous function space, a better solution will probably appear using random local mutation operation. But the situation differs much for optimizing discrete functions, where the simple random local search method will become inefficient in finding a neighbor suboptimum. Considering this, the discrete fireworks algorithm defines two special explosion operations. As aforementioned, they are called explosion operation I and explosion operation II, respectively.

13.2.2.1 Explosion Operation I

Similar to discrete particle swarm optimization (DPSO) where city swap is defined as basic operation, a kind of basic operation is also acquired in discrete fireworks algorithm. In concept, the interaction information between individuals in DFWA is to guide the global search process. Actually, there is no direct interaction between fireworks in FWA. As described before, the random local search mechanism is not suitable in discrete fireworks algorithm. Hence taking the properties of travelling salesman problem into consideration, we define the edge swap operation as the basic operation, which is equivalent to 2-opt local search method.

The 2-opt operation is shown in Fig. 13.2, where the edges—(a, b) and (c, d) are replaced by (a, c) and (b, d). This operation ensures that the result is still a feasible route. If a change reduces the route length, this operation is going to be accepted. Note that if they are substituted by (a, d) and (b, c), the original feasible route will be divided into two independent parts. If the route is described as x, the 2-opt operation can be completed by one sub-sequence reversion. For convenience, we define (a, b) as the edge between city c_a and city c_b and $x_{i,j}$ as the sub-sequence between ith city to jth city in this sequence.

Though explosion operation I is based on 2-opt method, there are two obvious differences between explosion operation I and the 2-opt local search method.

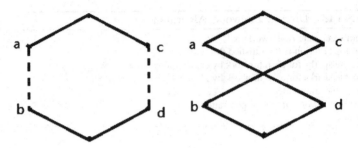

Fig. 13.2 The 2-opt local search operation (*left* the original route, *right* mutated route)

1. Explosion operation I is not as greedy as 2-opt method. During the searching process, explosion operation has the ability to accept a worse solution. Taking Fig. 13.2 for example, where $L_o = d(c_a, c_b) + d(c_c, c_d)$ denotes the original total length of (a, b) and (c, d), $L_m = d(c_a, c_c) + d(c_b, c_d)$ denotes the edge length after edge swap, we accept a worse solution with probability p_a:

$$p_a = \exp(-L_m/L_o * \theta),\ L_m > L_o, \tag{13.5}$$

where θ is a parameter controlling the probability of choosing a worse solution. According to the definition, the more similar L_m and L_o, the larger p_a is. Also the probability goes larger when θ decreases. The benefit of these two properties is to give an opportunity of jumping out of local optimum while reducing the useless mutations. Intuitively, if the selected cities are distant in the instance and L_m is much larger than L_o then this operation is likely to be invalid. Parameter θ can control the range of p_a in order to maintain an proper probability.
2. This operation can be applied to multiple binary operations. It is capable of smoothing the mutated route by additional 2-opt methods by which the probability of jumping out 2-opt local optimum increases and the probability of abandoning a potential solution decreases in the meantime. 2-opt local optimum is a kind of state where 2-opt method is unable to obtain a better solution.

The pseudocode is depicted by Algorithm 13.2, where 2-opt(c, k) indicates the 2-opt operation on edge (x_c, x_{c+1}) and (x_k, x_{k+1}). With all city iterated whose sequence index is 'k', 2-opt(c, k) for 'c' is called 2-opt optimization for 'c'. The 'rand' function randomly generates a real number in range [0, 1] and "randi(n)" randomly generates a integer between 1 and n.

Algorithm 13.2 Explosion operation I

1: Input: x : fireworks
2: Output: $spark$: spark generated by fireworks
3: $spark = x$
4: $z = randi(n)$ // n is the number of cities
5: $rp = randperm(n)$ // randperm randomly generates a permutation
6: **for** $i = 1 : n$, where $rp(i) \neq z$ **do**
7:    $a = z, b = z + 1, c = rp(i), d = rp(i) + 1$ is the indices of the sequence
8:    sort(a,b,c,d);
9:    $L_o = d(c_{x_a}, c_{x_b}) + d(c_{x_c}, c_{x_d}) L_m = d(c_{x_a}, c_{x_c}) + d(c_{x_b}, c_{x_d})$
10:    **if** $L_o > L_m$ **then**
11:      reverse $x_{b,c}$, return.
12:    **else**
13:      **if** $rand < p_a$ **then**
14:        reverse $x_{b,c}$
15:        %2-opt optimization for 'a'
16:        **for** k = 1:n **do**
17:          2-opt(a,k);
18:        **end for**
19:        %2-opt optimization for 'c'
20:        **for** k = 1:n **do**
21:          2-opt(c,k);
22:        **end for**
23:        return.
24:      **end if**
25:    **end if**
26: **end for**

13.2.2.2 Explosion Operation II

In the optimization procedure, the capacity of explosion operation I degrades as when the algorithm has already found a "good" local optimum. In that case, a more powerful operation is needed. Similar to explosion operation I, explosion operation II is based on 3-opt local search method and it changes 3 edges at one time.

The two possible 3-opt are shown in Fig. 13.3. In this figure, three edges, i.e., (a, b), (c, d) and (e, f), are removed and three other edges are created to form a new

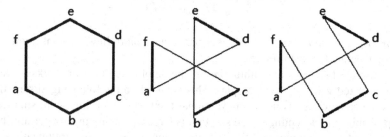

Fig. 13.3 Two possible 3-opt moves (*left* original tour, *right* mutated tours)

Algorithm 13.3 Explosion operation II

1: Input: fireworks x
2: Output: spark generated by fireworks $spark$
3: $spark = x$
4: $z_1 = randi(n)$, $z_2 = randi(n)$ // n is the number of cities
5: $rp = randperm(n)$, // randperm randomly generates a permutation with length 'n'
6: **for** $z_3 = 1 : n$, where $z_3 \neq z_1 and z_3 \neq z_2$ **do**
7:    $sort(z_1, z_2, z_3)$;
8:    $a = rp(z_1)$, $b = rp(z_1) + 1$, $c = rp(z_2)$, $d = rp(z_2) + 1$, $e = rp(z_3)$, $f = rp(z_3) + 1$
9:    **for** for 4 possible swap **do**
10:      calculate L_o and L_m
11:      **if** $L_o > L_m$ **then**
12:        accept and return.
13:      **else**
14:        **if** $rand < p_a$ **then**
15:          accept.
16:          2-opt optimization for 'a', 'b' and 'c'
17:          return.
18:        **end if**
19:      **end if**
20:    **end for**
21: **end for**

---

feasible route. If this operation reduces the total tour length, it is accepted. A route is described as city sequence $x = (x_1, x_2, x_3, \ldots, x_n)$ and 3-opt operation can be done within several subsequence reverse. Take the middle graph in Fig. 13.3, for example, reversing $x_{b,c}$ and $x_{d,e}$ is a possible 3-opt sequence operations. 3-opt operations has four possible swap in total, all enumeration would not be listed here for simplicity.

For explosion operation II, the same improvement in explosion operation I is applied. For more details, readers can refer to pseudocode Algorithm 13.3.

13.2.2.3 Sparks Number and Amplitude

The number of sparks is calculated as same as in former chapters. The spark number is:

$$s_i = M_e * \frac{1/(L_{max} - L(x^{(i)}) + \varepsilon)^2}{\sum_{j=1}^{n} 1/(L_{max} - L(x^{(i)}) + \varepsilon)^2} \tag{13.6}$$

where, $L_{max} = max\{L(x^{(i)})\}$, M_e is a parameter controlling the total number of sparks generated by the firework.

In discrete fireworks algorithm, we set the amplitude of all fireworks to be an identical fixed value (θ, for example). This is based on the following observation: θ has the same effect as in the conventional fireworks algorithm. With smaller θ, the probability of accepting worse solution is larger, resulting in stronger ability to escape the local optimum and find a better solution. In discrete optimization, the population fitness is believed to be close. The difference between individuals is tiny

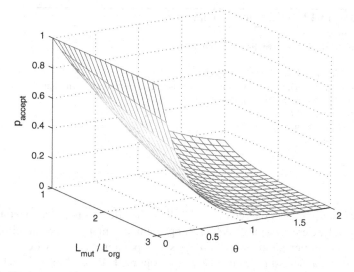

Fig. 13.4 The impact of θ and L_m/L_o for p_a

enough so that they should have identical probability to accept worse mutations. In addition, θ is a exponent factor for p_a. Figure 13.4 also illustrates that p_a is very sensitive to control parameter θ. Hence the amplitudes of fireworks are set to be same value.

13.2.3 Identical Mutation Operator

Explosion operator I and explosion operator II focus on the edge swaps, which is difficult to improve some special routes, such as the left side in Fig. 13.5. In this figure, the route is unable to be improved by 2-opt and 3-opt methods, but can be

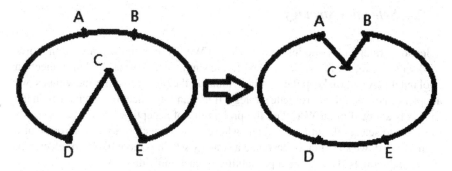

Fig. 13.5 2h-opt local search operation

Algorithm 13.4 Identical mutation operator

1: Input: fireworks x
2: Output: spark generated by fireworks $spark$
3: $spark = x$
4: $a = randi(n)$// n is the number of cities
5: $rp = randperm(n)$, // randperm randomly generates a permutation with length 'n'
6: **for** $k = 1 : n$ where $a \neq rp(k)$ **do**
7:     calculate L_o and L_m
8:     **if** $L_o > L_m$ **then**
9:         accept and return.
10:    **end if**
11: **end for**

---

done by inserting 'c' into edge (a, b). In FWA, the effect of mutation operator is to search locally. Here, we simply define 2h-opt method as mutation operator in discrete fireworks algorithm, aiming at optimizing single point in the route sequence. 2h-opt operation is one heuristic strategy between 2-opt method and 3-opt method, shown in Fig. 13.5 where edge (d, c), (a, b), and (c, e) are replaced by (a, c), (c, b), and (d, e) to reduce the route length.

In sequence format, two reversion operations are required to implement 2h-opt. Concretely, given sequence $(\ldots x_a, x_{a+1} \ldots x_{b-1}, x_b, x_{b+1} \ldots)$, where a, b denotes indices, in order to insert x_b into edge (x_a, x_{a+1}), we need to reverse sub-sequence $x_{a+1,c}$ at first and reverse $x_{a+2,c}$ then. This operation is called 2h-opt(a, b). If a is equal to n, then $x_{a+1} = x_1$.

In the consideration that mutation operation is mainly used for local search, the p_a is not defined for mutation operator which is equivalent to setting p_a to be 0. It means that mutation operation is unable to accept a worse solution. Thus we can randomly select one city 'a', go through all indices 'k' in the sequence to operate 2h-opt(a,k). Because the opportunity of being selected is same among all cities, we call this mutation operation as identical mutation operator. Specifically, the pseudocode described this procedure is given in Algorithm 13.4.

13.2.4 Selection Strategy

Selection strategy plays an important role in FWA. Similar to the conventional fireworks algorithm, the fireworks and sparks are simultaneously selected into selection pool in DFWA and we keep the best individual all the time. For other individuals, we adopt a strategy similar to roulette strategy. If the random selection mechanism in the EFWA is adopted in DFWA, then the probability of accepting "bad" sparks seems too large to select efficient individuals whose fitness are close to current optimum. In this regard, we specifically designed a strategy which is more likely to select high fitness individuals. The selected probability of each individuals is:

$$p_{sel}(\boldsymbol{x}^{(i)}) = \frac{1/(L(\boldsymbol{x}^{(i)}) - L_{\min} + \varepsilon)^2}{\sum_{j=1}^{n} 1/(L(x^{(j)}) - L_{\min} + \varepsilon)^2} \qquad (13.7)$$

where $L_{\min} = min\{L(\boldsymbol{x}^{(i)})\}$ is the smallest fitness, ε denotes the smooth parameter in case of the appearance of zero divisor. According to Eq. (13.7), the probability of individuals with lower fitness is larger. Comparing to the first power format, the second power format can increase the ratio of excellent sparks and hence increase the probability of being selected into next generation. We executed two experiments with random selection strategy and the one described above. The results demonstrates that the performance of random selection strategy is not very satisfactory which leads to 1 % more average error rate. Of course, there might be other better strategies rather than above discussions. We merely introduce one workable solution for selection strategy here.

13.2.5 Adaptive Strategy

In the dynamic fireworks algorithm (DynFWA), the best fireworks, i.e., core fireworks stated in original paper [14], is self-adaptive by the reward of previous generations. If a better solution is found, then the amplitude is increased by an amplification factor to enhance its global search ability. Otherwise, the amplitude is decreased by reduction factor to enhance local search ability. In the DFWA, the same idea can be adopted. Recall that the amplitude in DFWA is defined as the control parameter θ in p_a. If a better solution is obtained, we had better decrease p_a to accelerate local search, otherwise it means that the population is probably trapped in suboptimum, we had better to increase p_a to supplement local search by giving more chances to jump out of local optimum.

In the process of discrete optimization, there are much differences between local optimums. Hence DFWA may need more than one generation to obtain a better result which means it is necessary to observe the feedback for several consecutive generations and then make a reasonable adaption. In this version of DFWA, we count for 10 generations to see if a better solution is obtained. If not, increase amplitude. Once the current fireworks are better than the previous, decrease amplitude immediately. The specific strategy is similar to that in DynFWA. θ is properly confined within the interval [1,2]. The pseudocode is given in Algorithm 13.5.

13.3 Experimental Results and Analysis

At the first part of this section, we will discuss the issues of parameter settings and performance (runtime) analysis. The main topic followed in this section is to display the experimental results of the DFWA and the analysis. First, the experiments will be

---

Algorithm 13.5 Parameter θ self adaptive strategy

1: fireworks x_c
2: spark $\hat{x}_{best}$
3: amplification factor $Coef_a$
4: reduction factor $Coef_r$
5: $N_{un} = 0$ counts the unchanged times.
6: **if** $L(\hat{x}_{best}) < L(x_c)$ **then**
7:    $N_{un} = 0$;
8:    $\theta = \theta * Coef_a$;
9: **else**
10:    $N_{un} = N_{un} + 1$;
11:    **if** $N_{un} \geq 10$ **then**
12:       $\theta = \theta * Coef_r$;
13:       $N_{un} = 0$;
14:    **end if**
15: **end if**

---

executed in small test cases to verify if it has the capacity to solve simple travelling salesman problems. Then a middle-sized test case and some large test cases are employed. The test cases come from standard TSPLIB benchmark. All cases are symmetric TSPs. The platform for this experiment is windows 7-64 bit operation system with i3-370m CPU and matlab2013R software. The number of fireworks is 50, the maximal spark number M_e is 50 and the maximal iterations is identically set to 40,000. In the experiments, we had three different DFWA. DFWA-na is the discrete fireworks algorithm with p_a set to 0, i.e., the explosion operator is simply local search method. This version is not a standard discrete fireworks algorithm, solely for contrast. DFWA-nri denotes a version of standard DFWA without reboot mechanism, while DFWA-ri has the reboot mechanism.

13.3.1 Parameters Setting and Performance Analysis

Spark number has a huge impact on performance for fireworks algorithm. According to [15], pretty performance can be gained when fireworks number is set to 5 and M_e (Eq. 13.6) is set to 50. In the following, we will analyze the settings in discrete fireworks algorithm by two batches of experiments. In the first one, we tried to verify the acceptable fireworks number by fixing $M_e(50)$ and test case(d198). Figure 13.6 illustrates the consequence. Horizontal ordinate denotes fireworks number while vertical ordinate denotes the average error rate. It shows that a proper setting for fireworks number is 5.

In the second experiment, we analyze the effect of M_e. We fixed the fireworks number to be 5 in d198 test case. The result is in Fig. 13.7 where the horizontal ordinate is M_e and vertical ordinate means the error between global optimum. As we can see from this graph, the best performance is acquired by 50 maximal spark number. Note that the error rate difference between 30 and 50 is actually quite small,

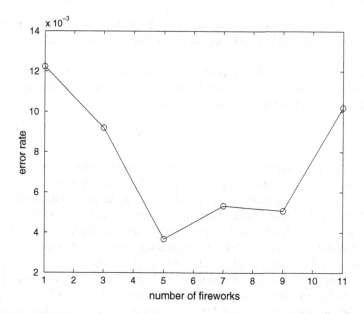

Fig. 13.6 The impact of fireworks number on performance

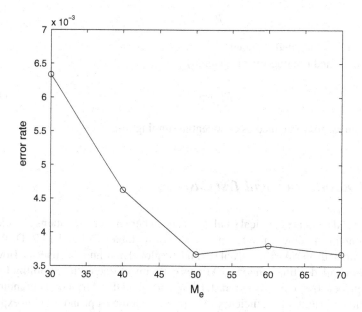

Fig. 13.7 The impact of maximal spark number on performance

only 0.3 %. Hence we can conclude that the influence of this parameter is not that vital as in conventional fireworks algorithm. Without loss of generalization, we set M_e to be 50.

Discrete fireworks algorithm conforms Markov model similar to conventional fireworks algorithm, because for the population in tth generation, the state of next generation is not related to the previous states($t' < t$):

$$P(s_{t+1}|s_t, s_{t_1} \ldots s_1) = P(s_{t+1}|s_t) \tag{13.8}$$

where s_t is the state of the t-th generation, $P(s_t)$ is the probability of being s_t in the tth generation. Due to the elite strategy, the discrete fireworks algorithm has absorbing property with which an algorithm would not discard the best solution once found. We call the Markov model with absorbing property as absorbing Markov process. According to fitness-level model, if there exists a certain state from which the probability of finding a better solution is exponentially small, this algorithm is supposed to spend more than polynomial time to find the global optimum. Assume there is a route that is the solution second to the global optimum, and it has a subsequence completely different from global optimum. The length of this subsequence is greater than 6 (the maximal edges changed by explosion operator II), then the opportunity of finding global optimum is less than:

$$p < C * p_a^{l_s/6}, \tag{13.9}$$

where C is an unrelated constant.

The expected running time is given by

$$E(t) > C * p_a^{-l_s/6}, \tag{13.10}$$

which can be approximated as exponential running time.

13.3.2 Results on Small Test Cases

oliver30 and att48 are classical small test cases where oliver30 contains 30 cities and att48 contains 48 cities. The results are shown in Tables 13.1 and 13.2. DFWA-na, DFWA-nri and DFWA-ri are all able to find the global optimum of oliver30. However, DFWA-na failed 7 times among 10 times experiment due to the disability to jump out 3-opt local optimum. By contrast, DFWA-nri and DFWA-ri found optimum in all runs which illustrates the efficiency of improvement for 2-opt and 3-opt in explosion operator. In att48, DFWA-na found suboptimums in all runs and DFWA-nri failed 5/10 times, but DFWA-ri gives the best solution every time. It indicates that reboot

Table 13.1 The results on oliver30, the data besides DFWA is from [16], the minimal route length is 420

oliver30	DFWA-na	DFWA-nri	DFWA-ri	Basic SA	Basic GA	Basic ACA
Average	421.4	**420**	**420**	437.6632	482.4671	447.3256
Best solution	420	**420**	**420**	424.9918	424.9918	440.8645
Worst solution	429	**420**	**420**	480.1452	504.5256	502.3694

Table 13.2 The results on att48, the data besides DFWA is from [16], the minimal route length is 33522

att48	DFWA-na	DFWA-nri	DFWA-ri	Basic SA	Basic GA	Basic ACA
Average	33878.5	**33608**	**33522**	34980	37548	41864
Best solution	33522	**33522**	**33522**	35745	36759	43561
Worst solution	34140	33700	**33522**	41864	34559	42256

mechanism may be helpful in optimizing small instances. Intuitively, owning to the lack of interactions between populations, it is expensive to repeatedly generate new population trying to leave locality. Reboot mechanism remits this disadvantage to some extent, avoiding the waste of computation resource.

13.3.3 Results on Medium Test Case

In the following, the experiment is based on d198, an instance with 198 cities. The other settings remain the same as above. Table 13.3 demonstrates the result on d198 test case, where the error rate of DFWA-ri is about 0.4 % and DFWA-nri is about 0.3 % slightly outperforms the Max Min Ant System (MMAS) 1.2 %, Ant Colony System(ACS) 1.7 %. It is necessary to point out that improved ant algorithm combined with local search strategy can gain the global optimum with high probability. Figure 13.8 shows the convergence curve of DFWA-nri and DFWA-na which verifies that the improvement in explosion operator is efficient in medium scale instances as well.

Table 13.3 Result on d198, the data besides DFWA is from [13], the minimal route length is 15780

Average value	DFWA-na	DFWA-nri	DFWA-ri	MMAS	ACS
d198	16334	**15838.6**	**15855**	15972.5	16054

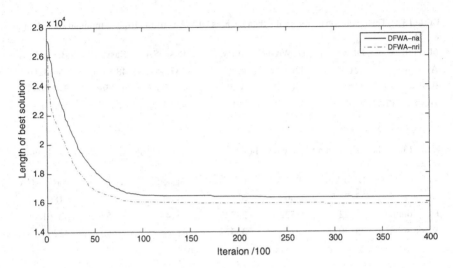

Fig. 13.8 The convergence curve of DFWA-na and DFWA-ri on d198

Table 13.4 Result in larger test cases, the data besides DFWA is from [10]

	DFWA-na	DFWA-nri	DFWA-ri	ACS	Optimum
pcb442	54351.8	52255.6	52341.3	51690	50778
att532	92317.5	89230	89237	88177.4	86729
rat783	9586.9	9198.1	9180.8	9066	8806

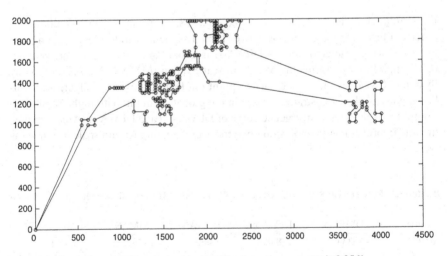

Fig. 13.9 A solution of d198 optimized by DFWA whose the error rate is 0.05 %

13.3.4 Results on Large Test Case

The result of larger test instances can refer to Table 13.4. The test set includes att532, pcb442, rat783. This table indicates that this version of discrete fireworks algorithm cannot compete with standard ant algorithm with large instances, there is much more room for progress (Fig. 13.9).

13.4 Comparison with Classical Algorithms

The experimental results displayed above demonstrate that the discrete fireworks algorithm can efficiently solve small instances with city number less than 100, outperforming the basic simulated annealing, basic genetic algorithm and basic ant algorithm in [16]. When solving medium cases, DFWA can obtain better solutions comparing to basic ant algorithms [10] owning to its excellent local search ability. However, the performance of discrete fireworks algorithm is limited in large-scale instances due to its lack of individual interactions. Compared with those algorithms combined with local search mechanisms such as genetic algorithm [11], Max Min Ant System [13], discrete particle swarm optimization method [9], this discrete version of FWA is still behind those classical algorithms.

In conclusion, this version of discrete fireworks algorithm has the following features:

1. DFWA can solve TSP in small instances with high probability and gain well results in optimizing medium scale cases.
2. DFWA has a strong local search ability. It combines three kinds of local search methods and some improvements are employed to enhance search capacity.
3. DFWA is based on the fundament of basic conventional fireworks algorithm where the interactions between individuals are limited. This weakness reveals when to solve large-scale problems.

In the future, we will mainly focus on the study of the interaction mechanism among individuals in the swarm for a sharp improvement in performance of DFWA.

13.5 Summary

Discrete FWA is just in its fiddle era, this chapter merely presents our first-trying discrete version of FWA and its application in TSP briefly. The discrete fireworks algorithm remains the basic framework of FWA and introduces some major changes in explosion operator, selection strategy and mutation operator respectively. In explosion operator, every firework is able to accept a worse solution and generate a spark with lower fitness, which is analogue to the mechanism of simulated annealing. However the controlling parameter θ changes with the feedback of optimization process rather than time. In addition, this version of discrete fireworks algorithm properly changes its behavior of the local search method to suit in the framework of FWA. As the experimental results indicate, discrete FWA is very effective for TSP, it can be applied to more discrete optimization problems potentially.

References

1. R.G. Bland, D.F. Shallcross, Large travelling salesman problems arising from experiments in X-ray crystallography: a preliminary report on computation. Oper. Res. Lett. **8**(3), 125–128 (1989)
2. R.G. Michael, D.S. Johnson, *Computers and Intractability: A Guide to the Theory of NP-completeness* (WH Freeman and Company, San Francisco, 1979)
3. H. Crowder, M.W. Padberg, Solving large-scale symmetric travelling salesman problems to optimality. Manag. Sci. **26**(5), 495–509 (1980)
4. M. Padberg, G. Rinaldi, Optimization of a 532-city symmetric traveling salesman problem by branch and cut. Oper. Res. Lett. **6**(1), 1–7 (1987)
5. D.S. Johnson, L.A. McGeoch, The traveling salesman problem: a case study in local optimization. Local Search Comb. Optim. **1**, 215–310 (1997)
6. H. Mhlenbein, M. Gorges-Schleuter, O. Krmer, Evolution algorithms in combinatorial optimization. Parallel Comput. **7**(1), 65–85 (1988)
7. H. Braun, On solving travelling salesman problems by genetic algorithms. *Parallel Problem Solving from Nature* (Springer, Berlin, 1991), pp. 129–133
8. M. Clerc, Discrete particle swarm optimization, illustrated by the traveling salesman problem. *New Optimization Techniques in Engineering* (Springer, Berlin, 2004), pp. 219–239
9. M.F. Tasgetiren, P.N. Suganthan, Q.-Q. Pan, A discrete particle swarm optimization algorithm for the generalized traveling salesman problem, in *Proceedings of the 9th Annual Conference on Genetic and Evolutionary Computation* (ACM, 2007), pp. 158–167
10. M. Dorigo, L.M. Gambardella, Ant colony system: a cooperative learning approach to the traveling salesman problem. IEEE Trans. Evol. Comput. **1**(1), 53–66 (1997)
11. B. Freisleben, P. Merz, A genetic local search algorithm for solving symmetric and asymmetric traveling salesman problems, in *Proceedings of IEEE International Conference on Evolutionary Computation* (IEEE, 1996), pp. 616–621
12. B. Freisleben, P. Merz, New genetic local search operators for the traveling salesman problem. *Parallel Problem Solving from Nature—PPSN IV* (Springer, Berlin, 1996), pp. 890–899
13. H.H. Hoos, T. Stutzle, Max min ant system. *Future Generation Computer Systems* (Elsevier, Paris, 2000)
14. S. Zheng, A. Janecek, J. Li, Y. Tan, Dynamic Search in Fireworks Algorithm, in *IEEE Congress on Evolutionary Computation (CEC)* (2014), pp. 3222–3229
15. Y. Tan, Y. Zhu, Fireworks algorithm for optimization, in *Advances in Swarm Intelligence* (Springer, Berlin, 2010), pp. 355–364
16. L. Fang, P. Chen, S. Liu, Particle swarm optimization with simulated annealing for TSP, in *Proceedings of the 6th WSEAS International Conference on Artificial Intelligence, Knowledge Engineering and Data Bases*, Corfu Island, Greece, February 2007, pp. 16–19

Chapter 14
Implementation of Fireworks Algorithm Based on GPU

In recent years, the graphics processing unit (GPU) has gained much popularity in general purpose computing, thanks to its low price and easy access. In this chapter, a very efficient FWA variant based on GPUs, so-called GPU–FWA for short, is introduced. GPU–FWA modifies the original FWA to suit the particular architecture of the GPU. It does not need special complicated data structure, thus making it easy to implement; meanwhile, it can make full use of the great computing power of GPUs. The key components of GPU–FWA are FWA search, attract-repulse mutation, and implementation which are elaborated in this chapter.

To make the chapter self-contained, a brief introduction of general purpose computing on GPUs (GPGPU) is presented first. Then we describe GPU–FWA in detail, followed by the empirical comparison of GPU–FWA with conventional FWA and popular PSO.

14.1 General Purpose GPU Computing

In the single-core CPU period, programmers have relied in enormous extension on the advances in hardware to accelerate their applications; as a new generation of processors is introduced, the same software just runs faster. However, due to energy-consumption and heat-dissipation issues, the increase of the clock frequency and the level of productive activities that can be performed in each clock period within a single CPU is significantly limited. Virtually, CPU vendors switched to models where multiple processor cores are used in each chip to increase the processing power. We have entered a multicore period and have to parallelize the legacy serial program to fully exploit the horsepower of the new generation of CPUs.

Over the last few years, driven by the insatiable demand for realtime, high-definition 3D graphics, the GPU has evolved into a massively parallel, many-core processor. The performance and capabilities of GPUs have been increasingly

© Springer-Verlag Berlin Heidelberg 2015

Y. Tan, *Fireworks Algorithm*, DOI 10.1007/978-3-662-46353-6_14

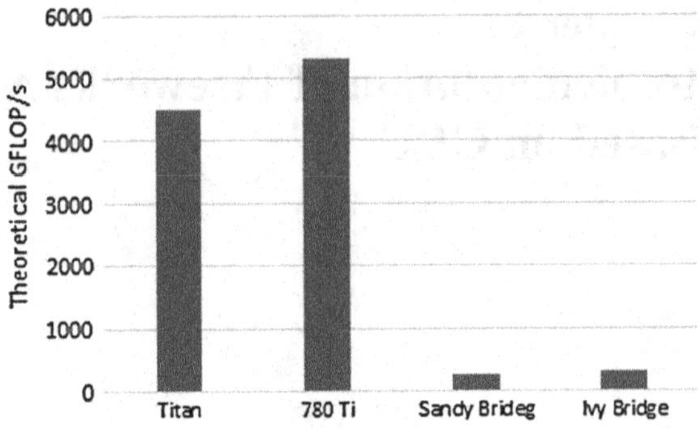

Fig. 14.1 Floating-point operations per second for the CPU and GPU (data from [3])

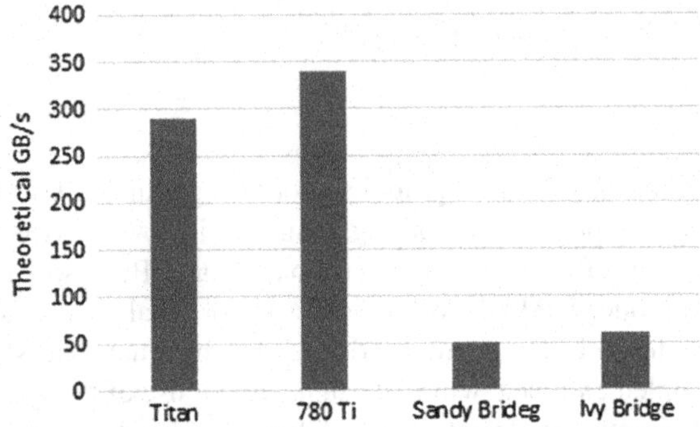

Fig. 14.2 Memory bandwidth for the CPU and GPU (data from [3])

improved. Today's GPU is not only a powerful graphics engine but is a highly parallel and programmable device that can be used for general purpose computing application. With its tremendous computational horsepower and very high memory bandwidth (as illustrated by Figs. 14.1 and 14.2), the GPU has become a significant part of modern mainstream, general purpose computing systems [1, 2].

14.1.1 Advantages of GPU Computing

The GPU was initially introduced especially for high-performance image and graphics processing, where computation-intensive and highly parallel computing is required. So the GPU is designed with many cores from its inception, thus quite effective in utilizing parallelism and pipelining. It takes many advantages over both single- and multicore CPUs [3].

Since the GPU is specialized for computation-intensive, high parallel graphics rendering, most of the transistors are devoted to data processing rather than data caching and flow control, as schematically illustrated by Fig. 14.3. The GPU is especially well-suited to address problems with high arithmetic intensity (the ratio of arithmetic operations to memory operations) where the same program is executed on many data elements in parallel.

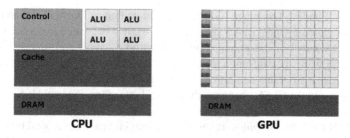

Fig. 14.3 The GPU devotes more transistors to data processing

14.1.2 Development Tools on GPU Hardware

At the very beginning of GPU computing in 2001, general purpose programming on GPU mainly relied on shading languages (SLs). A shading language is a C-like high language, which wrappers the low-level graphical rendering operators and is used to program code to manage the graphic processing.

A general computing problem is mapped to several shaders (a sequence of graphical rendering operations) that process in parallel, while the input and output data are stored as textures. The programming model of GPU for general purpose can be illustrated by Fig. 14.4.

A shader program operates on a single input element stored in the input registers, then it writes the extension result into the output registers. And this process is done in parallel by applying the same operations to all the data.

Among numerous shading languages, OpenGL Shading Language, DirectX High-Level Shader Language (HLSL), and Cg Programming Language are the most widely used ones.

Fig. 14.4 Shading language GPU computing model

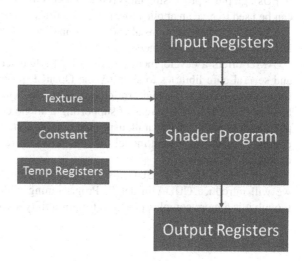

OpenGL Shading Language (GLSL) is a companion to OpenGL 2.0 and higher and part of the core OpenGL 4.3 specification [4]. It was created to give developers more direct control of the graphics pipeline without having to use assembly language or hardware-specific languages. Programs written by GLSL are compiled into OpenGL shader programs and must run through OpenGL APIs.

Analogous to the GLSL used with the OpenGL standard, DirectX HLSL is developed by Microsoft for cooperating with the Microsoft Direct3D API.

Cg (short for C for Graphics) is another shading language developed by NVIDIA in close collaboration with Microsoft. It is very similar to GLSL and HLSL, but unlike the former two shading languages dependent of specific APIs, it can works with both DirectX and OpenGL APIs.

While this GPU computing model can leverage the computing power of GPUs greatly, it faces several drawbacks. First, one needs not only to have a good knowledge of the problem to be solved, but also to possess intimate knowledge of graphics API and GPU architecture. Second, problems have to be expressed in terms of vertex coordinates, textures and shader programs, greatly increasing program complexity. Third, basic programming features such as random reads and writes to memory are not supported, which greatly restrict the flexibility of the programming model.

To address these problems, various programming platforms and technologies have been introduced. Instead of programming GPUs with graphical APIs, the programmer can now write programs in C or other high-level programming languages and target a general purpose and massively parallel processor.

Of all those developments, OpenCL and CUDA are two of the most prevalent GPU computing platforms available today.

OpenCL (Open Computing Language) is an open, royalty-free standard for cross-platform, parallel programming of modern processors. OpenCL is based on C language, and designed to enable the development of portable parallel applications for systems with heterogeneous computing devices consisting of central processing units (CPUs), graphics processing units (GPUs), DSPs and other processors. So, OpenCL can be used to give an application access to GPUs for non-graphical computing.

OpenCL implementations already exist on GPUs of AMD and NVIDIA, the two largest discrete GPU designers.

AMD offers a development toolkit, APP SDK (successor of the ATI Stream), and several core libraries to simplify the OpenCL programming on its GPU platform. NVIDIA's CUDA platform can also interact with OpenCL. OpenCL draws heavily on CUDA in the areas of supporting a single code base for heterogeneous parallel computing, data parallelism, and complex memory hierarchies. In fact, it employs a data parallelism model that has direct correspondence with the CUDA data parallelism model.

Compared to OpenCL, CUDA is a maturer platform for GPUs. In the following, we will introduce CUDA in detail. Programming with OpenCL is quite similar, though it is a more complex platform for portability's sake.

14.1.3 Compute Unified Device Architecture (CUDA)

Computing unified device architecture (CUDA), introduced by NVIDIA in November 2006, is a general purpose parallel computing platform and programming model, which leverages the parallel compute engine in NVIDIA GPUs to solve many complex computational problems in a more efficient way than on the CPU.

CUDA comes with a software environment that allows developers to use C as a high-level programming language, thus makes it easier for programmers to fully exploit the parallel feature of GPUs without an explicit familiarity with the GPU architecture.

14.1.3.1 Kernel

Kernel is a core concept of the CUDA programming model. A kernel is a function that explicitly specifies data parallel computations to be executed on a device (GPU) that operates as a co-processor to the host (CPU) running the program. When a kernel is launched on the GPU, it is executed by a batch of threads.

14.1.3.2 Thread Hierarchy

Threads are organized into independent blocks, and blocks in turn constitute a grid.

Threads can be identified by a set of intrinsic thread-identification variables (e.g., threadIdx, blockIdx, blockDim, GridDim). To help with complex addressing based on the thread, an application can also specify a block as a two or three-dimensional array of arbitrary size and identify each thread using a 2- or 3-component newIndex instead. For a two-dimensional block of size $D_x \times D_y$, the thread ID of newIndex (x, y) is $y * D_x + x$.

14.1.3.3 Memory Model

The memory model of CUDA is tightly related to its thread hierarchy. There are several kinds of memory spaces on the device:

- Read-write per-thread registers
- Read-write per-thread local memory
- Read-write per-block shared memory
- Read-write per-grid global memory
- Read-only per-grid constant memory
- Read-only per-grid texture memory

CUDA threads may access data from multiple memory spaces during their execution. Each thread has private registers and local memory. Each thread block has shared memory visible to all threads of the block. All threads have access to the same global memory. Shared memory has the same lifetime as the block, while the global, constant, and texture memory spaces are persistent across kernels.

There are also two additional read-only memory spaces accessible by all threads: the constant and texture memory spaces. They are little relevant to scientific computing, so we leave them out to interesting readers (refer to [3]).

14.1.3.4 Single-Instruction, Multiple-Thread (SIMT)

A CUDA-enabled GPU can have a scalable array of multithreaded streaming multiprocessors (SMs), which is roughly equivalent to CPU cores. Each SM can have certain number of scalar processors (i.e., Streaming Processors, SPs) with respect to the specific architecture.

When a CUDA program on the host CPU invokes a kernel grid, all blocks are distributed equally to the SMs with available execution capacity. The threads of a thread block execute concurrently on one multiprocessor in the entire execution period as a unit, and multiple-thread blocks can execute concurrently on one multiprocessor. As running blocks finish the execution, inactive blocks are launched on the vacated SMs.

To manage such a large amount of threads, it employs a unique architecture called single-instruction, multiple-thread (SIMT).

The multiprocessor creates, manages, schedules, and executes threads in groups of 32 parallel threads called warps. Individual threads composing a warp start together at the same program address, but they have their own instruction address counter and register state and are therefore free to branch and execute independently.

When a multiprocessor is given one or more thread blocks to execute, it partitions them into warps and each warp gets scheduled by a warp scheduler for execution. The way a block is partitioned into warps is always the same; each warp contains threads of consecutive, increasing thread IDs with the first warp containing thread 0.

A warp executes one common instruction at a time, so full efficiency is realized when all 32 threads of a warp agree on their execution path. If threads of a warp diverge via a data-dependent conditional branch, the warp serially executes each branch path taken, disabling threads that are not on that path, and when all paths complete, the threads converge back to the same execution path. Branch divergence occurs only within a warp; different warps execute independently regardless of whether they are executing common or disjoint code paths.

14.2 GPU–FWA

GPU–FWA [5] was proposed for the purpose of achieving the following goals:

- Good quality of solutions. The algorithm can find good solutions, compared to the state-of-the-art algorithms.

- Good scalability. As the problem gets complex, the algorithm can scale in a natural and decent way.
- Ease of implementation and usability, i.e., few control variables to steer the optimization. These variables should also be robust and easy to choose.

To meet these goals, several critical modifications to the original FWA are adopted to take benefit of this particular architecture. The pseudocode of GPU–FWA is depicted by Algorithm 14.1.

Like other swarm intelligence algorithms, GPU–FWA is an iterative algorithm. In each iteration, every firework does a local search independently. Then, an information-exchange mechanism is triggered to utilize the heuristic information to guide the search process. The mechanism should make a balance between exploration and exploitation.

As the algorithm is self-descriptive, what is left to be made clear is Algorithms 14.2 and 14.3. Below we will explain these two algorithms in detail, respectively.

Algorithm 14.1 GPU–FWA

1: Initialize n fireworks
2: calculate the fitness value of each fireworks
3: calculate A_i according to Eq. 2.3
4: **while** termination condition unsatisfied **do**
5:    **for** $i = 1$ to n **do**
6:       Search according to Algorithm 14.2
7:    **end for**
8:    Mutate according to Algorithm 14.3
9:    calculate the fitness values of the new fireworks
10:   update A_i according to Eq. 2.3
11: **end while**

Algorithm 14.2 FWA Search

1: **for** $i = 1$ to L **do**
2:    generate m sparks
3:    evaluate the fitness of each spark
4:    find the best spark with best fitness value, replace it with the current firework if it is better.
5: **end for**

14.2.1 FWA Search

In FWA, each firework generates certain number of sparks to exploit the nearby solution space. Fireworks with better fitness values generate more sparks with a smaller amplitude. This strategy aims to put more computational resources to the more potential position, thus making a balance between exploration and exploitation.

In FWA Search, this strategy is adopted, but in a "greedy" way, i.e., instead of a global selection procedure in FWA, each firework is updated by its current best spark. The mechanism exhibits an enhanced hill-climbing behavior search.

Each firework generates a fixed number of sparks. The exact number (m) of sparks is determined in accordance with the specific GPU hardware architecture. This fixed encoding of firework explosion is more suitable for parallel implementation on the GPUs.

As aforementioned in Sect. 14.1.3, within CUDA-enabled GPU, threads are scheduled by warp, which is nowadays 32 for all the CUDA-enable GPUs. Each warp is assigned certain number of stream processors (SPs). All threads in the same warp execute a common instruction at a time on these SPs. For the older generation Tesla architecture [6], the number is 8, and for Fermi architecture [7] is 16.

To avoid wastage of hardware resource, m should be multiple of number of SMs. But, it is unnecessary to pick m too large, as greater m is apt to over-exploit a certain position, while a better refined search can be achieved via running more explosions.

So as a rule of thumb, m should be 16 and 32 on GPUs of the Fermi architecture, and 8 or 16 on previous generation Tesla architecture. Thus the sparks of each firework can be generated by treads in a single warp, which, as aforementioned, need not any extra synchronization overhead.

Also, as can be seen from Algorithm 14.2, unlike FWA, in GPU–FWA, the fireworks do not exchange information in each explosion procedure, and the number of sparks for each firework generation is fixed.

Such a configuration takes many advantages.

Algorithm 14.3 Attract-Repulse Mutation

1: Initialize the new location: $\hat{\mathbf{x}}_i = \mathbf{x}_i$;
2: $s = U(1 - \delta, 1 + \delta)$;
3: **for** $d = 1$ to D **do**
4:   $r = rand(0, 1)$;
5:   **if** $r < \frac{1}{2}$ **then**
6:     $\hat{\mathbf{x}}_{i,d} = \hat{\mathbf{x}}_{i,d} + (\hat{\mathbf{x}}_{i,d} - \mathbf{x}_{best,d}) \cdot s$;
7:   **end if**
8:   **if** $\hat{\mathbf{x}}_{j,d} > x_{UB,d}$ or $\hat{\mathbf{x}}_{j,d} < x_{LB,d}$ **then**
9:     $\hat{\mathbf{x}}_{j,d} = x_{UB,d} + |\hat{\mathbf{x}}_{j,d} - x_{UB,d}|$ mod $(x_{UB,d} - x_{UB,d})$;
10:   **end if**
11: **end for**

First, global communications among fireworks need explicit synchronization, which implies a considerable overhead. By letting the algorithm to perform a given number of iterations without exchanging information, the running time can be reduced greatly.

Second, the number of sparks for each firework to generate is dynamically determined, the computation task must be assigned dynamically through the optimization procedure. As GPUs are inefficient at control operations, the dynamic computation assignment is apt to harm the overall performance of GPUs. By fixing the sparks number, we can assign each firework to a warp, this way, all sparks are synchronized implicitly without extra overhead.

The last but not the least, implemented the explosion in one block of threads, it can fully utilize the shared memory, thus, once the firework position and fitness are loaded from the global memory, no visit to the global memory is needed. The latency of visiting global memory can be reduced greatly.

14.2.2 Attract-Repulse Mutation

While the heuristic information is used to guide local search, other strategies should be taken to keep the diversity of the firework swarm. Keeping a diversity of the swarm is crucial for the success of optimization procedure.

In FWA, a Gaussian mutation is introduced to increase the diversity of the firework swarm. In this mutation procedure, m extra sparks are generated. To generate such a spark, first, a scaling factor g is generated from $G(1, 1)$ (Gaussian distribution with mean 1 and variance 1). For a randomly selected firework, the distance between each corresponding dimension of the firework and the current best firework is multiplied by g. Thus, the new sparks can be closer to the best firework or further away from it.

Similar to Gaussian mutation, in GPU–FWA, a mechanism called Attract-Repulse Mutation (AR-Mutation) is proposed to achieve this aim in an explicit way, as illustrated by Algorithm 14.3, where x_i depicts the ith firework, while x_{best} depicts the firework with the best fitness.

The philosophy behind AR-Mutation, is that, for non-best fireworks, they are either attracted by the best firework to "help" exploit the current best location or repulsed by the best firework to explore more space (see Fig. 14.5). In fact, the choice between "attract" and "repulse" reflects the balance between exploitation and exploration.

Despite Gaussian mutation is used in original FWA [8], various random distribution could be taken certainly. As uniform distribution is most straightforward and easiest, so uniform distribution is taken in the proposed algorithm.

Fig. 14.5 Schematic diagram of Attract–Repulse Mutation

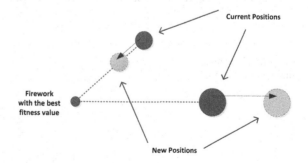

To theoretically analyze the AR-Mutation mechanism, the procedure can be simplified to a 1-order Markov chain. Given, $x_0 = 1$, the next state is generated by Eq. (14.1)

$$x_{t+1} = \alpha_t * x_t, \qquad (14.1)$$

where α_t subjects to uniform distribution between a and b, $0 < a < 1$ and $b > 1$.

Then the tth state can be expressed by the following equation:

$$x_t = \prod_{i=1}^{t} \alpha_i \cdot x_0, \qquad (14.2)$$

We can calculate the expected position,

$$E[x_t] = E\left[\prod_{i=1}^{t} \alpha_i\right] \cdot x_0 = \prod_{i=1}^{t} E[\alpha_i] \cdot x_0 = \prod_{i=1}^{t} E[\alpha] \cdot x_0 = A^t \cdot x_0, \qquad (14.3)$$

where $E[\alpha]$ is the expectation of α.

As can be seen from Eq. (14.3), if the expectation of α, i.e., A, is greater than 1, then x is expected to increase exponentially, otherwise, if A less than 1, x is expected to decay exponentially. Figure 14.6 plots a simulation result, where three traces subject to $U(0.9, 1.11)$ $(A = 1.005)$, $U(0.9, 1.1)$ $(A = 1)$, and $U(0.9, 1.09)$ $(A = 0.995)$, respectively. As the simulation showed, even a small disturbance on $A = 1$, the results tend to diverge to infinite or converge to 0, exponentially.

As for AR-Mutation, it means that fireworks are either "repulsed" to the bounds of feasible range or "attracted" to the current best position. Both conditions lead to prematurity and the loss of diversity.

To make sure that fireworks can "linger" around the search space more steadily, A should take 1. The distribution should be in the form of $s = U(1 - \delta, 1 + \delta)$, where $\delta \in (0, 1)$.

However, as the search range is limited, so δ should be taken with care, though A is set to 1.

As depicted in Fig. 14.7, from left to right, from top to bottom, δ takes 0.9 to 0.1, respectively. In the simulations, when $x > 100$, x is truncated to 10. x converges to 0

Fig. 14.6 $E[x]$ under different values of A

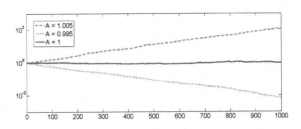

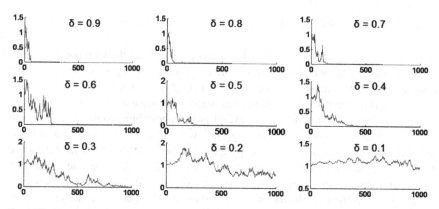

Fig. 14.7 Simulation results with different uniform distribution

with diverse speeds. As a tendency, greater δ corresponds to faster convergence, and vice versa. But what exact convergence speed is most suitable, is task-dependent. It relies on the landscape of the objective function and how many iterations the algorithm will run.

14.2.3 Implementation

The flowchart of GPU–FWA implementation on the platform of CUDA is as Fig. 14.8.

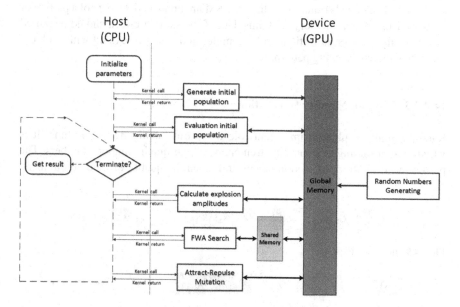

Fig. 14.8 The flowchart of the GPU–FWA implementation on CUDA

14.2.3.1 Thread Assignment

In the FWA search kernel, each firework is assigned to a single warp (i.e. 32 continuous threads). But, not all the threads in the warp are necessary to be used to execute computation. If the number of sparks is set to 16, then we use the former half-warp threads, or if the number is 32, all threads in the warp are used.

Such an implementation brings several advantages. First, since threads in the same warp are synchronized inherently, they will cut down the overhead of interspark communication. Second, by keeping each firework and their sparks in the same warp, the explosion process takes place in a single block, thus the shared memory can be utilized. As accessing to the shared memory is with much lower latency than global memory, the overall running time can be greatly reduced. Finally, as GPUs automatically dispatch block according to the computing and memory resources, it is easy for the proposed algorithm to extend with the scale of problem.

14.2.3.2 Data Organization

In implementation of GPU–FWA, the position and fitness values of each firework are stored in the global memory, while the the data of sparks are stored in the fast-accessed shared memory. For the purpose of coalescing global memory access [3], data is usually organized in an interleaving configuration (i.e., structure of arrays) [9, 10], as in Fig. 14.9. Here, we take the conventional way ,i.e., the data of the fireworks and sparks in both global and shared memory are stored in a continuous manner (i.e., array of structures, see Fig. 14.10). In our implementation, each firework occupies a single SM. The threads running on the same SM are up to load the data of a particular firework from global memory, and thus data of the same firework should be stored continuously. This organization is also simpler and easier to extend with problem scale than the interleaving pattern.

14.2.3.3 Random Number Generation

Random numbers play an important role in swarm intelligence algorithms. It is very time-consuming to generating tremendous, high-quality random numbers. The performance of the optimization heavily relies on the quality of random numbers.

Fig. 14.9 Interleaving storage

Fig. 14.10 Continuous storage

(Interested readers can refer to Chap. 3 for details). For our implementation, the efficient CURAND library [11] is used to generate high-quality random numbers on the GPU.

14.2.4 Empirical Analysis

The performance of GPU–FWA can be studied empirically. We compare GPU–FWA with both original FWA [8] and standard PSO [12].

14.2.4.1 Experimental Environment

The experiments were conducted on Windows 7 Professional x64 with 4G DDR3 Memory (1333 MHz) and Intel core I5-2310 (2.9, 3.1 GHz). The GPU used in the experiments is NVIDIA GeForce GTX 560 Ti with 384 CUDA cores. The CUDA runtime version is 5.0.

PSO is implemented according to [12] with a ring-topology and FWA according to [8] with minor modification as mentioned in Sect. 14.2.

In all simulations, each function was run 20 times independently. For GPU–FWA, in each running, 1000 iterations were executed. FWA and PSO executed the same number of function evaluations as GPU–FWA.

For GPU–FWA, the parameters are set as follows: $n = 48$, $L = 30$, $\delta = 0.5$. As in the experimental environment, the GeForce 560 Ti GPU has 12 CUDA cores, the number of fireworks should be the multiplication of 12 and big enough to avoid waste of computational power. 48 is adopted for the comparison of precision; when comparing the speedup, 72, 96, and 144 are also used.

So far, there is no theoretical rules on the criterion of the selection of L and δ. Some experiments are conducted to predetermine them. $L = 30$ and $\delta = 0.5$ performed quit well compared to various parameter settings ($L = 10, 20, 30, 40, 50$,

Table 14.1 Precision comparisons among GPU–FWA, FWA, and PSO

Fun	GPU–FWA		FWA		PSO	
	Avg.	Std.	Avg.	Std.	Avg.	Std.
f1	**1.31E–09**	1.85E–09	7.41E+00	1.98E+01	3.81E–08	7.42E–07
f2	1.49E–07	6.04E–07	9.91E+01	2.01E+02	**3.52E–11**	1.15E–10
f3	**3.46E+00**	6.75E+01	3.63E+02	7.98E+02	2.34E+04	1.84E+04
f4	**1.92E+01**	3.03E+00	4.01E+02	5.80E+02	1.31E+02	8.68E+02
f5	**7.02E+00**	1.36E+01	2.93E+01	2.92E+00	3.16E+02	1.11E+02
f6	−8.09E+03	2.89E+03	**−1.03E+04**	3.77E+03	−6.49E+03	9.96E+03
f7	1.33E+00	1.78E+01	**7.29E–01**	1.24E+00	1.10E+00	1.18E+00
f8	**3.63E–02**	7.06E–01	7.48E+00	7.12E+00	1.83E+00	1.26E+01

Table 14.2 *p*-values of t-test

	f1	f2	f3	f4	f5	f6	f7	f8
GPU–FWA versus FWA	1.00E–06	0.00E+00	0.00E+00	0.00E+00	0.00E+00	0.00E+00	5.16E–01	0.00E+00
GPU–FWA versus PSO	3.46E–01	1.21E–04	0.00E+00	2.15E–02	0.00E+00	6.50E–03	8.03E–01	1.21E–02

and $\delta = 0.1 \cdots 0.9$, as the limit of space, the results are omitted here). The total function evaluation time was $48 * 16 * 1000 = 768,000$.

For a fair comparison, all of the three algorithms were tested under the same scale. Here, by saying scale, we mean that the number of function evaluations that can be executed in parallel. For GPU–FWA, the scale in this experiment is 768, so PSO's swarm size is set as the same number. For FWA, as the firework number takes 64, and total spark number is 640 and number of gaussian sparks is 64.

14.2.4.2 Quality of Solutions

Sphere, Hyper-ellipsoid, Schwefel 1.2, Rosenbrock, Rastrigin, Schwefel, Griewangk, and Ackley functions ($f1 \sim f8$) were used as benchmark, see Appendix A.3 for the detailed configurations. The first three functions are unimodal functions, while others are multimodal functions.

All benchmark functions were optimized in 20 independent trails, and the average results and corresponding standard deviations are as Table 14.1.

Under the significance level of 0.01 (observe Table 14.2), it can be seen that GPU–FWA outperforms FWA on $f1 \sim f6$ and $f8$, it only lost to FWA on $f7$. PSO outperforms GPU–FWA on unimodal function $f2$, but fail to GPU–FWA on another unimodal function $f3$. GPU–FWA can get better results on multimodal functions $f4$, $f5$, $f6$, $f8$. In general, as far as the benchmark functions are concerned, we can see that GPU–FWA performs better than FWA and PSO.

14.2.4.3 Speedup Versus Swarm Size

Besides the precision of the solutions, speedup efficiency is another critical factor that has to be considered.

In order to observe the speedups GPU–FWA achieves in comparison with PSO and FWA, a series of experiments were conducted, where n is set respectively to 48, 72, 96, 144 for GPU–FWA. 1000 iterations are run, and the same function evaluation time under the same scale for PSO and FWA.

The running time (in seconds) and speedup with respect to Rosenbrock function is illustrated by Table 14.3. Figures 14.11 and 14.12 depict the speedup of all the eight benchmark functions with respect to the swarm size.

Table 14.3 Running time and speedup of Rosenbrock

n	FWA(s)	PSO(s)	GPU–FWA(s)	SU(FWA)	SU(PSO)
48	36.420	84.615	0.615	59.2	137.6
72	55.260	78.225	0.624	88.6	125.4
96	65.595	103.485	0.722	90.8	143.3
144	100.005	155.400	0.831	120.3	187.0

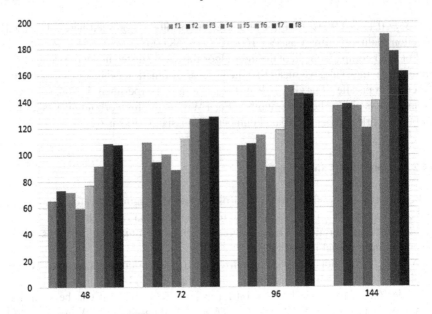

Fig. 14.11 Speedup versus FWA

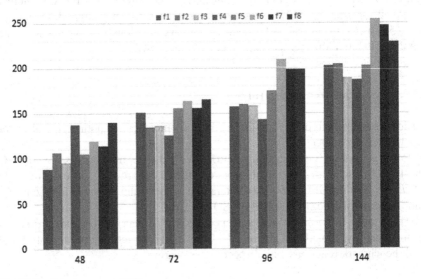

Fig. 14.12 Speedup versus PSO

GPU–FWA achieved a speedup as high as 180× with the scale of less than 200, in the meantime, the up-to-date GPU accelerated PSO achieve 200× fold speedup with the scale high up to 10,000 [10]. Thus GPU–FWA are more scalable than the conventional GPU-based PSO.

For extensive and deep analysis of swarm intelligence algorithms (SIAs) on GPU, please refer to [13] for further readings.

14.3 Summary

To take benefit of GPUs, GPU–FWA is proposed for optimization. GPU–FWA can fully leverage the great computing power of the GPU architecture, making it very well to parallel computation. It does not need special complicated data structures, thus making it easy to implement. As the problem scale goes great, it can extend in an easy and natural way. The new method requires few control variables, thus it is robust as well as easy to use.

Tested on suite of benchmark functions, it is demonstrated that the GPU–FWA outperforms FWA and the popular PSO in the quality of solution. Experimental results obtained a speedup up to 160× and 200× compared to CPU-based FWA and PSO, respectively, on an up-to-date CPU .

It can be concluded that GPU–FWA is a potential powerful tool for solving large-scale optimization problems on the massively parallel architecture.

References

1. J.D. Owens, M. Houston, D. Luebke, S. Green, J.E. Stone, J.C. Phillips, GPU computing. Proc. IEEE **96**(5), 879–899 (2008). ISSN: 0018-9219. doi:10.1109/JPROC.2008.917757
2. S. Stankovi, J. Astola, in *GPU Computing with Applications in Digital Logic*. Tampere International Center for Signal Processing. ed. by J. Astola, M. Kameyama, M. Lukac, R.S. Stankovi. Chap. An overview of miscellaneous applications of GPU computing (2012), pp. 191–215. ISBN: 978-952-15-2920-7
3. NVIDIA Corp., CUDA C Programming Guide, July 2013
4. M. Segal, K. Akeley, The OpenGL Graphics System: A Specification (Version 4.4). The Khronos Group Inc., July 2013
5. K. Ding, S. Zheng, Y. Tan, A GPU-based parallel fireworks algorithm for optimization, in *Proceeding of the Fifteenth Annual Conference on Genetic and Evolutionary Computation Conference*. GECCO (ACM, Amsterdam, 2013), pp. 9–16. ISBN: 978-1-4503-1963-8. doi:10.1145/2463372.2463377
6. NVIDIA Corp., NVIDIA GeForce 8800 GPU Architecture Overview. Technical report (2006)
7. NVIDIA Corp., NVIDIA's Next Generation CUDA^{TM} (2009)
8. Y. Tan, Y. Zhu, Fireworks algorithm for optimization, in *Advances in Swarm Intelligence* (Springer, Berlin, 2010), pp. 355–364
9. Y. Zhou, Y. Tan, GPU-based parallel multi-objective particle swarm optimization. Int. J. Artif. Intell. **7**(A11), 125–141 (2011)
10. V. Roberge, M. Tarbouchi, Parallel particle swarm optimization on graphical processing unit for pose estimation. WSEAS Trans. Comput. **11**(6), 170–179 (2012)
11. NVIDIA Corp., CURAND Library Programming Guide v5.5, July 2013
12. D. Bratton, J. Kennedy, Defining a standard for particle swarm optimization, in *Swarm Intelligence Symposium, SIS* (IEEE, Honolulu, 2007) pp. 120–127
13. Y. Tan, K. Ding, A survey on GPU-based implementation of swarm intelligence algorithms. IEEE Trans. Cybern. **45**(12), 1–14 (2015)

Part IV
Applications

Fireworks algorithm can be applied to many real-life applications that require to solve optimization algorithms. These applications can be generally transferred to single objective or multiple objective optimization problems, and thus fireworks algorithm can be applied to solve these problems easily and directly. In Chaps. 15–17, we verify how fireworks algorithms can be applied to various applications in different areas. These applications include related pattern recognition problems (nonnegative matrix factorization, document clustering, spam detection, and image recognition), complex model estimation problem (seismic inversion), and emerging swarm robotics searching problem. These applications sit in areas that differ greatly from each other and have different requirements for optimization algorithms. The fireworks algorithm can solve these problems successfully, which illustrates that FWA has great adaptiveness to different requirements in real-world applications.

Part IV
Applications

Chapter 15
FWA Application on Non-negative Matrix Factorization

This chapter presents the use of swarm intelligence algorithms for non-negative matrix factorization (NMF) [1]. The NMF is a special low-rank approximation which allows for an additive parts-based and interpretable representation of the data. Here, we present our efforts to improve the convergence and approximation quality of NMF using five different meta-heuristics based on swarm intelligence. Several properties of the NMF objective function motivate the utilization of meta-heuristics: this function is non-convex, discontinuous, and may possess many local minima. The proposed optimization strategies are twofold: On one hand, we present a new initialization strategy for NMF in order to initialize the NMF factors prior to the factorization; on the other hand, we present an iterative update strategy which improves the accuracy per runtime for the multiplicative update NMF algorithm.

15.1 Introduction

Low-rank approximations are utilized in several content based retrieval and data mining applications, such as text and multimedia mining, web search, etc. and achieve a more compact representation of the data with only limited loss in information. They reduce storage and runtime requirements, and also reduce redundancy and noise in the data representation while capturing the essential associations. The NMF leads to a low-rank approximation which satisfies non-negativity constraints [2]. NMF approximates a data matrix A by $A \approx WH$, where W, H and A are the NMF factors. NMF requires all entries in W, H and A to be zero or positive. Contrary to other low-rank approximations such as the singular value decomposition (SVD), these constraints force NMF to produce so-called additive parts-based representations. This is an impressive benefit of NMF, since it makes the interpretation of the NMF factors much easier than for factors containing positive and negative entries [3, 4].

The NMF is usually not unique if different initializations of the factors W and H are used. Moreover, there are several different NMF algorithms which all follow different strategies (e.g., mean squared error, least squares, gradient descent, ...)

© Springer-Verlag Berlin Heidelberg 2015
Y. Tan, *Fireworks Algorithm*, DOI 10.1007/978-3-662-46353-6_15

and produce different results. Mathematically, the goal of NMF is to find a (ideally the best) solution of an optimization problem with bound constraints in the form $min_{x \in \Omega} f$, where f is the nonlinear objective function of NMF and is usually not convex, discontinuous, and may possess many local minima [5].

Since meta-heuristic optimization algorithms are known to be able to deal well with such difficulties they seem to be a promising choice for improving the quality of NMF. Over the last decades nature-inspired meta-heuristics, including those based on swarm intelligence, have gained much popularity due to their applicability for various optimization problems [6]. They benefit from the fact that they are able to find acceptable results within a reasonable amount of time for many complex, large, and dynamic problems. Although they lack the ability to guarantee the optimal solution for a given problem (comparably to NMF), it has been shown that they are able to tackle various kinds of real-world optimization problems [7]. Meta-heuristics as well as the principles of NMF are in accordance with the law of sufficiency [8]: If a solution to a problem is good, fast, and cheap enough, then it is sufficient.

In this chapter, we present two different strategies for improving the NMF using five optimization algorithms based on swarm intelligence and evolutionary computing: particle swarm optimization (PSO), genetic algorithms (GA), fish school Search (FSS), differential evolution (DE), and fireworks algorithm (FWA). All algorithms are population based and can be categorized into the fields of swarm intelligence (PSO, FSS, FWA), evolutionary algorithms (GA), and a combination thereof (DE).

The goal is to find a solution with smaller overall error at convergence. The concepts of the two optimization strategies are the following: In the first strategy, meta-heuristics are used to initialize the factors for minimizing the NMF objective function prior to the factorization. The second strategy aims at iteratively improving the approximation quality of NMF during the first iterations.

15.1.1 NMF Research History

The work by Lee and Seung [2] is known as a standard reference for NMF. The original multiplicative update (MU) algorithm introduced in this article provides a good baseline than other algorithms (e.g., the alternating least squares algorithm [9], the gradient descent algorithm [10], ALSPGRAD [10], quasi Newton-type NMF [11], fastNMF, bayesNMF [12], etc. While the MU algorithm is still the fastest NMF algorithm per iteration and a good choice when a very fast and rough approximation is needed, ALSPGRAD, fastNMF, and bayesNMF have shown to achieve a better approximation at convergence compared to many other NMF algorithms [13].

15.1.1.1 NMF Initialization

Only few algorithms for non-random NMF initialization have been published. Wild et al. [14] used spherical-means clustering to group column vectors of as input. A similar technique was used in [15]. Another clustering-based method of structured initialization designed to find spatially localized basis images can be found in [11]. Boutsidis and Gallopoulos [16] used an initialization technique based on two SVD processes called non-negative double singular value decomposition (NNDSVD). Experiments indicate that this method has advantages over the centroid initialization in [14] in terms of faster convergence.

15.1.1.2 NMF and Meta-Heuristics

So far, only few studies can be found that aim at combining NMF and meta-heuristics, most of them are based on genetic algorithms (GAs). In [5], the authors have investigated the application of GAs on sparse NMF for microarray analysis, while [17] have applied GAs for boolean matrix factorization, a variant of NMF for binary data based on Boolean algebra. However, the methods presented in these studies are barely connected to the techniques presented in this article. In two preceding studies [18, 19], we have introduced the basic concepts of the proposed update strategies.

In this chapter, we extend our preliminary work in several ways by the following new contributions. We evaluate our methods on synthetic data which allows us to evaluate the proposed methods.

15.2 Low-Rank Approximations

Given a data matrix $A \in \mathbb{R}^{m \times n}$ whose n columns represent instances and whose m rows contain the values of a certain feature for the instances, most low-rank approximations reduce the dimensionality by representing the original data as accurate as possible with linear combinations of the original instances and/or features. Mathematically, A is replaced with another matrix A_k with usually much smaller rank. In general, a closer approximation means a better factorization. However, it is highly likely that in some applications specific factorizations might be more desirable compared to other solutions.

The most important low-rank approximation techniques are the singular value decomposition (SVD, [20]) and the closely related principal component analysis (PCA, [21]). Traditionally, the PCA uses the eigenvalue decomposition to find eigenvalues and eigenvectors of the covariance matrix $Cov(A)$ of A. Then the original data matrix can be approximated by $A_k := A Q_k$, with $[Q_k = [q_1, \ldots, q_k]]$, where $q_1, \ldots, q_k$ are the first k eigenvectors of $Cov(A)$. The SVD decomposes A into a product of three matrices such that $A = U \sum V^T$, where $\sum$ contains the singular values along the diagonal, and U and V are the singular vectors. The reduced rank SVD to A can be found by setting all but the first largest singular values equal to zero and using only the first k columns of U and V, such that $A_k := U_k \sum_k V_k^T$. Other well-known low-rank approximation techniques comprise factor analysis, independent components analysis, multidimensional scaling such as Fastmap or ISOMAP, or locally linear embedding (LLE), which are all summarized in [22].

Among all possible rank k approximations, the approximation A_k calculated by SVD and PCA is the best approximation in the sense that $||A - A_k||_F$ is as small as possible [23]. In other words, SVD and PCA give the closest rank k approximation of a matrix, such that $||A - A_k||_F \leq ||A - B_k||_F$, where B_k is any matrix of rank k, and $||.||_F$ is the Frobenius norm, which is defined as $(\sum |a_{ij}|^2)^{1/2} = ||A||_F$. However, the main drawback of PCA and SVD refers to the interpretability of the transformed features. The resulting orthogonal matrix factors generated by the approximation usually do not allow for direct interpretations in terms of the original features because they contain positive and negative coefficients [24]. In many application domains, a negative quantification of features is meaningless and the information about how much an original feature contributes in a low-rank approximation is lost. The presence of negative, meaningless components or factors may influence the entire result. This is especially important for applications where the original data matrix contains only positive entries, e.g., in text-mining applications, image classification, etc. If the factor matrices of the low-rank approximation were constrained to contain only positive or zero values, the original meaning of the data could be preserved better.

15.2.1 Non-negative Matrix Factorization (NMF)

The NMF leads to special low-rank approximations which satisfy these non-negativity constraints. NMF requires that all entries in A, W and H are zero or positive. This makes the interpretation of the NMF factors much easier and enables NMF a non-subtractive combination of parts to form a whole [2]. The NMF consists of reduced rank non-negative factors W and with H with $k \ll min\{m, n\}$ that approximate a matrix A by WH where the approximation WH has rank at most k. The nonlinear optimization problem underlying NMF can generally be stated as

$$min_{W,H} f(W, H) = min_{W,H} \frac{1}{2}||A - WH||_F^2 \tag{15.1}$$

The Frobenius norm $||.||_F$ is commonly used to measure the error between the original data A and the approximation WH but other measures such as the Kullback–Leibler divergence are also possible [2]. The error between A and WH is usually stored in a distance matrix $D = A - WH$ (cf. Fig. 15.2). Unlike the SVD, the NMF is not unique, and convergence is not guaranteed for all NMF algorithms. If they converge, then usually to local minima only (potentially different ones for different algorithms). Nevertheless, the data compression achieved with only local minima has been shown to be of desirable quality for many data mining applications [25]. Moreover, for some specific problem settings a smaller residual $D = A - WH$ (a smaller error) may not necessarily improve the solution of the actual application (e.g., classification task) compared to a rather coarse approximation. However, as analyzed

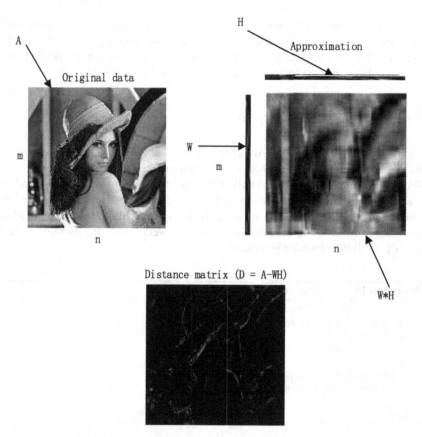

Fig. 15.1 Scheme of very coarse NMF approximation with very low-rank k. Although k is signif-icantly smaller than m and n, the typical structure of the original data matrix can be retained (note the three different groups of data objects in the *left*, *middle*, and *right part of A*)

in (Janecek and Gansterer 2010) a closer NMF approximation leads to qualitatively better classification results and turns out to achieve significantly more stable results (Fig. 15.1).

15.2.1.1 NMF Initialization

Algorithms for computing NMF are iterative and require initialization of the factors and NMF unavoidably converges to local minima, probably different ones for different initialization [16]. Hence, random initialization makes the experiments unrepeatable. A proper non-random initialization can lead to faster error reduction and better overall error at convergence. Moreover, it makes the experiments repeatable. Although the benefits of good NMF initialization techniques are well known in the literature, most studies use random initialization [16]. Since some initialization procedures can be rather costly in terms of runtime the trade-off between computational cost in the initialization step and the computational cost of the actual NMF algorithm need to be balanced carefully. In some situations, an expensive preprocessing step may overwhelm the cost savings in the subsequent NMF update steps.

15.2.1.2 General Structure of NMF

In the basic form of NMF (see Algorithm 15.1), W and H are initialized randomly and the whole algorithm is repeated several times (maxrepetition). In each repetition, NMF update steps are processed until a maximum number of iterations is reached (maxiter). These update steps are algorithm specific and differ from one NMF variant to the other. Termination criteria: If the approximation error drops below a predefined threshold, or if the shift between two iterations is very small, the algorithm might stop before all iterations are processed.

Algorithm 15.1 General structure of NMF algorithms.

1: given matrix $A \in R^{m \times n}$ and $k \ll min\{m, n\}$
2: **for** $rep = 1$ to $maxReptition$ **do**
3:    $W = rand(m, k)$
4:    $H = rand(k, n)$
5:    **for** $i = 1$ to $maxIter$ **do**
6:       perform algorithm specific NMF update steps
7:       check termination criterion
8:    **end for**
9: **end for**

15.2.1.3 Multiplicative Update (MU) Algorithm

To give an example of the update steps for a specific NMF algorithm, we provide the update steps for the MU algorithm in Algorithm 15.2. MU is one of the two original NMF algorithms presented in [2] and still one of the fastest NMF algorithms per iteration. The update steps are based on the mean squared error objective function and consist of multiplying the current factors by a measure of the quality of the current approximation. The divisions in Algorithm 15.2 are to be performed element-wise. ε is used to avoid division by zero.

Algorithm 15.2 Update steps of the multiplicative update algorithm.

1: $H = H. * (W^T A)./(W^T W H + \varepsilon)$
2: $W = W. * (A H^T)./(W H H^T + \varepsilon)$

15.3 Swarm Intelligence Optimization

Optimization techniques inspired by swarm intelligence (SI) have become increasingly popular and benefit from their robustness and exibility [7]. Swarm intelligence is characterized by a decentralized design paradigm that mimics the behavior of swarms of social insects, flocks of birds, or schools of fish. Optimization techniques inspired by swarm intelligence have shown to be able to successfully deal with increasingly complex problems [6]. In this article, we use five different optimization algorithms. Particle swarm optimization (PSO, [26]) is a classical swarm intelligence algorithm, while fish school search (FSS, [27]) and fireworks algorithm (FWA, [28]) are two recently developed swarm intelligence methods. These three algorithms are compared to genetic algorithm (GA, [29]), classical evolutionary algorithm, and differential evolution (DE, [30]), which shares some features with swarm intelligence but can also be considered as an evolutionary algorithm. For simplicity, the following only gives the details of FWA, for PSO, GA, DE, and FSS, the interested reader is referred to the references given above.

Algorithm 15.3 Pseudo code of the Fireworks Algorithm

1: Randomly initialize locations (x_i) of n fireworks
2: Repeat
3: Set off n fireworks respectively at the n locations;
4: Calculate number $\hat{s}_i$ and location of sparks for each x_i;
5: Generate $\hat{m}$ specific sparks, each for a randomly selected fireworks;
6: Keep best location and select $n - 1$ locations for next iteration;
7: Until termination (time, max. number of fitness evals., convergence,...)

The **Fireworks Algorithm** (Algorithm 15.3) is a novel swarm intelligence algorithm that is inspired by observing fireworks explosion. Two different types of explosion (search) processes are used in order to ensure diversity of resulting sparks, which are similar to particles in PSO or fish in FSS.

15.4 Improving NMF with Swarm Intelligence Optimization

Before describing our two optimization strategies for NMF based on swarm intelligence, we discuss some properties of the Frobenius norm [23]. We use the Frobenius norm (1.1) as NMF objective function (i.e., to measure the error between A and WH) because it offers some properties that are beneficial for combining NMF and optimization algorithms. The following statements about the Frobenius norm are valid for any real matrix. However, in the following we assume that refers to a distance matrix storing the distance (error of the approximation) between the original data and the approximation, $D = A - WH$. The Frobenius norm of a matrix $D \in \mathbb{R}^{m \times n}$ is defined as

$$||D||_F = \left(\sum_{i=1}^{min(m,n)} \sigma_i \right)^{1/2} = \left(\sum_{i=1}^{m} \sum_{j=1}^{n} d_{ij}^2 \right)^{1/2} \tag{15.2}$$

The Frobenius norm can also be computed row-wise or column-wise. The row-wise calculation is

$$||D||_F^{RW} = \left(\sum_{i=1}^{m} |d_i^r|^2 \right)^{1/2} \tag{15.3}$$

where $|d_i^r|$ is the norm of the ith row vector of D, i.e., $|d_i^r| = (\sum_{j=1}^{n} |r_j^i|^2)^{1/2}$, and r_j^i is the jth element in row i. The column-wise calculation is

$$||D||_F^{CW} = \left(\sum_{j=1}^{n} |d_j^c|^2 \right)^{1/2} \tag{15.4}$$

with $|d_j^c|$ being the norm of the jth column of D, i.e., $|d_j^c| = (\sum_{i=1}^{m} |c_i^j|^2)^{1/2}$, and c_i^j being the ith element in column j. Obviously, a reduction of the Frobenius norm of any row or any column of leads to a reduction of the total Frobenius norm D_F.

In the following, we exploit these properties of the Frobenius norm for the proposed NMF optimization strategies. While strategy 1 aims at finding heuristically optimal starting points for the NMF factors, strategy 2 aims at iteratively improving the quality of NMF during the first iterations. All meta-heuristics mentioned in Sect. 15.3 can be used within both strategies. Before discussing the optimization strategies, we illustrate the basic optimization procedure for a specific row (row l) of W in Fig. 15.2. This procedure is similar for both optimization strategies.

15.4.1 Parameters

Global parameters used for all optimization algorithms are upper/lower bound of the search space and the initialization, the number of particles (chromosomes, fish, ...), and maximum number of fitness evaluations. Parameter settings are discussed in Sect. 15.5. For all meta-heuristics, the problem dimension is equal to the rank k of the NMF. That is, if, for example, $k = 10$, a row/column vector with 10 continuous entries is returned by the optimization algorithms.

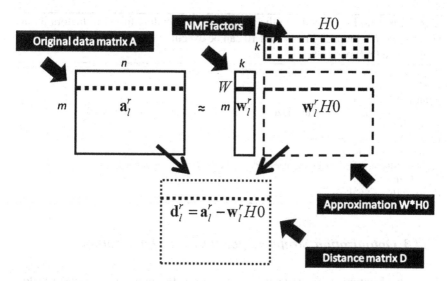

Fig. 15.2 Illustration of the optimization process for row l of the NMF factor W. The lth row of A (a_l^r) and all columns of $H0$ are the input for the optimization algorithms. The output is a row vector w_l^r (the lth row of W) which minimizes the norm of d_l^r, the lth row of the distance matrix D. The norm of d_l^r is the fitness function for the optimization algorithms (minimization problem)

15.4.2 Optimization Strategy 1 for Initialization

The goal of this optimization strategy is to find heuristically optimal starting points for the rows of W and the columns of H, respectively, i.e., prior to the factorization process. Algorithm 15.4 shows the pseudocode for the initialization procedure. In the beginning, $H0$ needs to be initialized randomly using a non-negative lower bound (preferably 0) for the initialization. In the first loop, W is initialized row-wise, i.e., row w_i^r is optimized in order to minimize the Frobenius norm of the ith row d_i^r of D, which is defined as $d_i^r = a_i^r - w_i^r H0$. Since the optimization of any row of W is independent to the optimization of any other row of W, all w_i^r can be optimized concurrently. In the second loop, the columns of H are initialized using the previously computed and already optimized rows of W, which need to be gathered beforehand (in line 7 of the algorithm). H is initialized column-wise, i.e., column h_j^c is optimized in order to minimize the Frobenius norm of the jth column d_j^c of D, which is defined as $d_j^c = a_j^c - W h_j^c$. The optimization of the columns of H can be performed concurrently as well.

Algorithm 15.4 Pseudocode for the initialization procedure for NMF factors W and H. The two for-loops in lines 4 and 10 can be executed concurrently. SIO = Swarm Intelligence Optimization.

1: given matrix $A \in R^{m \times n}$ and $k \ll min\{m, n\}$
2: $H0 = rand(k, n)$
3: **for** $i = 1$ to m **do**
4:     Use SIO to find w_i^r that minimizes $||a_i^r - w_i^r H||_F$, $(min||.||_F$ of row i of $D)$
5: **end for**
6: $W = [w_1^r; ...; w_m^r]$;
7: **for** $i = 1$ to n **do**
8:     Use SIO to find h_i^c that minimizes $||a_j^c - W h_j^c||_F$, $(min||.||_F$ of col j of $D)$
9: **end for**
10: $H = [h_1^c; ...; h_n^c]$;

15.4.3 Optimization Strategy 2 for Iterative Optimization

The second optimization strategy aims at iteratively optimizing the NMF factors W and H during the first iterations of the NMF. Compared to the first strategy not all rows of W and all columns of H are optimized instead the optimization is only performed on selected rows/columns. In order to improve the approximation as fast as possible we identify rows of D with highest norm (the approximation of this row is worse than for other rows of D) and optimize the corresponding rows of W. The same procedure is used to identify the columns of H, that should be optimized. Our experiments showed that not all NMF algorithms are suited for this iterative optimization procedure. For many NMF algorithms there was no improvement with respect to the convergence or a reduction of the overall error after a fixed number of iterations. However, for the multiplicative update (MU) algorithm which is one of the most widely used NMF algorithms this strategy is able to improve the quality of the factorization. Hence, Algorithm 15.5 shows the pseudocode for the iterative optimization of the NMF factors during the first iterations using the update steps of the MU algorithm. It can be seen that this update strategy is able to significantly reduce the approximation error per iteration for the MU algorithm. Due to the relatively high computational cost of the meta-heuristics, the optimization procedure is only applied in the first m iterations and only on c selected rows/columns of the NMF factors. Similar to strategy, the optimization of all rows of W are independent from each other (identical for columns of H), which allows for a parallel implementation of the proposed method. In the following, we describe the variables and functions (for updating rows of W) of Algorithm 15.5. Updating columns of H is similar to updating the rows of W.

- m: the number of iterations in which the optimization using meta-heuristics is applied
- c: the number of rows and/or columns that are optimized in the current iteration.
- Δc: the value of c is decreased by Δc in each iteration.

- $[Val, IX_W] = sort(norm(d_i^r),'descend')$: returns the values Val and the corresponding indices (IX_W) of the norm of all row vertors d_i^r of D in descending order.
- $IX_W = IX_W(1:c)$: returns only the first c elements of the vector IX_W.
- minimize $||a_l^r - w_i^r H||_F$.

Algorithm 15.5 Pseudocode for the iterative optimization for the multiplicative update algorithm. SIO = swarm intelligence optimization. The methods used in this algorithm are explained below.

1: **for** $iter = 1$ to $maxiter$ **do**
2:   $H = H. * (W^T A)./(W^T W H + \varepsilon)$
3:   $W = W. * (A H^T)./(W H H^T + \varepsilon)$
4:   **if** $iter < m$ **then**
5:     d_i^r is the ith row vector of D;
6:     $[Val, IX_W] = sort(norm(d_i^r),'descend')$
7:     $IX_W = IXW(1:c)$
8:     $\forall i \in IX_W$
9:     Use SIO to find w_i^r that minimizes $||a_i^r - w_i^r H||_F$;
10:
11:     $W = [w_1^r; ...; w_m^r]$;
12:     d_j^c is the jth column vector of D;
13:     $[Val, IX_H] = sort(norm(d_j^c),'descend')$
14:     $IX_H = IXH(1:c)$
15:     $\forall i \in IX_H$
16:     Use SIO to find h_i^c that minimizes $||a_j^c - Wh_j^c||_F$.
17:     $H = [h_1^c; ...; h_n^c]$;
18:     $c = c - \Delta c$;
19:   **end if**
20: **end for**

15.5 Experiment Setup

15.5.1 Software

All softwares are written in Matlab. We used only publicly available NMF implementations: multiplicative update (MU, Matlab's statistics toolbox since v6.2, nnmf()). ALS using projected gradient (ALSPG, [10]), BayesNMF and FastNMF [12]. Matlab code for NNDSVD (Sect. 15.1.1) is also publicly available [16]. Codes for PSO and DE were adapted from [31], and code for GA from the appendix of [29]. For FWA, we used the same implementation as in the introductory paper [28], and FSS was self-implemented following the algorithm provided in [27].

15.5.2 Hardware

All experiments were performed on a SUN FIRE X4600 M2 with eight AMD Opteron quad-core processors (32 cores overall) with 3.2 GHz, 2MB L3 cache, and 32GB of main memory (DDR-II 666).

15.5.3 Parameter Setup

The dimension of the optimization problem is always identical to the rank of the NMF (cf. Sect. 15.4). The upper/lower bound of the search space was set to the interval $[0, (4 * max(A))]$ and upper/lower bound of the initialization to $[0, max(A)]$. In order to achieve fair results which are not biased due to excessive parameter tuning, we used the same parameter settings for all data sets. These parameter settings were found by running a self-written benchmark program that tested several parameter combinations on randomly generated data. For some optimization strategies (PSO, FSS, and FWA), the recommended parameter settings from the literature worked fine. However, for GA and DE the parameter settings that were used in most studies in the literature did not perform very well. For GA, we found that a very aggressive (high) mutation rate highly improved the results. For DE, we observed a similar behavior and found that the maximum crossover probability (1) achieved the best results. For all experiments in this chaper, the following parameter settings were used:

1. GA: mutation rate of 0.5; selection rate of 0.65
2. PSO: G_{best} topology, $w = 0.8 c_1 = c_2 = 2.05$
3. DE: crossover probability (pc) set to upper limit 1
4. FSS: $step_{ind\_intial} = 1, step_{ind\_final} = 0.001 W_{scale} = 10$
5. FWA: number of sparks number is set to 10

15.5.4 Data Sets

We used the data set DS-RAND, which is a randomly created, fully dense 100×100 matrix in order to provide unbiased results.

15.5.5 Evaluation of Optimization Strategy 1

15.5.5.1 Initialization

Before evaluating the improvement of the NMF approximation quality as such, we first measure the initial error after initializing W and H (before running the NMF

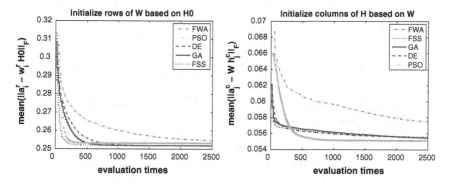

Fig. 15.3 *Left-hand side* average approximation error per row (after initializing rows of W). *Right-hand side* average approximation error per column (after initializing of H). NMF rank $k = 5$. Legends are ordered according to approximation error (*top* worst, *bottom* best)

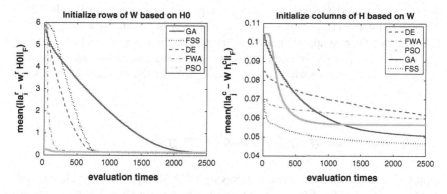

Fig. 15.4 Similar information as for Fig. 15.3, but for NMF rank $k = 30$

algorithm). Figures 15.3 and 15.4 show the average approximation error (i.e., Frobenius norm / fitness) per row (left) and per column (right) for data set DS-RAND.

The figures on the left side show the average (mean) approximation error per row after initializing the rows of W (first loop in Algorithm 15.4). The figures on the right side show the average (mean) approximation error per column after initializing the columns of H (second loop in Algorithm 15.4). The legends are ordered according to the average approximation error achieved after the maximum number of function evaluations for each figure (top = worst, bottom = best). When the NMF rank k is small (see Fig. 15.3, $k = 5$) all optimization algorithms except FWA achieve similar results. Except FWA, all optimization algorithms quickly converge to a good result. With increasing complexity (i.e., increasing rank k) FWA clearly improves its results, as shown in Fig. 15.4. The gap between the optimization algorithms is much bigger for larger rank k. Note that GA needs more than 2000 evaluations to achieve a low approximation error for initializing the rows of W. When initializing the columns of H, PSO, and GA suffer from their high approximation error during

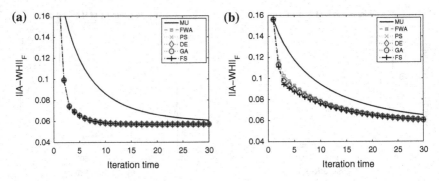

Fig. 15.5 Accuracy per Iteration when updating only the row of W, $m = 2$, $c = 20$

the first iterations, which is caused by the relatively sparse factor matrix W for PSO and GA. Although PSO is able to reduce the approximation error significantly during the first 500 iterations, FSS and GA achieve slightly better final results. Generally, FSS achieves the best approximation accuracy after the initialization procedure for large k. However, as shown later, the initial approximation error is not necessarily an indicator for the approximation quality of NMF or the resulting classification accuracy.

15.5.6 Evaluation of Optimization Strategy 2

Figure 15.5 shows the convergence curves for the NMF approximation using optimization strategy 2 for different values of rank k (data set DS-RAND). Due to the relatively high computational cost of the meta-heuristics we applied our optimization procedure here only on the rows of W, while the columns in H remained unchanged. Experiments showed that with this setting the loss in accuracy compared to optimizing both, W and H, is relatively small while the runtime can be increased significantly. m was set to 2 which indicates that the optimization is only applied in the first two iterations, and c was set to 20. As can be seen, the approximation error per iteration can be reduced when using optimization strategy 2. For small rank k (left side of Fig. 15.5), the improvement is significant but decreases with increasing values of k (see right side of Fig. 15.5). For larger k (larger than 10), the improvement over the basic MU is only marginal.

15.6 Summary

In this chapter, we presented two optimization strategies to improve the NMF using swarm intelligence-based optimization algorithms. In strategy, I use swarm intelligence algorithms to initialize the factors and prior to the factorization process of

NMF, the second strategy II aims at iteratively improving the approximation quality
of NMF during the first iterations of the factorization. Overall, five different optimization algorithms were used for improving NMF: particle swarm optimization (PSO),
genetic algorithms (GA), fish school search (FSS), differential evolution (DE), and
fireworks algorithm (FWA).

Both optimization strategies allow for efficiently computing the optimization of
single rows of W and/or single columns of H in parallel. The achieved results are
evaluated in terms of accuracy per runtime and per iteration, final accuracy after a
given number of NMF iterations, and in terms of the classification accuracy achieved
with the reduced NMF factors when applied to machine learning.

References

1. A. Janecek, Y. Tan, Swarm intelligence for non-negative matrix factorization. Int. J. Swarm Intell. Res. (IJSIR) **2**(4), 12–34 (2011)
2. D.D. Lee, H.S. Seung, Learning the parts of objects by non-negative matrix factorization. Nature **401**(6755), 788–791 (1999)
3. M.W. Berry, M. Browne, A.N. Langville, V. Paul Pauca, R.J. Plemmons, Algorithms and applications for approximate nonnegative matrix factorization. Comput. Stat. Data Anal. **52**(1), 155–173 (2007)
4. A.N. Langville, C.D. Meyer, R. Albright, J. Cox, D. Duling, Utilizing nonnegative matrix factorization for e-mail classification problems. *Survey of Text Mining III: Application and Theory* (Wiley, New York 2010), pp. 57–80
5. K. Stadlthanner, D. Lutter, F.J. Theis, E.W. Lang, A.M. Tom, P. Georgieva, et al., Sparse non-negative matrix factorization with genetic algorithms for microarray analysis, in *International Joint Conference on Neural Networks, IJCNN 2007* (IEEE, 2007), pp. 294–299
6. T. Blackwell, Particle swarm optimization in dynamic environments, in *Evolutionary Computation in Dynamic and Uncertain Environments* (Springer, Berlin, 2007), pp. 29–49
7. R. Chiong, *Nature-Inspired Algorithms for Optimisation*, vol. 193 (Springer, Berlin, 2009)
8. R.C. Eberhart, Y. Shi, J. Kennedy, *Swarm Intelligence* (Elsevier, Indianapolis, 2001)
9. P. Paatero, U. Tapper, Positive matrix factorization: a non-negative factor model with optimal utilization of error estimates of data values. Environmetrics **5**(2), 111–126 (1994)
10. C.-J. Lin, Projected gradient methods for nonnegative matrix factorization. Neural Comput. **19**(10), 2756–2779 (2007)
11. H. Kim, H. Park, Nonnegative matrix factorization based on alternating nonnegativity constrained least squares and active set method. SIAM J. Matrix Anal. Appl. **30**(2), 713–730 (2008)
12. M.N. Schmidt, H. Laurberg, Nonnegative matrix factorization with Gaussian process priors. Comput. Intell. Neurosci. **2008**, 3 (2008)
13. A. Janecek, S. Schulze Grotthoff, W.N Gansterer, LibNMF—a library for nonnegative matrix factorization. Comput. Inf. **30**(2), 205–224 (2011)
14. S. Wild, J. Curry, A. Dougherty, Improving non-negative matrix factorizations through structured initialization. Pattern Recognit. **37**(11), 2217–2232 (2004)
15. Y. Xue, C.S. Tong, Y. Chen, W.-S. Chen, Clustering-based initialization for non-negative matrix factorization. Appl. Math. Comput. **205**(2), 525–536 (2008)
16. C. Boutsidis, E. Gallopoulos, SVD based initialization: a head start for nonnegative matrix factorization. Pattern Recognit. **41**(4), 1350–1362 (2008)
17. V. Snel, J. Plato, P. Krmer, Developing genetic algorithms for boolean matrix factorization. Databases Texts **61**, (2008)

18. A. Janecek, Y. Tan, Using population based algorithms for initializing nonnegative matrix factorization. *Advances in Swarm Intelligence* (Springer, Berlin, 2011), pp. 307–316
19. A. Janecek, Y. Tan, Iterative improvement of the multiplicative update nmf algorithm using nature-inspired optimization, in *2011 Seventh International Conference on Natural Computation (ICNC)*, vol. 3 (IEEE, 2011), pp. 1668–1672
20. M.W. Berry, Large-scale sparse singular value computations. Int. J. Supercomput. Appl. **6**(1), 13–49 (1992)
21. I. Jolliffe, *Principal Component Analysis* (Wiley Online Library, 2005)
22. P.N. Tan, M. Steinbach, V. Kumar. *Introduction to Data Mining*. Pearson Education, Inc., London (2006)
23. M.W. Berry, Z. Drmac, E.R. Jessup, Matrices, vector spaces, and information retrieval. SIAM Rev. **41**(2), 335–362 (1999)
24. Q. Zhang, M.W. Berry, B.T. Lamb, T. Samuel, A parallel nonnegative tensor factorization algorithm for mining global climate data, in *Computational Science–ICCS 2009* (Springer, Berlin, 2009), pp. 405–415
25. A.N. Langville, C.D. Meyer, R. Albright, J. Cox, D. Duling, Initializations for the nonnegative matrix factorization, in *Proceedings of the Twelfth ACM SIGKDD International Conference on Knowledge Discovery and Data Mining* (Citeseer, 2006), pp. 23–26
26. D. Bratton, J. Kennedy, Defining a standard for particle swarm optimization, in *Swarm Intelligence Symposium, SIS 2007* (IEEE, 2007), pp. 120–127
27. C.J.A. Bastos Filho, F.B. de Lima Neto, A.J.C.C. Lins, A.I.S. Nascimento, M.P. Lima, Fish school search, in *Nature-Inspired Algorithms for Optimisation* (Springer, Berlin, 2009), pp. 261–277
28. Y. Tan, Y. Zhu, Fireworks algorithm for optimization, in *Advances in Swarm Intelligence* (Springer, Berlin, 2010), pp. 355–364
29. R.L. Haupt, S.E. Haupt, *Practical Genetic Algorithms* (Wiley, New York, 2004)
30. K. Price, R.M. Storn, J.A. Lampinen, *Differential Evolution: A Practical Approach to Global Optimization* (Springer, Berlin, 2006)
31. M.E.H. Pedersen, *SwarmOps: Black-Box Optimization in ANSI C* (Hvass Laboratories, Southampton, 2008)

Chapter 16
FWA Applications on Clustering, Pattern Recognition, and Inversion Problem

In this chapter, we will present the applications of fireworks algorithm for dealing with practical optimization problems including, document clustering, spam detection, image recognition, and seismic inversion problem [1]. The experimental results given herein suggest that fireworks algorithm is one of the most promising swarm intelligence algorithms in dealing with those practical problems.

16.1 Document Clustering

With the popularity of the Internet, people often post a lot of information online. The number of documents on the Web is growing rapidly. Analyzing and processing a huge number of document collections manually have become very unrealistic. Automatic documents processing has become the trend in the information age. The rapid development of natural language processing enables automatic document processing to step on to a new level. Document clustering is an important document processing task. It automatically organizes massive documents into clusters according to their topics for further usage.

Documents in the same clusters are expected to have the same topic, and documents in different clusters are expected to have different topics. Document clustering has been applied to many areas. For example, in the search engines, if the searching results are clustered into different clusters according to their different topics, people can choose to read documents in their interested topics. Therefore, people can quickly ignore the documents which they are not interested in, and can find information they need more quickly.

Document clustering generally contains two phases: feature extraction and clustering. Term frequency-inverse document frequency (TF-IDF) is a very common feature extraction method. Many classical clustering algorithms are based on partition or hierarchy. In recent years, some researchers have applied evolutionary algorithms such as genetic algorithms and particle swarm optimization algorithm to document clustering, and they have achieved promising results [2–4].

© Springer-Verlag Berlin Heidelberg 2015
Y. Tan, *Fireworks Algorithm*, DOI 10.1007/978-3-662-46353-6_16

As a new swarm intelligence algorithm, fireworks algorithm has showed excellent capabilities in function optimization [5–7]. And we can expect its excellent performance when applied to document clustering.

16.1.1 Document Features

Generally speaking, a machine learning algorithm requires the samples to be represented by vectors with a fixed dimension, and the vectors are used as input of the algorithm. Most clustering algorithms use this method to represent samples. The process of converting documents to vectors is called feature extraction.

After feature extraction, each document is represented by a feature vector. Each dimension of the vector is called a feature, which represents a particular property of the document. The multiple dimensions of feature vectors reflect the multiple properties of the documents. A good feature extraction method should be able to extract a variety of information of the original documents.

Vector space model (VSM) [8, 9] is a commonly used feature extraction method for documents. In this model, each word is regarded as a feature, and the feature value reflects the importance of the corresponding word.

16.1.1.1 Feature Selection

Vector space model regards each word as a feature. A collection of documents generally contains a larger number of words. Therefore, the number of features will be much. The subsequent calculations will spend a lot of computer resources. Meanwhile, not all features are beneficial to clustering, some of the redundant features may reduce the clustering effectiveness. It is necessary to select a subset of features according to a certain criterion. The word frequency is used as the criterion in this book.

The frequency of each word in all documents is calculated first. The most frequent N words are selected as features, and the remaining words are directly discarded. In the following experiment, the value of N is set as 2000.

16.1.1.2 TF-IDF Feature for Documents

TF-IDF is a common document feature extraction algorithm for document [10], which reflects the importance of a word in a document. TF-IDF represents the product of the term frequency (TF) and inverse document frequency (IDF). Term frequency of the ith word in the jth document is defined as follows:

$$tf_{ij} = n_{ij}/n_j \tag{16.1}$$

where n_{ij} represents number of the ith word in the jth document, and n_j represents the total number of words in the jth document.

Inverse document frequency of the ith words is defined as follows:

$$idf_i = log(D/d_i) \qquad (16.2)$$

where D represents the number of documents, and d_i represents the number of documents containing the ith word.

TF-IDF feature of the ith word in the jth document is defined as follows:

$$tfidf_{ij} = tf_{ij} * idf_i. \qquad (16.3)$$

16.1.1.3 Dimension Reduction

After the above processing each document is able to be represented as a high-dimensional vector. Using clustering algorithms directly on the high-dimensional vector will consume large computing resources. What is more, because some information of the high-dimensional vector is redundant, the machine learning algorithm applied on high-dimensional feature probably goes into overfitting. In such case the machine learning model learns too much about the redundant information of the training data, and ignores some essential characteristics of the training data for clustering [11].

Dimension reduction is a commonly used data processing technique. It maps data from high-dimensional space to low-dimensional space. Data in the low-dimensional space should reflect in the essential characteristics of the original data sufficiently, and should remove redundant information from the original data. Dimension reduction algorithms are often divided into two categories: linear dimensionality reduction and nonlinear dimensionality reduction.

This section uses a linear dimensionality reduction method called principal component analysis (PCA) to reduce the dimension of documents [12]. PCA first selects several orthogonal principal components from the high-dimensional space. The variance of data along the directions of these principal components should be the greatest. These principal components will be considered as the orthogonal bases of a low-dimensional space, and the original data is represented as low-dimensional vectors under such orthogonal bases. PCA retains the original data in several directions with the largest variance and discards other directions. We can see that it tries to keep the essential information of the original data and dismisses unimportant information.

16.1.1.4 Data Standardization

In many dataset the distribution of data over different dimensions have a variety of ranges. The data along some dimensions may be too large or too small. And the ranges along different dimensions may differ too much. This situation will decrease the performance of many machine learning algorithms.

Standardizing the data will unify the range of data along different dimensions. Each dimension is shrunk linearly to unify the range. After shrinking each dimension is with zero mean and unit standard deviation. Let x_{ij} denotes the jth dimension of the ith datum before standardization, and let $\hat{x}_{ij}$ represent the normalized data, i.e.,

$$\hat{x}_{ij} = (x_{ij} - \mu_j)/\sigma_j \tag{16.4}$$

where μ_j represents the mean of the jth dimension, and σ_j represents the standard deviation of the jth dimension.

16.1.2 Fireworks Algorithm for Document Clustering

Fireworks algorithm has shown excellent ability in optimization. Fireworks algorithm can be applied to document clustering by searching the optimal centroids of all the clusters. Therefore, the clustering problem is converted to optimization problem.

Fireworks algorithm searches the optimal centroids in the $M * F$-dimensional space, where M represents the number of document clusters, and F denotes the dimension of feature vectors. We denote the centroid of the ith cluster as $c_i, i = 1, 2, \ldots, M$, and x_i is an F-dimensional vector. A firework is defined as $<c_1, c_2, \ldots, c_M>$.

Once the centroid of each cluster is given, the documents can be categorized into clusters. The nearest centroid to each document is found and the document is categorized into the corresponding cluster. We choose the Euclidean distance to measure the distance between two vectors. That is, for vector $a = \{x_1, x_2, \ldots, x_p\}$ and vector $b = \{y_1, y_2, \ldots, y_p\}$, the distance between them is $d = \sqrt{\sum_{i=1}^{p} (x_i - y_i)^2}$.

The closer two vectors are, the smaller the value of d is, and vice versa. The Euclidean distance can be calculated easily and efficiently.

The clustering criteria intra-cluster distance is used as the fitness function of fireworks algorithm. The intra-cluster is calculated as follows:

$$J = \sum_i \sum_{j \in K_i} \left(\mathbf{x}_j - \mathbf{c}_i\right)^2 \tag{16.5}$$

where $\mathbf{c}_i$ represents the centroid of the ith cluster, and K_i represents the set of documents in the ith cluster, and $\mathbf{x}_j$ is the jth document in K_i.

The smaller the intra-cluster distance is, the more compact documents in the same cluster are, and the better clustering quality fireworks algorithm achieves.

Finally, we can use fireworks algorithm to find an optimal firework with minimal intra-cluster distance. And these fireworks are used as the centroids of document clusters.

Specific steps of using fireworks algorithm for document clustering are shown in Algorithm 16.1.

Algorithm 16.1 Fireworks algorithm for document clustering.

1: Randomly select N points in $M * F$-dimensional search space of N points as the initial fireworks. Each firework represents centroids of different document clusters.
2: Calculate the Euclidean distance between each document and each centroid. Each document is clustered into the cluster with the nearest centroid.
3: Get the fitness function of each fireworks by calculating the intra-cluster distance.
4: Explode and select some fireworks for the next generation. If the termination condition meets, go to step 5, or go to step 2.
5: Get the result of document clustering.

16.1.3 Experiment

The following experiment demonstrates the application of fireworks algorithm to document clustering.

20-newsgroup is a well-known English text mining data set [10], which contains messages from 20 newsgroups. Messages in each newsgroup are about one specific topic. These messages were clustered into four clusters by the fireworks algorithm, clustering results are shown in Fig. 16.1. Due to space limitations, Fig. 16.1 only shows some representative messages from 12 newsgroups. From Fig. 16.1 we can see that messages within the same clusters roughly have the same topic. For example, messages in the first row of the third column are mainly about politics, messages in the second row of the first column are mainly about religion, and messages in the third row of the third column are mainly about hardware. There are also a few messages with different topics in each cluster due to errors. But on the whole the fireworks algorithm can effectively organize documents into clusters according to their topics, which means that this is a good document clustering algorithm.

Traditional document clustering algorithms are easy to fall into local optimum while fireworks algorithm is effectively able to avoid local optimum, as a result, the fireworks algorithm can achieve a promising result when applied to document clustering, which lay a new way to document clustering.

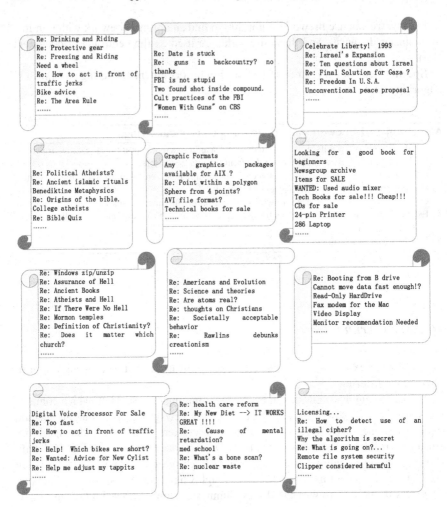

Fig. 16.1 Document clustering by fireworks algorithm on 20-newsgroup

16.2 Spam Detection

Spam, defined as Unsolicited Commercial Emails (UCE) or Unsolicited Bulk Emails (UBE) [13], has become a significant problem for both recipients and Internet Service Providers (ISPs). For recipients, coping with spam is time-consuming; furthermore, spam frequently contains images that recipients find offensive, or attached malicious programs that attack recipients' computers. For ISPs, large scale of spam is a considerable burden on their systems [14–17].

Many approaches have been proposed to handle the problem, in which intelligent approaches play an increasingly important role in antispam in recent years for their ability of self-learning and good performance. Spam detection is seen as a two-class categorization problem in intelligent approaches, which mainly contain feature extraction and classification phases. In previous researches, parameters of the spam detection approaches are set simply and manually. However, the manual setting might cause several problems. For instance, lack of prior knowledge may lead to improper parameter setting, repeated attempts of users cost overmuch human effort, and the inflexibility of the dataset-relevant parameters should also be taken into account. This section introduces application of FWA in parameter optimization of spam detection.

16.2.1 Problem Description

Intelligent spam detection approaches mainly contain two steps, feature extraction and classification. The former transforms email samples into specific representation form, and further utilized by the later to get the class. In this section, we set local concentration (LC) approach [18–20], which is one of the immune concentration-based feature extraction approaches [18–23], and support vector machine (SVM) technique [17, 24–28] as examples of feature extraction approach and classification algorithm, respectively, to introduce how to optimize parameters of spam detection approach with FWA (Fig. 16.2).

Figure 16.3 shows the intelligent spam detection model with LC and SVM, which mainly contains training phase and classification phase. In training phase, an email dataset, called training set, is preprocessed with tokenization and stemming, and large amount of words are obtained. These words are further selected according to their importance estimated by some term selection metrics. The selected words are

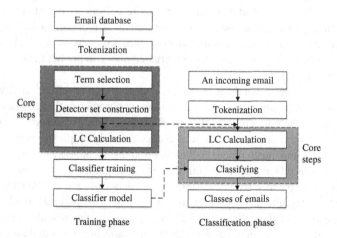

Fig. 16.2 Intelligent spam detection model with LC and SVM. Reprinted from Ref. [29], with kind permission from Springer Science+Business Media

Fig. 16.3 Framework of parameter optimization of spam detection with FWA. Reprinted from Ref. [29], with kind permission from Springer Science+Business Media

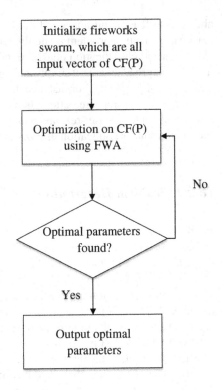

considered to be able to provide more useful information for email classification. Spam detector set and legitimate detector set are constructed based on these selected words. Concentration features of each email sample in the email dataset are calculated on the detector sets and concentration feature vectors are used to construct the classifier. In classification phase, an incoming email is preprocessed similarly and the concentration features are calculated to construct its feature vector. Finally, the feature vector is used by the classifier constructed above to predict the class label of the incoming email [28].

In the above model, all of the term selection, detector set construction, feature calculation, and classification involve some important parameters that can have apparent influence on the performance of spam detection. Terms with more information are selected by term selection to reduce computational complexity with relatively nonuseful words and interference on classification results of noisy terms. Percentage of term selection $m\%$ determines both the size and quality of the gene library for constructing detector sets. By calculating class tendency of each selected terms, spam detector set and legitimate detector set are constructed. Tendency threshold measures the absolute value of the difference of a term's posteriori probabilities appearing in two classes of emails. Concentration features of emails are calculated on each local area to construct the concentration vectors. Number of sliding windows N defines the partition granularity inside an email. Parameters of the classifier are also important for the detection performance.

Intelligent spam detection approach involves many parameters which immediately affect the performance of spam detection. Therefore, it is necessary to optimize the parameters of spam detection approach with FWA.

16.2.2 Optimization Principle

The classification problem that "whether an email is spam or a legitimate email" is here considered as an optimization problem, that is, to achieve the lowest error rate by finding the optimal parameter vector in the potential search space [29].

The optimal parameter vector $P^* = < F_1^*, F_2^*, \ldots, F_n^*, C_1^*, C_2^*, \ldots, C_m^* >$ composes of two parts: the first part is the feature calculation relevant parameters $< F_1^*, F_2^*, \ldots, F_n^* >$, and the second part is the classifier relevant parameters $< C_1^*, C_2^*, \ldots, C_m^* >$. Parameters $< F_1^*, F_2^*, \ldots, F_n^* >$ determine the performance of feature construction independently, and parameters $< C_1^*, C_2^*, \ldots, C_m^* >$ affect performance of the specific classifier. The optimal vector $P*$ is the vector whose cost function $CF(P)$ associated with classification achieves the lowest value, with

$$CF(P) = Err(P) \tag{16.6}$$

where $Err(P)$ is the classification error measured by cross-validation on the training set.

Vector P is the optimization vector whose performance is measured by $CF(P)$. Therefore, this optimization of concentrations can be formulated as follows: Finding $P^* = < F_1^*, F_2^*, \ldots, F_n^*, C_1^*, C_2^*, \ldots, C_m^* >$, so that

$$CF(P^*) = \min_{\{F_1, F_2, \cdots, F_m, C_1, C_2, \cdots, C_m\}} CF(P). \tag{16.7}$$

Different feature extraction methods hold different parameters and lead to different performances. For LC approach, specifically, there are term selection rate $m\%$, tendency threshold, and sliding windows number N. Different classifiers also hold different parameters. SVM-related parameters determine the location of optimal hyperplane for classification in feature space, including cost function value C and kernel function parameter γ, just to name a few.

Figure 16.3 shows the framework of parameter optimization of intelligent spam detection with FWA. Based on this framework, we further define two strategies for parameter optimization. For efficiency consideration, strategy 1 defines an independent validation set on training set, that is to divide the original training set into a new training set and a validation set. The fireworks are trained on the new training set and validated on the validation set independently, where each firework corresponds to a different parameter combination, until the optimal firework is obtained and used for detection on the testing set. For this strategy, fitness values of fireworks are calculated on the independent validation set in each cross-validation, so the computational complexity is relatively low.

For robustness consideration, strategy 2 randomly divides the original training set into ten parts instead of defining an independent validation set, and fitness values of fireworks are calculated by tenfold validation. In this case, each firework can get a comprehensive and integrated evaluation.

16.2.3 Experimental Results

Experiments were conducted on four benchmark corpora PU1, PU2, PU3, and PUA, using tenfold cross-validation to verify the effectiveness of FWA-based parameter optimization of spam detection. During the experiments, range of the term selection rate $m\%$ is set to [0, 1], range of the tendency threshold θ is set to [0, 0.5], and sliding window number N [1, 50]. SVM-related parameters $c \in (1, 100)$ and $\gamma \in (0, 20)$. Parameters involved in FWA are the same with original FWA.

Comparison of performance before and after parameter optimization with strategy 1 is shown in Fig. 16.4. It is obvious that the performance of spam detection is improved on all the corpora utilized. However, due to the limitation that the

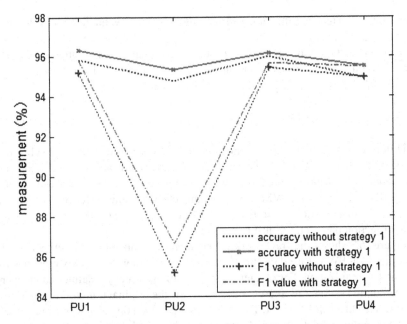

Fig. 16.4 Comparison of performance before and after parameter optimization with strategy 1. Reprinted from Ref. [29], with kind permission from Springer Science+Business Media

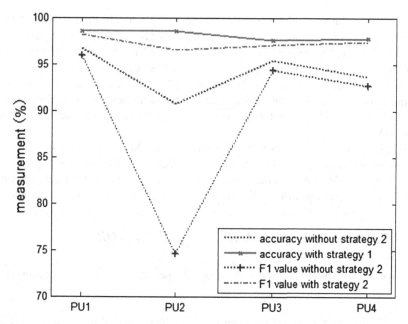

Fig. 16.5 Comparison of performance before and after parameter optimization with strategy 2. Reprinted from Ref. [29], with kind permission from Springer Science+Business Media

independent validation set cannot always describe the distribution feature of data in testing set precisely, the improvement of performance is somewhat limited.

Figure 16.5 shows the comparison of performance before and after parameter optimization with strategy 2, which randomly divides the original training set into ten parts instead of defining an independent validation set, and calculating fitness values of fireworks using tenfold validation. To some extent, strategy 2 solves the evaluation problem caused by different data distribution features in validation set and testing set. For experimental efficiency consideration, we randomly selected 20% samples from each corpus for experiments. As we can see in Fig. 16.5, performance of spam detection is improved substantially using strategy 2. In real world, spam filters are usually trained off-line, thus strategy 2 is also worth applying [29].

16.3 Image Recognition

For a recognition task, distance measure between patterns is one of the most important procedures [11, 30, 31]. Orientation coding-based methods have achieved high accuracy and speed in recognition, such as competitive code (CompCode) [32], palmprint orientation code (POC) [33, 34], and robust line orientation code (RLOC) [35]. In these methods, the commonly used strategy is to convolve the images with a number of filters, then, the orientation is recoded as orientation values by the dominate convolution results. For orientation coding-based patterns, OR_XOR (Hamming distance) and SUM_XOR (Angular distance) are usually used for measuring the distance. However, little work have been done to study the physical significance of these distances for orientation coding-based patterns. In [36], Guo et al. tried to use a unified distance measure for recognition, however, the physical meaning of the unified distances measure scheme is not explained. Moreover, the relation between *OR_XOR* and *SUM_XOR* is not clear.

In this section, a new unified distance measure (UDM) scheme is proposed [37]. To automatically compute the parameters introduced in UDM model, the population-based heuristic algorithms, particle swarm optimization (PSO), differential evolution (DE), and fireworks algorithm (FWA) [] are used for the determination of parameters.

16.3.1 Relation Between SUM_XOR and OR_XOR

Denoting $I(x, y)$ as the preprocessed image, the pattern image $P(x, y)$ is computed as

$$P(x, y) = arg \min_{\forall i \in [0, N-1]} I(x, y) \bigotimes G(x, y, \theta_i) \qquad (16.8)$$

where $\bigotimes$ is the convolution operation. Suppose $N = 6$, then $\theta_i = i\pi/6$. Each point in $P(x, y)$ is the orientation θ_i with the dominant response, and the distance between two pattern images $P(x, y)$, $Q(x, y)$ is sum of the distance of all corresponding points. Here, some basic rules are presented first. θ_a and θ_b are two corresponding points in $P(x, y)$ and $Q(x, y)$.

1. $d(\theta_a, \theta_b) = d(\theta_b, \theta_a)$ (Symmetry).
2. $d_0 = d(\theta_a, \theta_a) = 0$.
3. $d_1 = d(0, \pi/6) = d(\pi/6, \pi/3) = d(\pi/3, \pi/2) = d(\pi/2, 2\pi/3) = d(2\pi/3, 5\pi/6) = d(5\pi/6, 0)$. For any two orientations, the interval with $\pi/6$ should be equal.
4. $d_2 = d(0, \pi/3) = d(\pi/6, \pi/2) = d(\pi/3, 2\pi/3) = d(\pi/2, 5\pi/6) = d(2\pi/3, 0)$
   For any two orientations, the interval with $\pi/3$ should be equal.
5. $d_3 = d(0, \pi/2) = d(\pi/6, 2\pi/3) = d(\pi/3, 5\pi/6) = d(\pi/2, 5\pi/6)$
   For any two orientations, the interval with $\pi/2$ should be equal.

To represent the orientation, Kong et al. used a 3-bit ($log_2 6$) encode method for orientation coding [32], which encode the orientations $\{0, \pi/6, \pi/3, \pi/2, 2\pi/3, 5\pi/6\}$ as $\{000, 001, 011, 111, 110, 100\}$. Then *SUM_XOR* and *OR_XOR* distances between $P(x, y)$, $Q(x, y)$ are defined as follows:

$$D_{SUM\_XOR} = \frac{\sum_{y=1}^{M} \sum_{x=1}^{N} \sum_{i=1}^{3} P_i^b(x, y) \bigotimes Q_i^b(x, y)}{3 * M * N} \qquad (16.9)$$

$$D_{OR\_XOR} = 1/(M*N) * \sum_{y=1}^{M} \sum_{x=1}^{N} ((P_0^b(x,y) \oplus Q_0^b(x,y))|$$
$$(P_1^b(x,y) \oplus Q_1^b(x,y))|(P_2^b(x,y) \oplus Q_2^b(x,y))) \tag{16.10}$$

where M and N are the height and length of the pattern, $P_i^b(x,y)$, $Q_i^b(x,y)$ are the ith bit plane of P and Q. $\oplus$ denotes the bitwise exclusive OR (XOR), | denotes the bitwise OR. Based on the five basic rules, denote by a_i, $i \in [0,3]$ the number of pixels where the distance is d_i, $i \in [0,3]$, (16.9) and (16.10) can be rewritten as follows:

$$D_{SUM\_XOR} = \frac{k_0 * a_0 + k_1 * a_1 + k_2 * a_2 + k_3 * a_3}{M*N} \tag{16.11}$$

$$= \frac{1/3 * a_1 + 2/3 * a_2 + 3/3 * a_3}{M*N} \tag{16.12}$$

$$D_{SUM\_XOR} = \frac{k_0 * a_0 + k_1 * a_1 + k_2 * a_2 + k_3 * a_3}{M*N} \tag{16.13}$$

$$= \frac{1 * a_1 + 1 * a_2 + 1 * a_3}{M*N}. \tag{16.14}$$

Here, k_i, $i \in 0,1,2,3$ can be seen as the cost of wrong decision, and $k_0 = 0$. For a 6 orientation filter, k_i denotes the cost that the decision of θ_l while the ground truth is $\theta_{(l+i)mod6}$ or $\theta_{(l-i)mod6}$.

16.3.2 The Proposed Unified Distance Measure Scheme

In [38], Kong pointed out that SUM_XOR outperforms OR_XOR in PolyU palm-print databases. Compare (16.11) and (16.13), the difference between OR_XOR and SUM_XOR is that the using of different coefficients, which denotes the cost of wrong decision. Based on the comparison, we define the unified distance measure scheme between patterns $P(x,y)$ and $Q(x,y)$ as

$$D_u = \frac{k_1 * a_1 + k_2 * a_2 + k_3 * a_3}{M*N}. \tag{16.15}$$

For a recognition task, the performance index are usually equal error rate (EER). For a two-choice decision task, equal error rate is the false reject rate which equals the false accept rate. The EER is scale invariant with the distributions. Thus, the unified distance can be rewritten as

$$D_u = \frac{k_1}{M*N} * \left(a_1 + \frac{k_2}{k_1} * a_2 + \frac{k_3}{k_1} * a_3 \right) \tag{16.16}$$

$$D'_u = K_1 * a_1 + K_2 * a_2 + K_3 * a_3. \tag{16.17}$$

For the unified distance shown in (16.17), there are three parameters (K_1, K_2, K_3) to be determined, in which $K_1 = 1$. *SUM_XOR* and *OR_XOR* are two special cases, which take $K_2 = 2$, $K_3 = 3$ and $K_2 = 1$, $K_3 = 1$, respectively.

The optimization problem can generally be stated as

$$\min eer = \min_{x=[K_2, K_3] \in \Omega} f(x), \tag{16.18}$$

where $f : \mathbb{R}^2 \to \mathbb{R}$ is a nonlinear function used for the computation of EER values, and Ω is the feasible region. For this optimization problem, to automatically determine the parameters of the unified distance measure model in (16.17), the heuristic algorithms, particle swarm optimization [39], differential evolution [40], and firework algorithm [5] are used for the optimization, which are named as PSO-UDM, DE-UDM, FWA-UDM, respectively. Details about each heuristic algorithm can be found in the references provided below.

Particle Swarm Optimization (PSO, [39, 41]) mimics the process of the search for food of flocks. The particles (flocks) in the swarm move under the guidance of the cognitive information and social information.

Differential Evolution (DE, [40]) maintains a population of candidate solutions while it also creates some new candidate solutions by combining existing ones according to its formulae. Then it keeps the candidate solution which has the best score or fitness on the optimization problem.

Fireworks Algorithm (FWA, [5]) Inspired by the explosion of fireworks in the night sky, a firework explodes and generates the sparks in the nearby space can be seen as the search of an optimization problem.

16.3.3 Experiment

To validate the performances of the proposed PSO-UDM, DE-UDM, and FWA-UDM, three databases, the artificial data set, the PolyU palmprint database, and the collected finger-vein database are used in [37] and this chapter only chooses the PolyU palmprint database to present the performance. POC, RLOC, and CompCode are used for pattern extraction. The EER is compared to evaluate the performance. In the experiments, the search range of K_2 and K_3 are both set to [0, 6].

To reduce the influences of shift and rotation of the patterns, in the matching part, a shift of [−2, 2] in X-axis and [−2, 2] in Y-axis is taken, and the minimal distance under a certain K_2, K_3 is recorded as the distance between two patterns.

The swarm size for each algorithm is set to 50, and the maximum evaluation times is 5000.

1. PSO, $velocity_{max} = 20$, $c_1 = c_2 = 2$, $\omega = 0.5$, $\alpha = 1$
2. DE, the rest of parameters are same as [40]
3. FWA, the rest of parameters are same as [5]

Table 16.1 Experimental results of EER and parameters of K_2, K_3 on the PolyU palmprint database

Distance	POC			RLOC			CompCode		
	EER	K_2	K_3	EER	K_2	K_3	EER	K_2	K_3
OR_XOR	2.2143E-3	1	1.0	4.2664 E-4	1.0	1.0	2.3652E-3	1.0	1.0
SUM_XOR	1.9425E-3	2	3.0	4.2873 E-4	2.0	3.0	2.8809E-3	2.0	3.0
PSO-UDM	**1.7894E-3**	2.79628	2.12560	**3.9649E-4**	1.76602	2.06041	**2.1216E-3**	0.71403	1.01558
DE-UDM	1.7912E-3	1.90490	2.0377	4.2904E-4	3.10227	3.35624	2.3385E-3	0.835902	1.48357
FWA-UDM	1.7907E-3	2.30518	1.87876	3.9739E-4	1.57043	1.46151	2.1539E-3	0.70509	0.85458

16.3.3.1 Experiments Using the PolyU Palmprint Database

Experimental Design: The PolyU palmprint database [42] is designed by The Hong Kong Polytechnic University, which contains palmprint images from 500 individuals. Each individual provides 12 images, each 6 images one time. Therefore, the database contains 6000 images from 500 individuals. The size of all the images is 128×128.

First, the images are preprocessed for noises reduction and patterns are extracted by POC, RLOC, and CompCode algorithms. Then the three distances are computed. There are totally 17,997,000 match times with 33,000 genuine match time and 17,964,000 imposters. To reduce the computation load, the images are downsampled.

Experimental Results: Table 16.1 lists the results of EER, parameters of K_2, K_3 introduced in the unified distance model. From the results, it can be concluded that the using of population-based algorithms for parameters optimization for the unified distance model can achieve lower EER than OR_XOR and SUM_XOR on the palmprint database. Among the three methods, PSO-UDM, DE-UDM, and FWA-UDM, PSO-UDM has the lowest EER which can reduce the EER by up to 7.88 % (1.94250E-03 to 1.78942E-03) for POC, 3.31 % (4.10080E-04 to 3.96492E-04) for RLOC, and 10.30 % (2.36520E-03 to 2.12160E-03) for CompCode, respectively.

16.3.3.2 Performance Comparison

We investigate the relation between hamming distance and angular distance, and based on the study, a unified distance is proposed. To validate the proposed distance methods, we test the orientation algorithms on one databases for recognition. Experimental results of EER shown in Table 16.1, suggest that the proposed PSO-UDM, DE-UDM, and FWA-UDM gain great advantages compared with the OX_XOR and SUM_XOR distance measure schemes. Among the three population based heuristic algorithms, PSO-UDM and FWA-UDM gains the better performance compared with DE-UDM while PSO-UDM and FWA-UDM nearly have the same performance.

16.4 Seismic Inversion Problem

Seismic inversion is a critical and challenging task in geophysics. It is widely used for oil/gas exploration and development. In recent decade, intelligence optimizations such as GA and PSO [17], are applied to the inversion problem and achieve a great success. In this section, we try to apply FWA to inversion problem, and we will see that FWA outperforms PSO, GA, and its variants [7, 16, 24, 43].

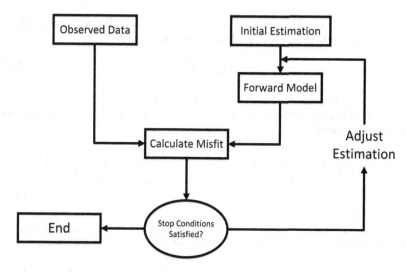

Fig. 16.6 Flowchart of inversion

16.4.1 Problem Statement

In Geophysics, inversion methods are usually used for exploring the structure of the Earth.

In general, the target of an inversion problem is to find a parameterized model that can fit the observed data as best. It is based on the forward problem, i.e., for a given set of parameters, the expected observation can be calculated. The flowchart of inversion is depicted by Fig. 16.6.

There exist many inversion models, here only reflectivity model is discussed. The seismic wave can be regarded as light-like wave. When the seismic wave reaches the boundary between two medium, both reflection and refraction happen. The seismic wave can be observed after many such reflections and refractions. On these hypotheses, we can set up the geographical reflectivity model [44].

Given a point and its horizontal and vertical distances between the seismic source denoted as r and z, respectively.

$$\Phi_0(r, z, t) = \frac{1}{R} F\left(t - \frac{R}{\alpha_1}\right), \qquad (16.19)$$

where $R = r^2 + z^2$ t denotes time and α_1 wave velocity.

Fourier transform can be expressed in calculus as

$$\Phi_0(r, z, \omega) = \bar{F}(\omega) \int_0^{\inf} \frac{k}{jv_1} J_0(kr) exp(-jv_1 z) dk, \qquad (16.20)$$

where, $\bar{F}(\omega)$ is the Fourier transform of $F(t)J_0(kr)$ order 0 Bessel function of first kind, j imagery unit, k the number of horizontal wave, and

$$v_1 = (k_{\alpha_1}^2 - k^2)^{1/2}, \tag{16.21}$$

where $k_{\alpha_1} = \frac{\omega}{\alpha_1}$ denotes the numbers of vertical waves.

Generally, after the refraction of m layers, the seismic wave can be calculated in frequency domain as

$$\bar{\Phi}_1(r, z, \omega) = \bar{F}(\omega) \int_0^{\inf} \frac{k}{jv_1} J_0(kr) P_d(\omega, k) \times exp\left[-j\left(\sum_{i=1}^{m-1} h_i v_i + \left(z - \sum_{i=1}^{m-1} h_i\right)v_m\right)\right] dk, \tag{16.22}$$

where $P_d(\omega, k)$ is the product of each layer's down-wave transmission coefficient.

In the reflection interface, the generated P-wave can be calculated as

$$\bar{\Phi}_2(r, z, \omega) = \bar{F}(\omega) \int_0^{\inf} \frac{k}{jv_1} J_0(kr) P_d(\omega, k) \tilde{R}_{pp}(\omega, k) \times exp\left[-j\left(\sum_{i=1}^{m-1} h_i v_i + \left(z - \sum_{i=1}^{m-1} h_i\right)v_m\right)\right] dk, \tag{16.23}$$

where $\tilde{R}_{pp}(\omega, k)$ is complex reflection.

Fluctuations through m layers' reflection back to the original layer can be calculated as follows:

$$\bar{\Phi}_3(r, z, \omega) = \bar{F}(\omega) \int_0^{\inf} \frac{k}{jv_1} J_0(kr) P_d(\omega, k) \tilde{R}_{pp}(\omega, k) P_u(\omega, k)$$
$$\times exp\left[-j\left(2\sum_{i=1}^{m-1} h_i v_i + \left(z - \sum_{i=1}^{m-1} h_i\right)v_m\right)\right] dk, \tag{16.24}$$

where $P_u(\omega, k)$ is the product of all layers' up-wave transmission coefficients.

Thus, we can calculate the P-wave and S-wave received on the ground as

$$\bar{\Phi}_4(r, z, \omega) = \bar{F}(\omega) \int_0^{\inf} \frac{k}{jv_1} J_0(kr) P_d(\omega, k) \tilde{R}_{pp}(\omega, k) P_u(\omega, k) r_{pp}(\omega, k)$$
$$\times exp\left[-j\left(2\sum_{i=1}^{m-1} h_i v_i + \left(z - \sum_{i=1}^{m-1} h_i\right)v_m\right)\right] dk, \tag{16.25}$$

$$\bar{\Phi}_5(r, z, \omega) = \bar{F}(\omega) \int_0^{\inf} \frac{k}{jv_1} J_0(kr) P_d(\omega, k) \tilde{R}_{ps}(\omega, k) P_u(\omega, k) r_{ps}(\omega, k)$$
$$\times exp\left[-j\left(2\sum_{i=1}^{m-1} h_i v_i + \left(z - \sum_{i=1}^{m-1} h_i\right)v_m\right)\right] dk, \tag{16.26}$$

where, r_{pp} and r_{ps} denote the P–P and P–S reflection ratios, respectively.

The generated fluctuation is

$$\bar{u} = \frac{\partial \bar{\Phi}_3}{\partial r} + \frac{\partial \bar{\Phi}_4}{\partial r} - \frac{\bar{\Psi}}{\partial z}, \tag{16.27}$$

$$\bar{w} = \frac{\partial \bar{\Phi}_3}{\partial z} + \frac{\partial \bar{\Phi}_4}{\partial z} + \frac{\bar{\Psi}}{\partial r} + \frac{\bar{Psi}}{r}. \tag{16.28}$$

Here, let $z = 0$, then we get the expected synthesized wave. Based on this forward model, we can solve the inversion problem as illustrated by Fig. 16.6.

16.4.2 Experiment and Analysis

To justify the feasibility of FWA for inversion problems, simulation experiment was conducted. In the simulation, GA, NGA, PSO as well as FAW were applied to find the structure parameters (density etc.). The results are presented by Figs. 16.7 and

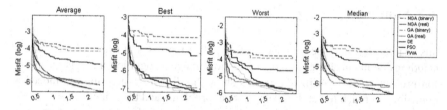

Fig. 16.7 Convergency curve of optimization

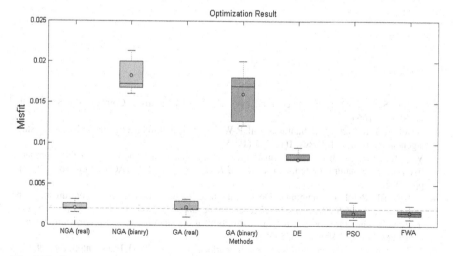

Fig. 16.8 Comparison of optimization results

Table 16.2 Comparison of optimization results

Parameters	1	2	3	4	5	6	7	8	9
Real value	4	6	6.25	6.95	8.1	1.5	10	20	18
NGA	3.955	5.970	6.222	6.92	1 8.111	1.478	8.278	20.560	1.903
DE	3.830	6.102	6.282	7.02	2 8.066	1.725	11.055	19.741	18.612
PSO	3.907	5.983	6.237	6.96	9 8.099	1.347	9.908	20.312	17.993
FWA	4.043	5.998	6.242	6.94	6 8.095	1.521	9.945	19.849	18.079

16.8. As can be seen, PSO and FWA outperform GA and NGA, and FWA achieves the best results.

The inversion result is very close to the real condition (see Table 16.2).

The whole results show that FWA can inherit the rapid convergence characteristics of swarm intelligence algorithms. It also has the same ability of searching global optimization as GA and PSO so that it can get better results.

16.5 Summary

This chapter described several successful application cases of fireworks algorithm, including document clustering, spam detection, image recognition, and seismic inversion problem. All experimental results show that fireworks algorithm is indeed a promising approach to handle such practical problems and their like.

In the future, the performance of FWA is believed to have better performance while the run time is smaller, and could be applied to deal with a lot of application problems.

References

1. Y. Tan, Swarm robotics: collective behavior inspired by nature. J. Comput. Sci. Syst. Biol. (JCSB) **6**, e106 (2013)
2. G. Jones, A. Robertson, C. Santimetvirul, P. Willett, Non-hierarchic document clustering using a genetic algorithm. Inf. Res. **1**(1), 1–1 (1995)
3. V.V. Raghavan, K. Birchard, A clustering strategy based on a formalism of the reproductive process in natural systems, in *ACM SIGIR Forum*, vol. 14(2). (ACM, New York, 1979), pp. 10–22
4. X. Cui, T.E. Potok, P. Palathingal, Document clustering using particle swarm optimization, in *Proceedings 2005 IEEE Swarm Intelligence Symposium. SIS 2005*, (IEEE, 2005), pp. 185–191
5. Y. Tan, Y. Zhu, Fireworks algorithm for optimization, in *Advances in Swarm Intelligence* (Springer, Berlin, 2010), pp. 355–364
6. Y. Tan, S. Zheng, Research progress on fireworks algorithm. CAAI Trans. Intell. Syst. **9**(10), 1–17 (2014)

7. Y. Tan, C. Yu, S. Zheng, K. Ding, Introduction to fireworks algorithm. Int. J. Swarm Intell. Res. (IJSIR) **4**(4), 39–70 (2013)
8. B.S. Everitt, S. Landau, M. Leese, D. Stahl, *Cluster Analysis*. Wiley series in Probability and Statistics (Wiley, New York, 2011). http://books.google.com.hk/books?id=w3bE1kqd-48C. ISBN: 9780470978443
9. S.C. Gu, Y. Tan, X. He, Recentness biased learning for time series forecasting. Inf. Sci. **237**, 29–38 (2013)
10. T. Joachims, A probabilistic analysis of the Rocchio algorithm with TFIDF for text categorization. Technical report, DTIC Document (1996)
11. J. Wang, Y. Tan, Efficient euclidean distance transform algorithm of binary images in arbitrary dimensions. Pattern Recognit. **46**(1), 230–242 (2013)
12. I. Jolliffe, *Principal Component Analysis* (Wiley Online Library, New York, 2005)
13. L.F. Cranor, B.A. LaMacchia, Spam! Commun. ACM **41**(8), 74–83 (1998)
14. F. Research, Spam, spammers, and spam control: a white paper by ferris research. Technical report (2009)
15. Y. Tan, Y. Zhu, Advances in anti-spam techniques. CAAI Trans. Intell. Syst. **5**(3), 189–201 (2010)
16. Y. Tan, Particle swarm optimizer algorithms inspired by immunity-clonal mechanism and their application to spam detection. Int. J. Swarm Intell. Res. **1**(1), 64–86 (2010)
17. J. Zhang, Y. Tan, L. Ni, C. Xie, Z. Tang, Hybrid uniform distribution of particle swarm optimizer. IEICE Trans. Fundam. Electron. Commun. Comput. Sci. **E93-A**(10), 1782–1791 (2010)
18. G. Ruan, Y. Tan, A three-layer back-propagation neural network for spam detection using artificial immune concentration. Soft Comput. **14**(2), 139–150 (2010)
19. Y. Zhu, Y. Tan, Extracting discriminative information from E-mail for spam detection inspired by immune system, in *2010 IEEE Congress on Evolutionary Computation (CEC)*. (IEEE, 2010), pp. 1–7
20. Y. Zhu, Y. Tan, A local-concentration-based feature extraction approach for spam filtering. IEEE Trans. Inf. Forensics Secur. **6**(2), 486–497 (2011)
21. Y. Tan, C. Deng, G. Ruan, Concentration based feature construction approach for spam detection, in *International Joint Conference on Neural Networks, 2009. IJCNN 2009*. (IEEE, 2009), pp. 3088–3093
22. G. Mi, P. Zhang, Y. Tan, A multi-resolution-concentration based feature construction approach for spam filtering, in *The 2013 International Joint Conference on Neural Networks (IJCNN)* (IEEE, 2013), pp. 1–8
23. G. Mi, P. Zhang, Y. Tan, Feature construction approach for email categorization based on term space partition, in *The 2013 International Joint Conference on Neural Networks (IJCNN)* (IEEE, 2013), pp. 1–8
24. J. Zhang, Y. Tan, L. N, C. Xie, Z. Tang, AMT-PSO: an adaptive magnification transformation based particle swarm optimizer. IEICE Trans. Fundam. Electron. Commun. Comput. Sci **E94-D**(4), 786–797 (2011)
25. Y. Tan, J. Wang, A support vector network with hybrid kernel and minimal Vapnik-Chervonenkis dimension. IEEE Trans. Knowl. Data Eng. **26**(2), 385–395 (2004)
26. H. Drucker, D. Wu, V.N. Vapnik, Support vector machines for spam categorization. IEEE Trans. Neural Netw. **10**(5), 1048–1054 (1999)
27. S.C. Gu, Y. Tan, X.G. He, Discriminant analysis via support vectors. Neurocomputing **73**(10–12), 1669–1675 (2010)
28. Y. Tan, G.C. Ruan, Uninterrupted approaches for spam detection based on SVM and AIS. Int. J. Comput. Intell. Pattern Recognit. (IJCIPR) **1**(1), 1–26 (2014)
29. W. He, G.Mi, Y. Tan, Parameter optimization of local-concentration model for spam detection by using fireworks algorithm. Advances in Swarm Intelligence (Springer, Berlin 2013), pp. 439–450
30. X. Huang, Y. Tan, X.G. He, An intelligent multi-feature statistical approach for discrimination of driving conditions of hybrid electric vehicle. IEEE Trans. Intell. Transp. Syst. **12**(2), 453–456 (2011)

31. Y. Tan, J. Wang, Recent advances in finger vein based biometrics techniques. CAAI Trans. Intell. Syst. **6**(6), 471–482 (2011)
32. A.W.K. Kong, D. Zhang, Competitive coding scheme for palmprint verification, in *2004 Proceedings of the 17th International Conference on Pattern Recognition. ICPR 2004*, vol. 1. (IEEE, 2004), pp. 520–523
33. X. Wu, K. Wang, D. Zhang, Palmprint authentication based on orientation code matching. *Audio-and Video-Based Biometric Person Authentication* (Springer, Berlin, 2005), pp. 83–132
34. S.C. Gu, Y. Tan, X.G. He, Laplacian smoothing transform for face recognition. Sci. China (Information Science) **53**(12), 2415–2428 (2010)
35. W. Jia, D.S. Huang, D. Zhang, Palmprint verification based on robust line orientation code. Pattern Recognit. **41**(5), 1504–1513 (2008)
36. Z. Guo, W. Zuo, L. Zhang, D. Zhang, A unified distance measurement for orientation coding in palmprint verification. Neurocomputing **73**(4), 944–950 (2010)
37. Z. Shaoqiu, Y. Tan, A unified distance measure scheme for orientation coding in identification, in *2013 IEEE Congress on Information Science and Technology,* (IEEE, 2013), pp. 979–985
38. A.W.K. Kong, Palmprint identification based on generalization of iriscode. Ph.D. thesis, University of Waterloo (2007)
39. D. Bratton, J. Kennedy, Defining a standard for particle swarm optimization, in *IEEE Swarm Intelligence Symposium, 2007. SIS 2007.* (IEEE, 2007), pp. 120–127
40. R. Storn, K. Price, Differential evolution-a simple and efficient heuristic for global optimization over continuous spaces. J. Glob. Optim. **11**(4), 341–359 (1997)
41. J. Kennedy, R. Eberhart, Particle swarm optimization, in *Proceedings IEEE International Conference on Neural Networks*, vol. 4 (IEEE, 1995), pp. 1942–1948
42. Polyu palmprint database. URL: http://www.comp.polyu.edu.hk/~biometrics/MultispectralPalmprint/MSP.htm
43. Y. Tan, J. Wang, Nonlinear blind separation using higher-order statistics and a genetic algorithm. IEEE Trans. Evol. Comput. **5**(6), 600–612 (2001)
44. D. Hampson, B. Russell, B. Bankhead et al., Simultaneous inversion of pre-stack seismic data. SEG Tech. Progr. Expand. Abstr. **24**, 1633–1637 (2005)

Chapter 17
Group Explosion Strategy for Multiple Targets Search in Swarm Robotics

Swarm robotics is an emerging research area combining swarm intelligence and robotics. Thanks to the recent achievements in optimization problem using swarm intelligence, searching problems in swarm robotics have attracted a large number of researchers. In searching problems, a swarm of robots searches for multiple targets in the environment without knowing any prior knowledge about the targets. This progress is quite similar with that of optimization problems in many aspects. Moreover, in most of the swarm robotics searching problems so far, some kinds of fitness functions are introduced for guiding the search of the swarm. This makes it a natural advantage to introduce swarm intelligence algorithms into swarm robotics. In this chapter, inspired by the fireworks algorithm, the group explosion strategy (GES) is proposed for searching multiple targets in swarm robotics. In the GES model, the whole swarm is divided into several groups. Robots in a group are spatially adjacent within the sensing range of each other. The swarm searches and collects targets in the environment without prior knowledge. Different groups do not intersect directly and their search for targets is parallel and independent. Through certain strategies, groups that run into each other will be re-arranged into new groups with possibly different members and search directions. In this way, inter-group cooperation can emerge in the swarm. The simulation results indicate that the proposed method with GES in this chapter shows great advantage against the comparison algorithm inspired from PSO.

17.1 Introduction

Swarm robotics has achieved significant progress benefiting from the development of artificial intelligent [1, 2]. Many potential applications exist for the deployment of a swarm of robots [3], especially those require large amount of agents and time or are dangerous to human being, e.g., foraging, surveillance, monitoring and search, and rescue operations [1, 4]. In general, these applications can be regarded as search-and-explore tasks in unknown environments. Therefore, searching strategy is an important challenge for swarm robotics researchers.

© Springer-Verlag Berlin Heidelberg 2015
Y. Tan, *Fireworks Algorithm*, DOI 10.1007/978-3-662-46353-6_17

Controlling a swarm of robots is still a challenge in the robotic area despite its fast development. Robots in the swarm should have as limited functions or abilities as possible, including motion ability, energy storage, sensing, communication, and computation capability due to their size, power constraints, cost, and maintenance issues. Thus, cooperation plays the most important role in the swarm robotics control strategies to distribute and share resources across the swarm to complete the task [5]. PSO [6], inspired from birds flocking, is the most common swarm intelligence algorithm introduced for motivating swarm robotics for its simplicity and similarity with real robots [7]. Doctor et al. [8] are one of the first to use PSO for multi-robot searching though they mainly focus on optimizing the model parameters. Pugh and Martinoli [9] and their follow-up work [10] designed an effective searching algorithm inspired from PSO modified with various topologies. Hereford and Siebold [11] developed a distributed particle swarm optimization algorithm and used it for real robots in a physically embedded version. Xue et al. [12] presented their PSO application for robots in target searching with a parallel asynchronous control strategy.

However, there still remain many problems when adopting PSO in swarm robotics searching applications, such as the disadvantages of PSO as well as differences between optimization problems and searching applications which cannot be neglected. These problems can be named: large amount of random movements, trapping in local minimal, speed limitation [13], and others. To solve these problems, some researchers divide the swarm into sub-groups for better cooperation to accelerate the searching progress. Xue et al. [14] introduced a mechanism for predicating target positions using information from at least three neighbors. Couceiro et al. [15] proposed a RDPSO that involves dynamic sub-grouping of the whole swarm. However, the main problem of this research is that the robots in the swarm require global communication for arranging sub-groups, which is normally unavailable for large-scale outdoor applications. Therefore, new strategies capable of solving these problems should be proposed.

Inspired from the explosion phenomenon in fireworks algorithm, a swarm robotics searching strategy, referred to as "group explosion strategy" [16], is proposed in this chapter. In the proposed method, the entire swarm is divided into sub-groups in a self-organizing way and each group searches for the targets independently. Robots maintain the group structure and may split into smaller groups during the search. The strategy provides both intra-group cooperation and inter-group cooperation within the swarm. The sub-grouping strategy can overcome the problems mentioned above under limited sensing constraints.

17.2 Problem Statement

In this problem, a swarm of robots is applied to solve the problem of searching multiple targets in obstructive environment. The swarm has no prior information about the environment and targets. The problem is simulated in a computer program and time is divided into discrete iterations. For every iteration, each individual first retrieves information from environment and their neighbors, and then moves according to the sensing results. Maximum speed of robots is restricted so that robots' movements are guaranteed to complete before next iteration which is not too long. In real-life applications, all robots in the swarm are not restricted to share the same iteration cycle.

Fitness functions in swarm robotics are usually developed to measure the distance between robot and the target(s), such as adopting Euclidean distance directly [5], using olfaction measurements [17] or some type of potential functions [18]. However, these functions are usually continuous and the robots can easily converge to the target position with gradient descent methods. In real applications, on-board sensors of swarm robotics should be as cheap and simple as possible and may not detect the fitness values in such high accuracies limited abilities. Instead, sensors provide a limited number of rounded sensing results. Therefore, discrete fitness functions, rather than continuous one, are taken into consideration in this chapter.

The problem is quite similar with the searching problems in other researches [14, 15, 19], except that the fitness values detected by the robots are discrete. The details of the problem are defined as follows: there exists m targets in the environment and information of a target can only be sensed as fitness values, which are discrete values inversely proportional to the distance from the target, as shown in Fig. 17.1. The aim of this problem is to search and collect the targets. It takes 10 iterations for an individual to collect a target and the cooperation of multiple individuals collecting one target in the same time can accelerate this progress.

Each target has a randomly generated fitness ranging from F_{Max}-2 to F_{Max}, where F_{Max} is a pre-defined constant and is set to 20 in our experiments. Considering sensitivity and errors in real-time application, fitness values sensed by the robots are

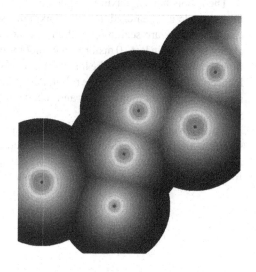

Fig. 17.1 A screenshot of the problem at the beginning of the simulation. *Red circles* stand for the targets. The *background color* illustrates fitness of that position. Robots and obstacles are not illustrated in this figure (Color figure online)

discrete values ranging from F_{Max} to 0. A target disappears when it is collected by the swarm and its fitness can be no longer sensed.

Each target is shaped as a circle with a radius of $Size_t$. The target is identified as found if a robot overlaps with the ring and the robot can start collect the target at the position. A ring with a same radius of $Size_t$ outside the target has the same fitness as the target. The fitness reduces by 1 when the distance from the target is increased by $2Size_t$ until the fitness drops to 0. This indicates that fitness values are shaped as a ring with width of $2Size_t$ with increased radius and reduced fitness value. When fitness of several targets overlap, the largest value is adopted. Value of constant $Size_t$ is set to 10.

A swarm of n autonomous mobile robots is used to solve this problem. The robots are designed to be as simple as possible and modeled as squares able to move freely in environment. The swarm has no leader or unique IDs and share no common coordinate systems or global position systems. Each robot can sense the fitness of its current position and has a limited sensing range to detect the relative positions of their neighbor robots. They have a limited memory of past states of themselves. Each robot normally does not communicate explicitly with other robots except when sharing current fitness to their neighbors. Each individual executes the same algorithm but acts independently and asynchronously from others.

Each robot has the ability to sense neighbor robots within the range of $4Size_t$. Since no global positioning system is available, the positions are relative which can be detected without direct communications with the help of an infrared sensor and angle transducer. If F_{Max} is quite small, the robots can detect neighbors' fitness values without direct communication through colored lights equipped on the robot. Otherwise, just like the situation in this chapter, robots share their fitness values to all their neighbors through direct communications or other strategies which is not focused in this chapter.

The robots have a maximum speed of $2Size_t$ per iteration so that they can past a fitness grade in one iteration at the maximum speed and react quickly to the fitness changes if they are searching at the right directions. The individuals also have the ability to maintain last 10 history states including past position and the corresponding fitness values. Positions in the history are relative positions updated according to the local coordinating system of the robot. Past states cannot be shared among the swarm since complex communications and localizations are required for sharing the past states.

Static obstacles are also introduced in the problem. Each obstacle is regarded as a square with different sizes. Collisions are detected every iteration in the simulation and any robots that run into an obstacle are considered as broken and will be removed from the swarm permanently while the obstacle remains still in the environment.

17.3 Group Explosion Strategy

17.3.1 Introducing Fireworks Explosion into Swarm Robotics

In this section, the group explosion strategy (GES) designed for searching multiple targets is explained in detail. In GES model, the whole swarm is divided into several groups. Robots in a group are spatially adjacent, i.e., within the sensing range of each other. Different groups do not intersect directly and their search for targets is parallel and independent. However, through certain strategies, groups that run into each other will be re-arranged into new groups with possibly different members and search directions. In this way, inter-group cooperation can emerge in the swarm.

Explosion strategies are introduced into GES for searching multiple targets. The comparison of an explosion process and the corresponding actions in GES is shown in Fig. 17.2 which are inspired from the fireworks algorithm for optimization problems in [20], since both algorithms are inspired from explosion phenomenon. An explosion starts from a point and generates several sparks with different distances around the initial point. The center point can be regarded as a robot and the sparks exploded are the neighbor robots. These robots as a whole aggregates into a group, randomly distributed around the center robot just like the sparks. In each iteration, robots process sensing data and decide their movements. In this way, a new explosion center is selected and the group explodes again in the next iteration to search near the new center.

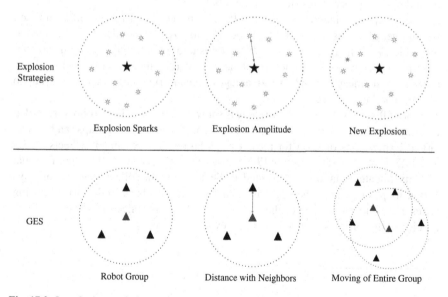

Fig. 17.2 Introducing explosion strategies into swarm robotics

17.3.2 Framework of Group Explosion Strategy

A group of robots can converge to the target much more quickly than a single robot with the help of intra-group cooperation, since the trends of the fitness in the environment are clearer within the group than a single individual. The more the robots in a group, the quicker the group can converge to a target, yet resulting in fewer targets, and the swarm is searching simultaneously since the number of groups is reduced. If the sizes of groups become too large, searching efficiency of the entire swarm declines instead. Therefore, a balanced group size should be adopted and the swarm can thus take the advantage of quick convergence from intra-group cooperation as well as searching several targets in parallel as inter-group cooperation. With a carefully designed strategy, the swarm can search and collect the targets more efficiently.

In GES model, robots first retrieve information from the environment and their neighbors, then calculate their new movements of this iteration (referred as velocity) according to the sensing data, and carry out the move before next iteration. The state of this iteration is stored in history before robots actually move towards the new positions. In this way, the history of each robot contains 10 past states of the robots excluding their current states.

Velocity of robot i at every iteration consists two components: grouping component $G(i)$ and history component $H(i)$. $G(i)$ controls the robots' behavior relevant to grouping and is calculated according to the current fitness of the robot and states of its neighbor including their relative positions and fitness values. $H(i)$ is computed from the past history states stored in the robot.

The grouping behavior of a robot differs with the size of its current group, i.e., the number of robots in its neighborhood. Group size is controlled by a pre-defined threshold β_G. When the size exceeds the threshold, robots in the group try to split this group into two smaller ones to balance the search efficiency. Otherwise, robots try to maintain the group and take advantage of the information shared in the group to benefit searching progress.

A brief flow chart of the GES strategy is shown in Fig. 17.3. Each robot is regarded as a finite-state machine with three states: group search, split groups, and collect target. Expressions of $G(i)$ for these two situations are explained in Sects. 17.3.3 and 17.3.4, respectively. Section 17.3.5 gives the expression of $H(i)$ and the final velocity update equation is presented in Sect. 17.3.6. In collect state, robot stays still at its position until the target is collected and goes back to the two searching states according to the size of the group. Several robots collecting same target can accelerate this process as mentioned in previous section.

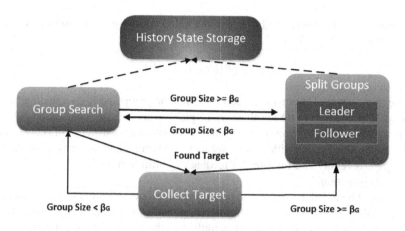

Fig. 17.3 Flow chart of GES

17.3.3 Searching in Groups

When the size of group is beyond the threshold β_G, the group tries to search for a target through intra-group cooperation. Following the strategies in fireworks explosion, searching in next iteration should take place in the area around the current best position found in the group. Therefore, the searching strategy of the group is to move the group center towards the best position within the group. The best position is selected as the current position of the robot with highest fitness value among all the robots in the group. Historical positions of any robots are not taken into account, since the communication overload for exchanging historical states is too large. In this way, the group should steadily converge to a target much quicker than only one robot.

The group center is calculated as the centroid of all the robots in the group. The group center $C(i)$ for robot i is calculated as follows:

$$C(i) = \frac{\sum_{j \in N(i)} P(j) + P(i)}{|N(i)| + 1}, \tag{17.1}$$

where $P(i)$ is the current position of robot i and $N(i)$ is the collection consisting of all neighbor robots of robot i, i.e., all other robots within the group.

It can be easily seen from Eq. (17.1) that although robots do not exchange their positions with each other directly, they can calculate the same centroid of the group if no noise and error is considered. Even with little noise and error from the sensors, the result could be quite similar and should not affect the cooperative scheme.

Given the group center, the robot can compute the grouping component $G(i)$ used in the velocity update equation. Before calculating, the robots in the group should exchange the fitness information among the group with the help of direct communication or colored lights due to the hardware design of the robots. $G(i)$ is computed using Eq. (17.2).

$$G(i) = (P(b) - C(i)) * R_S, \qquad (17.2)$$

where robot b is the robot with the best fitness in the group and R_S is a scaling factor randomly selected from three values with equal possibilities: $1 - \beta_S$, 1, and $1 + \beta_S$. β_S is used to adjust the shape of the group by controlling the distance of robots from the group center. Robots with various distances from the center can provide the group with great diversity and meaningful feedbacks in different scales for choosing the searching direction.

It should be noted in GES that robots do not explicitly tell their neighbors which robot has the best fitness. Instead, they exchange their fitness with all their neighbors. Each robot computes the best robot according to the fitness values it receives independently. Therefore, robots in the group may choose different best robots if noise or error occurs. However, the sensing distance is twice the time of the maximum speed of the robot, so it could be possible that the robot still remains in the group. If a robot goes out of the group, it will start searching the targets itself and rejoin a group if encounters one. The most possible place for rejoining a group is near a target, since robots always try to converge into a target even if a single robot without group.

17.3.4 Splitting Groups

When group size exceeds the threshold β_G, the strategy for the robots in the group is splitting the large group into two smaller ones. Robots with the best two fitness, denoted as $L1$ and $L2$, are selected as two leaders for the new groups. Without loss of generality, it is assumed that fitness of $L1$ is not worse than $L2$. To separate the two groups apart, two opposite directions are selected for the new groups. Two leaders repulse with each other and try to be as far away as possible. The repulsive vector is calculated in Eq. (17.3):

$$\begin{aligned} R(L1) &= (P(L1) - P(L2)) * \beta_R \\ R(L2) &= (P(L2) - P(L1)) * \beta_R \end{aligned}, \qquad (17.3)$$

where β_R is a pre-defined coefficient for repulsing the two leaders $L1$ and $L2$.

As for other robots in the group, each of them selects a leader to follow independently and randomly. Each of $L1$ and $L2$ is given a weight for selecting, the higher the weight, the more change a leader is selected since a leader with higher fitness is more possible to find a target. Besides, the leaders are repulse to each other and thus the robot $L1$ is quite possibly being repulsed towards a target. The weights of two leaders are assigned based on their fitness in Eq. (17.4) and the possibility for selecting leaders is calculated in Eq. (17.5):

$$w(L1) = F(L1) - F(L2) + \beta_P, \; w(L2) = \beta_P, \qquad (17.4)$$

$$P_{L1} = \frac{w(L1)}{w(L1) + w(L2)}, \quad P_{L2} = \frac{w(L2)}{w(L1) + w(L2)}, \tag{17.5}$$

where function F indicates the current fitness of the robot; $w(L1)$, $w(L2)$, P_{L1}, and P_{L2} are the weights and possibilities of two leaders, respectively; and β_P is a constant to balance the weights and fixed to 1 in this chapter.

Since the sensing range of a robot is double of size of a fitness value can occupy, the value of $F(L1) - F(L2)$ is restricted to be zero or one. This guarantees that the difference between the two weights is not too large that makes the new group sizes very unbalanced.

The grouping components of the leaders and other robots can be now calculated using Eq. (17.6):

$$G(i) = \begin{cases} R(i) & , \quad i = L1, L2 \\ R(i) + (P(l) - P(i)) * R_S, & \text{otherwise} \end{cases}, \tag{17.6}$$

where l is the leader robot i has selected. The leaders just repulse each other to separate the groups. Other robots use the same repulse vector with the leader and the new groups move apart as a whole to separate with each other. Besides separating, distances between robots and their leaders are also changed by the factor R_S which is the same as mentioned in previous section.

17.3.5 History State Storage

History component $H(i)$ is independent with the grouping situations and is computed based only on the stored states in robot's history. In GES, only ten latest history states (position and fitness) are maintained. When robots search in a wrong direction and depart the targets (fitness observed is decreasing), they can get back to route if making full use of the history components as computed in Eq. (17.7):

$$H(i) = (P(i) - h(i)) * r, \tag{17.7}$$

where $h(i)$ is the position of the history state with the best fitness and r is a random number uniformly distributed within the range $[0.4, 0.8]$.

If several history states share the same fitness, the most recent position is selected. Note that when counting $h(i)$, current position is also taken into account to make sure that the robot is not attracted by a worse position. Thus, the history component is set to be 0 if fitness of current position is better than all the states in history and will not contribute to the velocity update of the robot as the history state can provide no positive information to guide the search.

17.3.6 Velocity Update Equation

After calculating the two components $G(i)$ and $H(i)$, the velocity of robot i at iteration t, denoted as $V_t(i)$, is updated as follows:

$$V_t(i) = \begin{cases} G(i) + H(i), & \|G(i)\| > 0 \\ H(i) + R_p, & \|G(i)\| = 0 \wedge \|H(i)\| > 0, \\ V_{t-1}(i), & \|G(i)\| = 0 \wedge \|H(i)\| = 0 \end{cases} \qquad (17.8)$$

where R_P is a randomly generated unit vector and $V_{t-1}(i)$ is the velocity of the last iteration.

In the equation, velocity update strategy is different if $\|G(i)\|$ is 0. Since no grouping action is taking place when $\|G(i)\| = 0$, only the history component remains in the equation. This will lead to a vibration around the history best position if the robot gets stuck in an area with same fitness value. A small random vector R_P is introduced to avoid such situation. R_P is an unit vector while $H(i)$ is normally quite large, so the movement of the robot is not too stochastic.

The situation that both $G(i)$ and $H(i)$ are 0 usually occurs when the robot is the best of the group and fitness is improving in recent iterations, so the robot just remains its searching direction as the previous iteration.

17.3.7 Obstacle Avoidance

Obstacles are also considered in GES. Obstacle avoidance is computed stand alone after the velocity update and is not used in the $V_{t-1}(i)$ in Eq. (17.8). Since obstacle avoidance is not the main concern of this problem, a simple avoiding scheme is used for both GES and the baseline algorithm (introduced in Sect. 17.4.1). Each robot checks if it will run into any obstacles with the velocity $V_t(i)$. If so, it adds a small repulsive force perpendicular to $V_t(i)$ from the obstacle to make sure that a collision will not happen. This simple scheme can provide acceptable performance for obstacle avoidance from the simulation results in Sect. 17.4.4.

17.4 Simulation Results and Discussions

In this section, comparison between GES and a searching algorithm inspired from standard PSO is presented. We first introduce the baseline algorithm briefly and present the simulation results of several experiments. Two methods are simulated in a self-built simulation platform [21] and tested under different situations. The first experiment is to validate the GES to see if it is capable of solving the problem. Various population sizes and number of targets are considered in this experiment.

The second experiment is the scalable experiment which has the same condition as the first one except the scales of swarm size, and a number of target and map sizes are enlarged. Obstacles are introduced in the last experiment and results are compared with the same situation without obstacles.

In most of the experiments, the stop criteria are to collect a certain percent of the targets and the essential standard for judging the performance is the total iteration used. For the problem definition, it can be easily seen that fitness becomes inadequate as large areas with 0 fitness appear in the environment as targets are collected or in large maps. This means that the search of a swarm becomes harder as the experiment goes.

All the experiments in this section use the same environment setup: 20 randomly generated maps are used and each method is repeated for 20 times in the map of size $500 * 500$. Average results of these 400 runs are presented in results. Parameters of the two methods are tuned in advance with 10 robots and 20 targets.

17.4.1 Baseline Algorithm

In the experiment, the baseline algorithm uses the same strategy of RPSO in [15]. Each robot acts as a particle and the spacial-based topology of the robots for calculating gbest is adopted. However, RPSO requires a large communication range for the swarm which is restricted in the problem we consider, thus a little modification is introduced. In case the robot vibrates in an area, the small random vector R_P in GES is introduced if both pbest and gbest are the current position. Except the velocity update strategy, baseline algorithm shares the same scheme with GES for better comparison, such as obstacle avoiding and history state updating.

17.4.2 Validation Experiment

The first experiment validates the GES method and compares the iterations used for two methods to collect half and all the targets in the environment, respectively. The experiment is simulated in a small-scale setup: map size is $500 * 500$, population n ranges from 5 to 25, and the number of targets m varies from 10 to 30.

The results are shown in Tabel 17.1. In the table, results for collecting half targets and all targets are presented in different columns. "Iteration" stands for the iterations used to collect the targets and is the most important criterion to evaluate the performance. "Collect" stands for the targets collected in the "Half Target" situation as several targets may be collected in the same iteration. "Distance" stands for the averaged moving distance for each robot in the entire searching process, and "Time" indicates the average simulation time on PC in milliseconds which is used to evaluate the computation overload. The last two columns indicate the ratio of iterations between GES and RPSO.

Table 17.1 Validation results

m	n	GES						RPSO						GES/RPSO	
		Half target		All target				Half target		All target				Iteration	
		Iteration	Collect	Iteration	Distance	Time		Iteration	Collect	Iteration	Distance	Time		Half (%)	Full (%)
10	5	162.6	5.0	369.66	3285.16	238.2		178.7	5.0	763.67	3184.45	243.5		91.0	48.4
	10	133.4	5.0	271.09	2414.35	242.4		145.1	5.0	558.92	2456.72	250.9		91.9	48.5
	15	124.1	5.0	233.11	2066.17	246.4		132.2	5.0	487.40	2200.03	259.2		93.9	47.8
	20	123.4	5.0	215.86	1905.82	252.7		124.1	5.0	457.59	2094.18	266.8		99.4	47.2
	25	125.7	5.0	206.81	1819.75	259.9		122.3	5.0	424.48	1973.48	275.5		102.7	48.7
20	5	273.8	10.0	574.06	4951.14	992.9		277.1	10.0	826.82	4151.13	997.1		98.8	69.4
	10	191.5	10.0	388.22	3372.75	1000.1		226.7	10.0	600.58	3335.97	1000.8		84.5	64.7
	15	159.6	10.0	318.80	2774.25	1001.2		201.8	10.0	529.61	3054.10	1014.0		79.1	60.2
	20	152.4	10.1	285.26	2481.51	1010.1		190.3	10.0	487.58	2885.94	1024.9		80.1	58.5
	25	155.1	10.1	273.29	2378.51	1017.0		185.6	10.0	471.03	2822.60	1037.4		83.5	58.0
30	5	339.8	15.0	728.57	6167.18	2265.0		336.7	15.0	972.59	5095.47	2256.0		100.9	74.9
	10	217.5	15.0	457.15	3891.14	2267.2		255.6	15.0	676.36	3934.55	2273.1		85.1	67.6
	15	188.1	15.0	373.94	3202.25	2268.1		237.0	15.0	595.55	3657.55	2275.6		79.3	62.8
	20	171.2	15.0	327.33	2811.04	2272.7		221.4	15.0	550.29	3479.02	2293.6		77.4	59.5
	25	175.3	15.1	315.63	2723.57	2291.4		203.7	15.1	507.25	3260.95	2307.8		86.1	62.2

From the table, it can be easily seen that GES dominates RPSO in the performance regardless of population and number of targets, especially when collecting all targets. GES can complete most of the missions with only 70–90 and 50–70 % of the iterations than that of RPSO in "Half" and "All" situations, respectively. It also shows that GES has a smaller total moving distance which means that GES is more energy efficient than RPSO. Considering the smaller iterations, GES utilizes the maximum speed much better than RPSO which is one of the important reasons why GES has a better efficiency. In the table, GES also has a shorter time which indicates that the GES strategy requires less computation resources than RPSO. This is important for swarm robotics applications since the robots normally do not have powerful computation abilities.

The results also show that GES outperforms RPSO more when population size is larger. This indicates that the strategy introduced in GES shows great ability in cooperation among robots and can accelerate the searching process when there are more robots in the environment. When the number of targets increases, the average iteration GES used to collect one target decreases quicker than RPSO which also indicates that the strategy is taking full advantage of cooperation among robots.

17.4.3 Scalable Experiment

The main aim of this experiment is to see if the proposed GES strategy can scale to problems with larger number of robots, targets, or map sizes. In this experiment, environment is setup much bigger than previous experiments: map size is $1000 * 1000$, n ranges from 10 to 50, and m varies from 10 to 100. Both methods use the same parameter with the previous experiments. Our results showed that the trends of two algorithms are quite the same than in previous sections. GES has great advantages when population is large. It can also be induced that the advantages are not that large when the number of targets is increased. The main reason is that the map size is not large enough, so the environment is full of fitness values even when only 25 % of targets remains. As shown in previous section, advantage of GES is smaller when fitness is adequate.

17.4.4 Obstacle Avoidance Experiment

In this experiment, we test the performance of the proposed algorithm in environments with small static obstacles. Obstacles are randomly distributed in the environment and the number ranges from 0 to 100. Performances in obstacle environments are compared with that of non-obstacle environment. In this experiment, the changes of the two numbers, population m and number of targets n, are tested in a more conscientious way, with a step of 2 instead of 5 and 10 in previous experiments. Three main criteria are considered in this experiment: number of surviving robots,

"Iteration," and "Distance." The last one indicates the remaining robots after the simulation and is used to judge how many collisions were taking place. The stop criteria are set as collecting 75 % of the targets.

Our results showed that GES shows advantages in avoiding the obstacles than RPSO even with the same obstacle avoidance strategy. There are possibly two reasons for such results. The first one is that GES takes much less iterations than RPSO and thus has a lower possibility of encountering obstacles. The second one may be the cooperation schemes in GES. Although the strategy does not include a direct cooperation of obstacle avoidance, the swarm do benefit from the cooperation. The robots that successfully avoided any obstacles usually have lower fitness values in the group since they took a longer way round. Therefore, the groups will possibly move towards other directions and thus avoid the obstacles indirectly.

The trends of "Iteration" and "Distance" are very similar for both GES and RPSO strategies. Robots take shorter moves per iteration to avoid more obstacles which leads to larger iterations. However, it is obvious that GES has a much better stability than RPSO when the environment setups change, as the points in the figure of GES are much more compact. Stability can be also indicated from the trend of the two algorithms as RPSO is more sloped. If comparing the mean value of all the results with same number of obstacles, two strategies are quite the same when few obstacles exist, but GES becomes better when the number of obstacles increases. Considering the big differences of "Iteration" and "Distance" from the results of previous experiments, GES definitely outperforms RPSO in environments with obstacles.

17.5 Summary

Inspired by the Fireworks Algorithm, a group explosion strategy (GES) for searching multiple targets was proposed, which is applied to the multiple targets searching problem on a self-built simulation platform. The swarm searches and collects targets in the environment without prior knowledge. Several tests are run to evaluate how GES performs in various aspects including stability, robustness, and flexibility. Simulation results demonstrate that GES shows great efficiency when fitness is either adequate or inadequate in the environment. GES also shows good stability in obstructive and large-scale environments. These results indicate that the GES strategy has great ability in cooperating robots to accomplish the tasks.

References

1. Y. Tan, Z.Y. Zheng, Research advance in swarm robotics. Def. Technol. **9**(1), 31–62 (2013)
2. Q. Tang, P. Eberhard, Cooperative motion of swarm mobile robots based on particle swarm optimization and multibody system dynamics. Mech. Based Des. Struct. Mach. **39**(2), 179–193 (2011)

3. R. Grabowski, L.E. Navarro-Serment, P.K. Khosla, An army of small robots. Sci. Am. **18**, 34–39 (2008)
4. Y. Tan, Swarm robotics: collective behavior inspired by nature. J. Comput. Sci. Syst. Biol. (JCSB) (2013)
5. K. Derr, M. Manic, Multi-robot, multi-target particle swarm optimization search in noisy wireless environments, in *2009 2nd Conference on Human System Interactions. HSI'09* (IEEE, 2009), pp. 81–86
6. R. Eberhart, J. Kennedy, A new optimizer using particle swarm theory, in *1995 Proceedings of the Sixth International Symposium on Micro Machine and Human Science, MHS'95* (IEEE, 1995), pp. 39–43
7. D. Gong, C. Qi, Y. Zhang, M. Li, Modified particle swarm optimization for odor source localization of multi-robot, in *2011 IEEE Congress on Evolutionary Computation (CEC)*. (IEEE, 2011), pp. 130–136
8. S. Doctor, G.K. Venayagamoorthy, V.G. Gudise, Optimal PSO for collective robotic search applications, in *2004 Congress on Evolutionary Computation, CEC2004*, vol. 2 (IEEE, 2004), pp. 1390–1395
9. J. Pugh, A. Martinoli, Multi-robot learning with particle swarm optimization, in *Proceedings of the Fifth International Joint Conference on Autonomous Agents and Multiagent Systems* (ACM, 2006), pp. 441–448
10. J. Pugh, A. Martinoli, Inspiring and modeling multi-robot search with particle swarm optimization, in *2007 IEEE Swarm Intelligence Symposium, SIS 2007* (IEEE, 2007), pp. 332–339
11. J. Hereford, M. Siebold, Multi-robot search using a physically-embedded particle swarm optimization. Int. J. Comput. Intell. Res. **4**(2), 197–209 (2008)
12. S. Xue, J. Zhang, J. Zeng, Parallel asynchronous control strategy for target search with swarm robots. Int. J. Bio-Inspired Comput. **1**(3), 151–163 (2009)
13. J.G. Li, J. Yang, S.G. Cui, L.H. Geng, Speed limitation of a mobile robot and methodology of tracing odor plume in airflow environments. Procedia Eng. **15**, 1041–1045 (2011)
14. S. Xue, Y. Zan, J. Zeng, Z. Xue, J. Du, Group decision making aided PSO-type swarm robotic search, in *2012 International Symposium on Computer, Consumer and Control (IS3C)* (IEEE, 2012), pp. 785–788
15. M.S. Couceiro, R.P. Rocha, N.M.F. Ferreira, A novel multi-robot exploration approach based on particle swarm optimization algorithms, in *2011 IEEE International Symposium on Safety, Security, and Rescue Robotics (SSRR)* (IEEE, 2011), pp. 327–332
16. Z. Zheng, Y. Tan, Group explosion strategy for searching multiple targets using swarm robotic, in *IEEE Congress Evolutionary Computation* (2013), pp. 821–828
17. A. Marjovi, J. Nunes, P. Sousa, R. Faria, L. Marques, An olfactory-based robot swarm navigation method, in *2010 IEEE International Conference on Robotics and Automation (ICRA)* (IEEE, 2010), pp. 4958–4963
18. H.E. Espitia, J.I. Sofrony, Path planning of mobile robots using potential fields and swarms of Brownian particles, in *2011 IEEE Congress on Evolutionary Computation (CEC)*. (IEEE, 2011), pp. 123–129
19. Q. Tang, P. Eberhard, A PSO-based algorithm designed for a swarm of mobile robots. Struct. Multidiscip. Optim. **44**(4), 483–498 (2011)
20. Y. Tan, Y. Zhu, Fireworks algorithm for optimization, in *Advances in Swarm Intelligence* (Springer, Berlin, 2010), pp. 355–364
21. Z. Zheng, Y. Tan, An indexed K-D tree for neighborhood generation in swarm robotics simulation. *Advances in Swarm Intelligence*. Lecture Notes in Computer Science, vol. 7929 (Springer, Berlin, 2013), pp. 53–62

Postscript

This book does not end here, neither does the research on FWA. On the contrary, it seems to me the age of FWA has just begun, giving the bulk of challenges before us waiting to be dealt with. Considering the length of the book, I have to come to a temporary stop in the introduction of FWA. Stop for a deeper thinking. Regarding FWA research, what does the future hold in store for it, what are the most urgent problems to be answered? All of these problems haunt me and I think a tentative stop is for a better unveiling of the mystery.

After five years of FWA research, it has already become an effective and efficient swarm intelligence optimization algorithm whose performance catches up with or even exceeds most state-of-the-art swarm intelligence optimization algorithms, and becomes a new efficient tool for solving complex optimization problems. Nevertheless, FWA is still in the stage of infancy. Many areas are yet to be explored and novel ideas are increasingly emerging.

The future research of FWA may happen in the following areas:

1. Theoretical analysis. Including global convergence, computational efficiency, parameter setting, etc.
2. Performance improvement. One of the most important topics is to establish efficient cooperation mechanisms in FWA.
3. Hybrid algorithms. Combination of FWA with other SI algorithms.
4. Problem solving in the scenario of big data. As big-data application becomes popular and important, research on FWA for big data is a must-have topic.
5. Dynamic optimization problems such as dynamic object or target tracking and multi-targets tracking in search of swarm robots.
6. Applications in a variety of real-world problems.

Till now from accomplishing the manuscript of the book, the CIL team at Peking University has made several significant progresses on FWA study, including how to fully utilize the acquired/informed objective information assisting the swarm's future search, i.e., raising information utility and how to sufficiently enhance information sharing about each firework as well as cooperative evolutionary ability so as to fully

© Springer-Verlag Berlin Heidelberg 2015
Y. Tan, *Fireworks Algorithm*, DOI 10.1007/978-3-662-46353-6

bring the cooperative mechanism of a swarm into play for enhancing the solving capability of the swarm.

In the meantime, we have also made some beneficial contributions to pushing the frontiers of FWA research and organizing several forums and platforms for academic opinion exchanges.

First of all, I had organized a special issue in the International Journal of Swarm Intelligence Research (IJSIR) named "Developments and Applications of Fireworks Algorithm," which received widespread attention. We received a large number of research papers on FWA. After a strict and vigorous reviewing process, finally only five high-quality research papers were accepted. By the way, this special issue will appear in IJSIR-2015, Vol. 6, no. 2, which shows the epitome of FWA researches at present.

Second, we have organized and hosted a special session on FWA at IEEE CEC'2015 at Sendai, Japan, in which six papers are thoroughly discussed on this session that further indicates the hotness and prosperity of FWA researches. Furthermore, there are a few FWA research papers to be reported in regular sessions of The Sixth International Conference on Swarm Intelligence held in conjunction with The Second BRICS Congress on Computational Intelligence (ICSI-CCI'2015) which was successfully held during June 25–29, 2015, in BICC, Beijing, China.

Finally, other research institutes and universities both domestic and international have also published a number of research papers featuring FWA. I have received numerous emails and phone calls from scholars interested in the development and future prospects of FWA, and deeply felt their enthusiasm and desires for exploring FWA further.

It is shown from all of these works and papers that FWA is harvesting acknowledgment and attention from researchers and practitioners, and FWA's increasing computing efficacy is paving the road for more applications. FWA has widespread applications in various fields attracting more and more researchers to devote themselves to the research of FWA.

I believe that as more and more researchers join forces, research on FWA will be propelled to a new height, turning it, no doubt, into a mainstream swarm intelligence algorithm in the near future.

Appendix A
Benchmark Suites

A.1 Benchmark Suite 1

A.1.1 Sphere

A.1.1.1 Expression

$$f(x) = \sum_{i=1}^{D} x_i{}^2 \tag{A.1}$$

A.1.1.2 Parameters

Range: $[-100, 100]$, Optimum point: 0^D, Optimum value: 0, Dimension: 30

A.1.1.3 Image

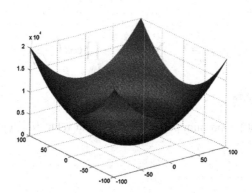

© Springer-Verlag Berlin Heidelberg 2015
Y. Tan, *Fireworks Algorithm*, DOI 10.1007/978-3-662-46353-6

A.1.2 Rosenbrock

A.1.2.1 Expression

$$f(x) = \sum_{i=1}^{D-1} \left(100(x_{i+1} - x_i{}^2)^2 + (x_i - 1)^2\right) \qquad \text{(A.2)}$$

A.1.2.2 Parameters

Range: $[-30, 30]$, Optimum point: 1^D, Optimum value: 0, Dimension: 30

A.1.2.3 Image

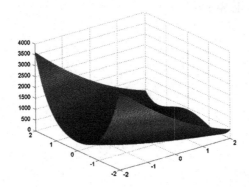

A.1.3 Griewank

A.1.3.1 Expression

$$f(x) = 1 + \sum_{i=1}^{D} \frac{x_i{}^2}{4000} + \prod_{i=1}^{D} \cos\left(\frac{x_i}{\sqrt{i}}\right) \qquad \text{(A.3)}$$

A.1.3.2 Parameters

Range: $[-600, 600]$, Optimum point: 0^D, Optimum value: 0, Dimension: 30

A.1.3.3 Image

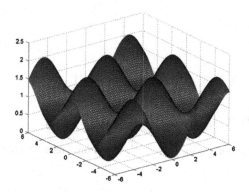

A.1.4 Rastrigin

A.1.4.1 Expression

$$f(x) = \sum_{i=1}^{D} (x_i^2 - 10\cos(2\pi x_i) + 10) \qquad (A.4)$$

A.1.4.2 Parameters

Range: $[-5.12, 5.12]$, Optimum point: 0^D, Optimum value: 0, Dimension: 30

A.1.4.3 Image

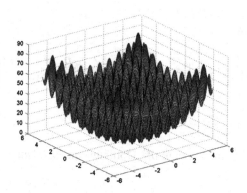

A.1.5 Schwefel's Problem 1.2

A.1.5.1 Expression

$$f(x) = \sum_{i=1}^{D} \left(\sum_{j=1}^{i} x_j \right)^2 \tag{A.5}$$

A.1.5.2 Parameters

Range: $[-100, 100]$, Optimum point: 0^D, Optimum value: 0, Dimension: 30

A.1.5.3 Image

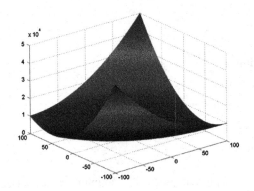

A.1.6 Ackley

A.1.6.1 Expression

$$f(x) = -20 \exp\left(-0.2 \sqrt{\frac{1}{D} \sum_{i=1}^{D} x_i^2} \right) - \exp\left(\frac{1}{D} \sum_{i=1}^{D} \cos(2\pi x_i) \right) + 20 + e \tag{A.6}$$

A.1.6.2 Parameters

Range: $[-32, 32]$, Optimum point: 0^D, Optimum value: 0, Dimension: 30

A.1.6.3 Image

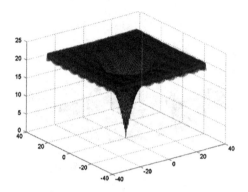

A.1.7 Penalized

A.1.7.1 Expression

$$f(x) = 0.1 \left\{ \sin^2(3\pi x_1) + \sum_{i=1}^{D-1} (x_i - 1)^2 [1 + \sin^2(3\pi x_{i+1})] \right.$$

$$\left. + (x_D - 1)^2 [1 + \sin^2(2\pi x_D)] \right\} + \sum_{i=1}^{D} \mu(x_i, 5, 100, 4), \quad \text{(A.7)}$$

where

$$\mu(x_i, a, k, m) = \begin{cases} k(x_i - a)^m & x_i > a \\ 0 & a \le x_i \le a \\ k(-x_i - a)^m & x_i < -a \end{cases} \quad \text{(A.8)}$$

A.1.7.2 Parameters

Range: $[-50, 50]$, Optimum point: 1^D, Optimum value: 0, Dimension: 30

A.1.7.3 Image

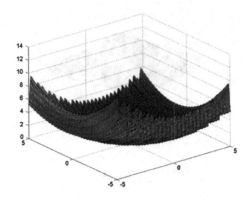

A.1.8 Six-Hump Camel-Back

A.1.8.1 Expression

$$f(x) = 4x_1^2 - 2.1x_1^4 + \frac{x_1^6}{3} + x_1 x_2 - 4x_2^2 + 4x_2^4 \tag{A.9}$$

A.1.8.2 Parameters

Range: $[-5, 5]$, Optimum point: $(-0.08983, 0.7126)$, $(0.08983, -0.7126)$, Optimum value: -1.032, Dimension: 2

A.1.8.3 Image

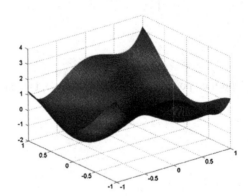

A.1.9 Goldstein-Price

A.1.9.1 Expression

$$f(x) = [1 + (x_1 + x_2 + 1)^2(19 - 14x_1 + 3x_1{}^2 - 14x_2 + 6x_1x_2 + 3x_2{}^2)]$$
$$\cdot [30 + (2x_1 - 3x_2)^2(18 - 32x_1 + 12x_1{}^2 + 48x_2 - 36x_1x_2 + 27x_2{}^2)]$$

$$(A.10)$$

A.1.9.2 Parameters

Range: $[-2, 2]$, Optimum point: $(0, -1)$, Optimum value: 3, Dimension: 2

A.1.9.3 Image

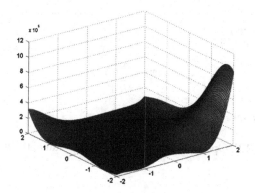

A.1.10 Schaffer's F6

A.1.10.1 Expression

$$f(x) = \frac{\sin^2\sqrt{x_1^2 + x_2^2} - 0.5}{[1 + 0.001(x_1^2 + x_2^2)]^2} + 0.5 \qquad (A.11)$$

A.1.10.2 Parameters

Range: $[-100, 100]$, Optimum point: 0^D, Optimum value: 0, Dimension: 2

A.1.10.3 Image

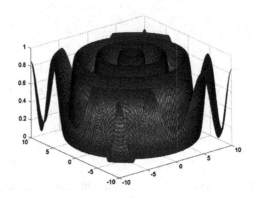

A.1.11 Axis Parallel Hyper Ellipsoid

A.1.11.1 Expression

$$f(x) = \sum_{i=1}^{D} i x_i^2 \tag{A.12}$$

A.1.11.2 Parameters

Range: $[-5.12, 5.12]$, Optimum point: 0^D , Optimum value: 0, Dimension: 30

A.1.11.3 Image

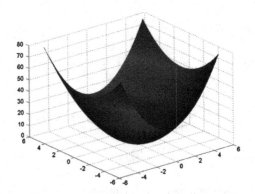

A.1.12 Rotated Hyper Ellipsoid

A.1.12.1 Expression

$$f(x) = \sum_{i=1}^{D} \left(\sum_{j=1}^{i} x_j^2 \right)^2.$$

(A.13)

A.1.12.2 Parameters

Range: $[-65.536, 65.536]$, Optimum point: 0^D, Optimum value: 0, Dimension: 30.

A.1.12.3 Image

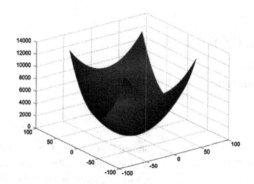

Table A.1 Test functions in Chap. 2

Number	Function name
1	Sphere
2	Rosenbrock
3	Griewank
4	Rastrigin

Table A.2 Test functions in Chap. 6

Number	Function name
1	Sphere
5	Schwefels Problem 1.2
2	Rosenbrock
6	Ackley
3	Griewank
4	Rastrigin
7	Penalized
8	Six-Hump Camel-Back
9	Goldstein-Price
10	Schaffers F6
11	Axis Parallel Hyper Ellipsoid
12	Rotated Hyper Ellipsoid

A.2 Benchmark Suite 2

Table A.3 25 Benchmark functions in CEC2005 [1]

	Number	Function name
Unimodal functions	1	Shifted Sphere Function
	2	Shifted Schwefel's Problem 1.2
	3	Shifted Rotated High Conditioned Elliptic Function
	4	Shifted Schwefel's Problem 1.2 with Noise in Fitness
	5	Schwefel's Problem 2.6 with Global Optimum on Bounds
Basic functions	6	Shifted Rosenbrock's Function
	7	Shifted Rotated Griewank's Function without Bounds
	8	Shifted Rotated Ackley's Function with Global Optimum on Bounds
	9	Shifted Rastrigin's Function
	10	Shifted Rotated Rastrigin's Function
	11	Shifted Rotated Weierstrass Function
	12	Schwefel's Problem 2.13
Expanded functions	13	Expanded Extended Griewank's plus Rosenbrock's Function (F8F2)
	14	Shifted Rotated Expanded Scaffer's F6

(continued)

Table A.3 (continued)

	Number	Function name
Hybrid composition functions	15	Hybrid Composition Function
	16	Rotated Hybrid Composition Function
	17	Rotated Hybrid Composition Function with Noise in Fitness
	18	Rotated Hybrid Composition Function
	19	Rotated Hybrid Composition Function with a Narrow Basin for the Global Optimum
	20	Rotated Hybrid Composition Function with the Global Optimum on the Bounds
	21	Rotated Hybrid Composition Function
	22	Rotated Hybrid Composition Function with High Condition Number Matrix
	23	Non-Continuous Rotated Hybrid Composition Function
	24	Rotated Hybrid Composition Function
	25	Rotated Hybrid Composition Function without Bounds

A.3 Benchmark Suite 3

Table A.4 28 Benchmark functions in CEC2013 [2]

	Number	Function name
Unimodal functions	1	Sphere Function
	2	Rotated High Conditioned Elliptic Function
	3	Rotated Bent Cigar Function
	4	Rotated Discus Function
	5	Different Powers Function
Basic multimodal functions	6	Rotated Rosenbrock's Function
	7	Rotated Schaffers F7 Function
	8	Rotated Ackley's Function
	9	Rotated Ackley's Function
	10	Rotated Griewank's Function
	11	Rastrigin's Function
	12	Rotated Rastrigin's Function
	13	Non-Continuous Rotated Rastrigin's Function

(continued)

Table A.4 (continued)

	Number	Function name
	14	Schwefel's Function
	15	Rotated Schwefel's Function
	16	Rotated Katsuura Function
	17	Lunacek Bi_Rastrigin Function
	18	Rotated Lunacek Bi_Rastrigin Function
	19	Expanded Griewank's plus Rosenbrock's Function
	20	Expanded Scaffer's F6 Function
Composition functions	21	Composition Function 1 ($N = 5$, Rotated)
	22	Composition Function 2 ($N = 3$, Unrotated)
	23	Composition Function 3 ($N = 3$, Rotated)
	24	Composition Function 4 ($N = 3$, Rotated)
	25	Composition Function 5 ($N = 3$, Rotated)
	26	Composition Function 6 ($N = 5$, Rotated)
	27	Composition Function 7 ($N = 5$, Rotated)
	28	Composition Function 8 ($N = 5$, Rotated)

Note N is the number of the functions that are combined

A.4 Benchmark Suite 4

Table A.5 30 Benchmark functions in CEC2014 [3]

	Number	Function name
Unimodal functions	1	Rotated High Conditioned Elliptic Function
	2	Rotated Bent Cigar Function
	3	Rotated Discus Function
Simple multimodal functions	4	Shifted and Rotated Rosenbrock's Function
	5	Shifted and Rotated Ackley's Function
	6	Shifted and Rotated Weierstrass Function
	7	Shifted and Rotated Griewank's Function
	8	Shifted Rastrigin's Function
	9	Shifted and Rotated Rastrigin's Function
	10	Shifted Schwefel's Function
	11	Shifted and Rotated Schwefel's Function
	12	Shifted and Rotated Katsuura Function
	13	Shifted and Rotated HappyCat Function

(continued)

Table A.5 (continued)

	Number	Function name
	14	Shifted and Rotated HGBat Function
	15	Shifted and Rotated Expanded Griewank's plus Rosenbrock's Function
	16	Shifted and Rotated Expanded Scaffer's F6 Function
Hybrid functions	17	Hybrid Function 1 ($N = 3$)
	18	Hybrid Function 2 ($N = 3$)
	19	Hybrid Function 3 ($N = 4$)
	20	Hybrid Function 4 ($N = 4$)
	21	Hybrid Function 5 ($N = 5$)
	22	Hybrid Function 6 ($N = 5$)
Composition functions	23	Composition Function 1 ($N = 5$)
	24	Composition Function 2 ($N = 3$)
	25	Composition Function 3 ($N = 3$)
	26	Composition Function 4 ($N = 5$)
	27	Composition Function 5 ($N = 5$)
	28	Composition Function 6 ($N = 5$)
	29	Composition Function 7 ($N = 3$)
	30	Composition Function 8 ($N = 3$)

Note N is the number of the functions that are combined

Appendix B
Resources

B.1 Internet Resources

1. Fireworks Algorithm Research Forum: http://www.cil.pku.edu.cn/research/fwa/index.html
2. Computational Intelligence Laboratory of Peking University: http://www.cil.pku.edu.cn
3. Source Codes of Fireworks Algorithm and its Variants: http://www.cil.pku.edu.cn/research/fwa/resources/index.html
4. 2014 International Conference on Swarm Intelligence Competition on Single Objective Optimization(ICSI-2014-BS) http://www.ic-si.org/competition/
5. IEEE CEC 2014 Competition Benchmark Functions http://www.ntu.edu.sg/home/EPNSugan/index_files/CEC2013/CEC2013.htm
6. IEEE CEC 2013 Competition Benchmark Functions: http://www.ntu.edu.sg/home/EPNSugan/index_files/CEC2014/CEC2014.htm

B.2 Organizations

1. IEEE Computational Intelligence Society: http://cis.ieee.org/
2. ACM SIGEVO: sigevo-members@ACM.ORG
3. World Federation on Soft Computing: http://www.softcomputing.org/
4. IEEE Systems, Man, and Cybernetics Society: http://www.ieeesmc.org/

B.3 Journals

1. International Journal of Swarm Intelligence Research
2. International Journal of Artificial Intelligence

© Springer-Verlag Berlin Heidelberg 2015
Y. Tan, *Fireworks Algorithm*, DOI 10.1007/978-3-662-46353-6

3. International Journal of Swarm Intelligence
4. IEEE Transactions on Evolutionary Computation
5. IEEE Transactions on Cybernetics
6. Softcomputing
7. Applied Softcomputing
8. International Journal of Computational Intelligence and Pattern Recognition
9. CAAI Transactions on Intelligent Systems

B.4 Conferences

1. International Conference on Swarm Intelligence(ICSI): http://www.ic-si.org
2. IEEE Symposium on Swarm Intelligence
3. IEEE Congress on Evolutionary Computation (CEC)
4. IEEE International Conference on System, Man and Cybernetics (IEEE SMC)
5. IEEE-INNS, International Joint Conference on Neural Networks (IJCNN)
6. IEEE World Congress on Computational Intelligence
7. ACM The Genetic and Evolutionary Computation Conference (GECCO)

References

1. P.N. Suganthan, N. Hansen, J.J. Liang, K. Deb, Y.-P. Chen, A. Auger et al., Problem definitions and evaluation criteria for the CEC 2005 special session on real-parameter optimization. KanGAL Report 2005005 (2005)
2. J.J. Liang, B.Y. Qu, P.N. Suganthan, A.G. Hernndez-Daz, Problem definitions and evaluation criteria for the CEC 2013 special session on real-parameter optimization (2013)
3. J.J. Liang, B.Y. Qu, P.N. Suganthan, Problem definitions and evaluation criteria for the CEC 2014 special session and competition on single objective real-parameter numerical optimization. Technical report 201311. Computational Intelligence Laboratory, Zhengzhou University, Zhengzhou China and Technical Report, Nanyang Technological University, Singapore (2013)

Index

© Springer-Verlag Berlin Heidelberg 2015
Y. Tan, *Fireworks Algorithm*, DOI 10.1007/978-3-662-46353-6

Printed in the United States
By Bookmasters